I0037032

Chitosan-Based Adsorbents for Wastewater Treatment

Edited by
Abu Nasar

Chitosan is a natural amino polymer. It is eco-friendly, biocompatible, biodegradable, cost-effective, easily available and has high potential to be utilized as an adsorbent. Because of their excellent chelating power, chitosan-based adsorbents have a very high ability to tightly bind the pollutants present in contaminated water and wastewater. Different heavy metals and toxic dyes can be effectively removed.

Chapter 1 deals with the chemical, physical, physicochemical and mechanical properties of chitosan and chitosan-based materials. Adsorption data on the removal of heavy metals and different dyes have been compiled. Chapter 2 covers the utilization of chitosan and its derivatives for the adsorptive removal of mercury from water and wastewater. Chapter 3 describes novel chitosan-based nanocomposites for dye removal applications. Chapter 4 discusses the effect of different chitosan modifications on its structure and specific surface area. Chapter 5 covers the applications of chitin and chitosan-based adsorbents for the removal of natural dyes from wastewater. Chapter 6 highlights the adsorptive treatment of textile effluents using chemically modified chitosan as adsorbents. Chapter 7 reviews the applications of chitosan-based adsorbents for the removal of arsenicals. Chapter 8 centers on the adsorption capacity enhancement of chitosan by chemical modification. Chapter 9 focuses on the smart use of surfactants for the modification of chitosan and some other biomaterials and their subsequent use for the removal of contaminants from aqueous solutions. Chapter 10 reviews the use of chitosan-based nanocomposites as adsorbents for the removal of dyes from wastewater. Chapter 11 describes the preparation of uniformly distributed platinum nanoparticles decorated with graphene oxide-chitosan by employing a microwave-assisted method. The nanocomposite can be used for the removal of dye from aqueous solution.

Keywords

Wastewater Treatment, Adsorbent, Chitosan, Adsorption Capacity of Chitosan, Heavy Metal Removal from Wastewater, Dye Removal from Wastewater, Chitosan-Based Nanocomposites, Removal of Arsenicals from Wastewater, Surfactants for Chitosan, Surfactants for Biomaterials

Chitosan-Based Adsorbents for Wastewater Treatment

Edited by

Abu Nasar

Department of Applied Chemistry, Z.H. College of Engineering and Technology, Aligarh Muslim University, Aligarh – 202002, India

Copyright © 2018 by the authors

Published by **Materials Research Forum LLC**
Millersville, PA 17551, USA

All rights reserved. No part of the contents of this book may be reproduced or transmitted in any form or by any means without the written permission of the publisher.

Published as part of the book series
Materials Research Foundations
Volume 34 (2018)
ISSN 2471-8890 (Print)
ISSN 2471-8904 (Online)

Print ISBN 978-1-945291-74-6
ePDF ISBN 978-1-945291-75-3

This book contains information obtained from authentic and highly regarded sources. Reasonable efforts have been made to publish reliable data and information, but the author and publisher cannot assume responsibility for the validity of all materials or the consequences of their use. The authors and publishers have attempted to trace the copyright holders of all material reproduced in this publication and apologize to copyright holders if permission to publish in this form has not been obtained. If any copyright material has not been acknowledged please write and let us know so we may rectify in any future reprint.

Distributed worldwide by

Materials Research Forum LLC
105 Springdale Lane
Millersville, PA 17551
USA
http://www.mrforum.com

Manufactured in the United States of America
10 9 8 7 6 5 4 3 2 1

Table of Contents

Preface

Many industries, such as textile, paper, plastics, dyestuffs, chemical, metallurgical, refineries, tanneries, etc. consume a large amount of water, and ultimately generate a substantial volume of wastewater containing toxic pollutants such as dyes, heavy metals, pesticides and other chemicals. The contaminated industrial effluents are one of the major sources of water pollution. Thus treatment of industrial wastewater is essential before its disposal. A number of different techniques such as electrodialysis, ion exchange, photocatalysis, nanofiltration membranes, electroflotation, electrokinetic/electrooxidation, coagulation-flocculation, reverse osmosis, ozonation, anaerobic-aerobic, adsorption, etc. are available to decontaminate wastewater. However, each technique has its own merits and demerits. Among them, adsorption technique has been considered to be preferable due to a number of factors such as reduced cost, the simplicity of the method, flexibility and ease of design, insensitivity to toxic pollutants and lower interference from diurnal variations. However, the choice of suitable adsorbent was always a challenging task. The conventional commercial adsorbents like activated carbon and silica gel are highly efficient but expensive and associated with regeneration problem. Thus the adsorbents made from natural sources attracted prominent attention.

Chitosan is a natural amino polymer obtained from deacetylation of chitin, which is richly available in nature. In fact, chitosan is the second most abundant biopolymer after cellulose and constitutes the structural component of mollusks, insects, crustaceans, fungi, algae, crab, shellfish, shrimp, lobster, crawfish, krill, etc. Chitosan is natural, renewable, eco-friendly, biocompatible, biodegradable, cheap and easily available and has high potential to be utilized as an adsorbent. Because of its excellent chelating power, chitosan has a very high ability to tightly bind the pollutants present in contaminated water. The waste, released from the processing of krill, shellfish, crabs, and shrimps may be deacetylated to chitosan, which can thus be efficiently utilized as a cost-effective adsorbent. In recent years, chitosan has received significant attention due to its specific benefits such as high adsorption capacity, abundance, and cost-effectiveness. Different heavy metals and dyes have been reported to be effectively removed by chitosan-based adsorbents.

The main objective of the book was to compile and present relevant information on the modification and utilization of chitosan-based adsorbents for the elimination of toxic dyes and heavy metals from wastewater. The present book is the outcome of contributions of 30 researchers from the international scientific community. It consists of

11 chapters. **Chapter 1** deals with the chemical, physical, physicochemical and mechanical properties of chitosan and chitosan-based materials. Applications and market trends of these materials have been thoroughly discussed. The adsorption data on the removal of heavy metals and different dyes have been compiled. **Chapter 2** provides a review on the current development on the utilization of chitosan and its derivatives for the adsorptive removal of mercury from water and wastewater. **Chapter 3** describes the preparation of monodisperse palladium nanoparticles decorated chitosan-graphene oxide as a nonadsorbent. The utilization of this nanocomposite for the effective removal of methylene blue dye from the aqueous solution has been explained. **Chapter 4** discusses the effect of chitosan modification on its structure and specific surface area. The different chitosan modification methods such as preparation of hydrogel beads from powdered chitosan, cross-linking of beads using epichlorohydrin or glutaraldehyde and conditioning of beads using NaHSO₄ have also been discussed. **Chapter 5** covers the applications chitin and chitosan-based adsorbents for the removal of natural dyes from wastewater. **Chapter 6** highlights the adsorptive treatment of textile effluents using chemically modified chitosan as adsorbents. **Chapter 7** reviews the environmental applications of chitosan-based adsorbents in the light of the removal of arsenicals. **Chapter 8** offers a thorough review on the enhancement of adsorption capacity of chitosan by chemical modification. **Chapter 9** focuses on the role of anionic surfactant (sodium dodecyl sulfate) for enhancing the removal capacity of chitosan. The chapter summarizes recent researches focusing on the smart use of surfactant for the modification of chitosan and some other biomaterials and their subsequent use for the removal contaminants from aqueous solutions. **Chapter 10** describes the recent developments in the field of chitosan-based nanocomposites as adsorbents for the removal of dyes from wastewater. **Chapter 11** deals with the preparation of uniformly distributed platinum nanoparticles decorated with graphene oxide-chitosan by employing a microwave-assisted method. The nanocomposite was used for the removal of dye from aqueous solution.

I am grateful to all contributing authors/coauthors for their appreciable contributions to the book. I would like to express my gratitude to Materials Research Forum LLC (USA) for bringing out the book in the present form. I am also thankful to the Chairperson, Department of Applied Chemistry, Z.H. College of Engineering & Technology, Aligarh Muslim University, Aligarh for providing the necessary facilities.

Abu Nasar

Editor

Chitosan-Based Adsorbents for Wastewater Treatment, Ed. Abu Nasar   Materials Research Forum LLC
Materials Research Foundations **34** (2018) 1-28   doi: http://dx.doi.org/10.21741/9781945291753-1

Chapter 1

Chitosan: Structure, Properties and Applications

Volkan Ugraskan[1], Abdullah Toraman[1], Afife Binnaz Hazar Yoruç[2*]

[1]Department of Chemistry, Yildiz Technical University, Turkey

[2]Department of Metallurgical and Materials Engineering, Yildiz Technical University, Turkey

*afife.hazar@gmail.com

Abstract

Chitosan is a deacetylated form of chitin which is the second most renewable biopolymer in nature. Chitosan has many features like biocompatibility, biodegradability, non-toxic properties, adsorption and functioning properties which make it promising materials in many applications. This chapter focused on chemical, physical, physicochemical and mechanical properties of chitosan and chitosan-based materials first of all, right after their commercial applications and also, in conclusion, the adsorption capacity and adsorbent applications of chitosan and its derivatives. Finally, new trends and fields about chitosan-based materials were present.

Keywords

Biopolymer, Chitin, Chitosan, Structure, Properties, Applications, Market Trends

Contents

1.   Introduction

Chitin, which is an ancient biopolymer has been studying since 1884. This biopolymer is produced by a large number of living organisms and is the most abundant polymer after cellulose in the world. Chitin is a hard, non-elastic, and nitrogenous polysaccharide. There are several derivatives of chitin and chitosan constitutes as the most important one. Chitosan has a similar to cellulose structure. Instead of the hydroxyl groups in the cellulose, the amine groups are present in the structure.

Chitosan, the acetylated form of chitin is soluble in acidic solution compared to non-soluble chitin. The fact that chitosan is soluble according to the chitin gives it many advantages. It can be used in many applications, especially in food, cosmetics, agriculture, medicine, paper, and textile.

The most important peculiarity of chitosan is the presence of the modifiable locations in its chemical structure. Because of this feature, chitosan can be modified with grafting and crosslinking processes may be converted to derivatives with superior properties. With grafting reactions, the addition of extra functional groups on chitosan increases the number of adsorption sites and hence the adsorption capacity and allows the use of chitosan in adsorbent applications.

This chapter focused on chemical, physical, physicochemical and mechanical properties of chitosan and chitosan-based materials first of all, right after their commercial applications and also, in conclusion, the adsorption capacity and adsorbent applications of chitosan and its derivatives. Finally, new trends and fields about chitosan-based materials are presented.

2.   Chitosan

Chitin is a polysaccharide which is bound to the β-1,4 bond of N-acetyl-D-glucosamine monomers and is the most abundant after cellulose in nature [2]. The main ingredient of crustaceans, such as crabs and shrimps, as well as the cell walls of mushrooms, contains lots of chitin molecules. From these organisms, chitin is extracted after several steps including deproteinization and demineralization.

Chitosan is obtained as a result of deacetylation of the chitin. In order to produce chitosan, the deacetylation degree of chitin must reach 50%. At the desired degree of deacetylation, the chitin is rendered soluble in the aqueous acidic medium. Dissolution occurs by protonation of the $-NH_2$ function on the C-2 position of the D-glucosamine repeat unit, and the polysaccharide is converted to a polyelectrolyte in the acidic medium. Chitosan is a natural cationic polymer. Because it is soluble in aqueous solutions, the solution is used extensively in different applications such as gels or films and fibers [3].

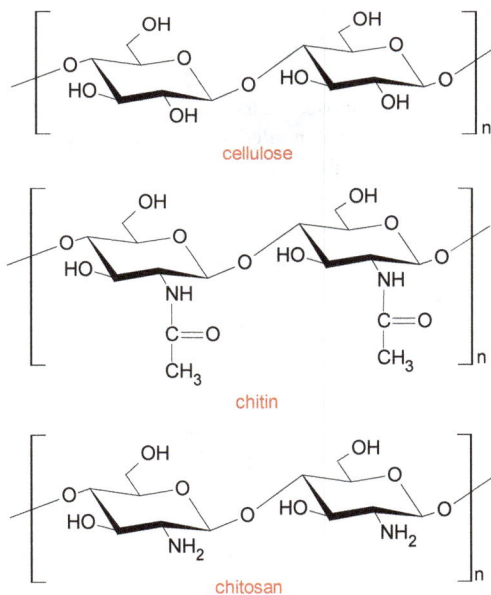

Figure 1 Structure of cellulose, chitin, and chitosan.

2.1 Structure of chitosan

Chitosan is semi-crystalline in solid form [4]. There are many studies on chitosan in polymorph structure in the literature [5]. Chitosan is structurally similar to cellulose (Fig. 1). Instead of the hydroxyl groups in cellulose, there is a groups of amines in chitosan. Unlike cellulose, it contains between 5 and 8% nitrogen. The amine group in the chitosan is the result of deacetylation of the acetamide groups in the chitin, but there always remain some acetamide groups which are not deacetylated in the structure [6].

However, the presence of hydroxyl groups on each repeat unit and the amine groups on the deacetylated unit are highly active compared to the chitin. These active groups modify the mechanical and physical properties of chitosan [7]. In addition, the presence of these groups provides biologically functional facilitation. For this reason, biocompatible, biologically active, non-toxic and good adsorption properties of chitosan make this material a very favorable biomaterial [8].

2.1.1 N-acetylation degree

During deacetylation of the chitin, the acetyl groups in the structure are converted to free amine groups by hydrolysis. N-acetylation degree (DD) is in proportion to the 2-acetamido-2-deoxy-D-glucopyranose, 2-amino-2-deoxy-D-glucopyranose units. In other words, DD is the ratio of deacetylated units to acetylated ones. This ratio has a significant effect on the solubility of the chitin and the solution properties. Chitosan is an N-deacetylated nontoxic derivative of chitin. In the chitin, the acetylated units are dominant (the acetylation level is typically 0.90). Chitosan is a fully or partially N-deacetylated derivative, less than 0.35 fully or partially N-deacetylated derivative [3,7].

Moreover, this step determines DD varies with temperature, time, and amount of sodium hydroxide (NaOH) used in the process. DD value affects the adsorption capacity of chitosan. High DD value usually results from the presence of high amounts of amino groups and may enhance the dye adsorption capacity of chitosan as a result of the protonation process. DD can be used for characterization of molecular weight, crystallinity, distribution of amine groups except for adsorption [9].

Characterization of a chitosan sample requires the determination of mean deacetylation. The degree of deacetylation can be determined by potentiometric titration, IR, elemental analysis, UV, ^{1}H NMR and ^{13}C NMR [3, 10-14].

The deacetylation level can be calculated using the IR technique. The percentage of acetylated amine group is calculated according to equation 1. The deacetylation degree in the infrared spectrum of the chitosan is calculated as 87% in Fig 2.

N-acetyl % = 100 −(A1652/A3446) x 100/1.33   (1)

A: Absorbance

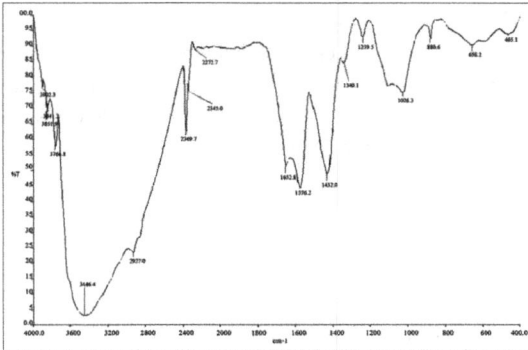

Figure 2 Infrared spectrum of chitosan [15].

Another method to calculate the DD is the 1H NMR technique. In Equation 2, DD can be calculated. In Figure 3, DD is calculated as 97% [12].

$$DD = \left[1 - \left(\frac{1/3 I_{CH_3}}{1/6 I_{H_2 - H_6}} \right) \right] \times 100 \qquad (2)$$

I: Peak of a proton (ppm)

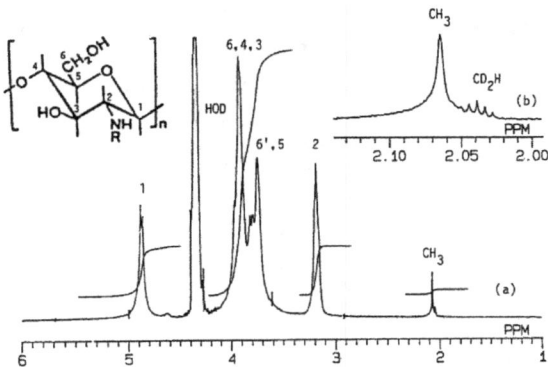

Figure 3 NMR spectrum of chitosan [12].

2.1.2 Molecular weight

The molecular weight distribution of chitosan is obtained using HPLC [16]. The average molecular weight is also determined by light scattering [17]. The fastest and simplest method of determining the molecular weight is carried out using the Mark-Houwink equation with a viscometer (Equation 3) [18].

$$[\eta]=KM_v^a \tag{3}$$

$[\eta]$: Viscosity (cP)

Mv: Molecular weight (g/mol)

K and a: Constants for temperature and solvent-solvent system

2.1.3 Mechanical properties of chitosan

Chitosan is found in various forms. Chitosan films have poor mechanical strength. For this reason, the use of these films is limited. In particular, it is desirable to have films with pharmacologically active coatings that have improved mechanical properties for the application of chitosan films or for the protection of seeds. Therefore, a composite form of chitosan films can increase its mechanical properties (Table 1).

Table 1 Mechanical properties of chitosan films.

Film Samples	Tensile Strength (MPa)	Elongation at Break (%)	Reference
Chitosan-poly(ethylene oxide) films	0.011±0.2	33±3	[19]
Neat chitosan	37.7±4.5	49.9±5.6	[20]
Chitosan-0.8 carbon nanotube film	74.9±4.8	19.5±3.3	[20]
Chitosan film	70.3	6.2	[21]
Cross-linked chitosan	93.3	5.1	[21]

2.2 Chitosan production

Chitosan is the result of deacetylation of chitin, a polysaccharide formed by β-1,4 bonds of N-acetyl-D-glucosamine monomers. The main source is seashells, insects, and mushrooms [22]. There is about 20-30% chitin in seashells [2]. Chitin production stages from fungi and seashells are given in Fig. 4.

Depending on the source, there are two allomorphic structures, α, and β. These two forms can be distinguished by infrared, NMR spectroscopy and X-ray diffraction. α-Chitin is the most commonly found form [23]. It is seen on the walls of fungus and yeast cells, on lobster and crab tendons, on shrimp shells and also on insect cuticle [24]. The rarer β-chitin is found in the squamous-pencil proteins, pogonophorans and tubers synthesized by vestimethetic worms. Examination of the two isomorphic crystallographic structures reveals that there are two antiparallel molecules in the α-chitin and only one antiparallel molecule in the β-chitin [25, 26].

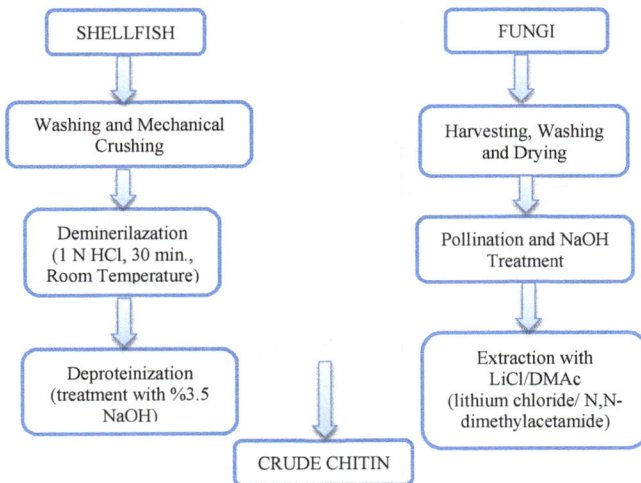

Figure 4 Chitin production stages from fungi and seashells [1].

The nitrogen content of the chitin varies from 5% to 8%, depending on the degree of deacetylation, while the nitrogen in the chitin is mostly in the form of primary aliphatic amino groups. For this reason, chitosan reacts with typical amines and the N-acylation and Schiff reaction are the most important. Chitosan derivatives are easily obtained under mild conditions [7].

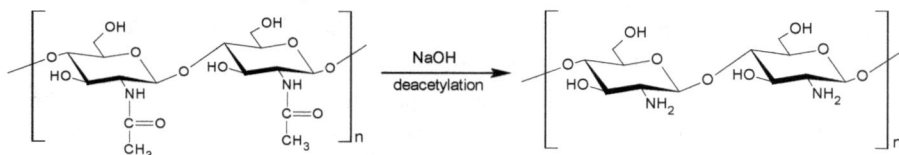

Figure 5 Production of chitosan.

There are many methods of the production of chitosan from chitin such as alkaline treatment at high temperature, and intermittent washing with water, using water-miscible organic solvents and enzymatic N-deacetylation. Commercially, chitosan is produced by N-deacetylation with a high concentration sodium hydroxide solution of chitin in high temperature and pressure (Fig. 5). Recently, the method used in the production of chitosan is done by microwave. This process, which is done by microwave, takes considerably shorter than other methods [27]. In recent years, microwave heating has attracted considerable interest because it is much faster than conventional heating. Especially for the modification of chitosan [28].

3.   Applications and market trends

Chitosan has many features like biocompatibility, biodegradability, non-toxic properties, adsorption and functioning properties which make it promising materials in many applications [7]. The high nitrogen ratio (%6,89) in the structure ensures the reactive amino and hydroxyl groups to chelate with metal ions [7, 29]. Over and above with hemostatic, fungistatic, spermicidal, antitumor, anticholesteremic, accelerating bone formation and central nervous system depressant etc. biological properties, application area can be expanded [29]. Chitosan has wide use in food, agriculture, medical, cosmetics, biotechnology and purification applications [2].

In Grand View Research's report dated January 2017 it is stated that in 2014, chitosan market value accounted for 1.06 billion and rose to 3.19 billion in 2015. The market for chitosan is expected to reach USD 17.84 billion in 2025. It is also stated that chitosan demand in wastewater treatment was valued at over USD 220 million in 2015 and it is predicted that the use of chitosan will increase by 17% for water treatment processes [30]. In the case of Global Market Insights report September 2016, a market value will hit USD 7.90 billion in 2024. The report states that water treatment applications in 2015 are above 30% of chitosan use [31]. Finally, it is stated in Global Industry Analysis data that the use of chitin and chitosan will reach 155.5 tons in 2022 [32].

Generally, the applications of chitosan according to the sectors are shown in Table 2.

Table 2 Chitosan applications and specific uses.

Application Area	Specific Uses	Reference
Agriculture	Antimicrobial Agent Additive Seed Coat Fertilizer Component	[33-38]
Food Industry	Thickener Animal Feed Component Waste Food Recycling Filtration and Removal Hypocholesterolemic Agent	[39-45]
Cosmetics	Hair Styling Agent Moisturizer Hydrating Agent Deodorizing Agent Viscosity Modifying Agent	[46-50]
Biomedical and Pharmaceutical	Wound Band and Bandage Hemostatic Agent Contact Lenses Drug Delivery Material Wound-Bone Healing Material	[51-62]
Purification	Adsorbent	[63-74]

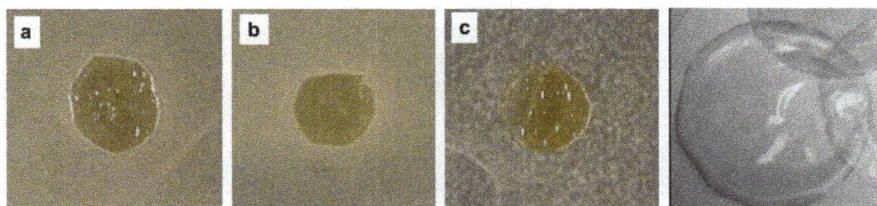

Figure 6 Inhibitory effect of the chitosan-starch film against (a) E. coli, (b) S. aureus, and (c) B. Subtilis [75, 76]

3.1   Food applications

Packaging is one of the major applications in the food sector. Compared to other biopolymers used in packaging applications, chitosan has advantages like showing an antibacterial effect as well as incorporation ability with functional groups of mineral or vitamin molecules. Lately, researchers focused on production and optimization of chitosan/starch film as a potential application in food packaging. The studies have proven antibacterial properties of the films (Figure 6) [75].

By dint of its good physicochemical properties, short-time biodegradability, biological function, biocompatibility, antimicrobial and antifungal activities and non-toxic properties, chitosan shows significant potential for packaging applications in the food industry. By the way, chitosan has interesting properties such as gas and aroma barriers as well as excellent film-forming capacity under dry coating conditions. These features offer new opportunities to preserve food quality and increase the shelf life of products [77, 78].

3.2   Agricultural applications

Chitosan has a great importance in environmentally friendly products used for plant defense in agricultural applications. The chitosan molecule triggers the defense response within the plant and creates physical and chemical precautions against invading pathogens.

The use of chitosan has generally been used as a fertilizer in seeds, leaves, fruits and vegetables, and as a controlled chemical release to protect plants against microorganisms and to stimulate plant growth [79].

Chitosan effectiveness against plant pathogens was investigated by researchers. Chien and Chou [80] found that Cytosporina sp. isolate 75 mg/L was completely inhibited by chitosan, whereas the second isolate of the same gender was not affected by 1000 mg/L. As a result of this experiment, it was reported that the antifungal activity of chitosan depends on the species, concentration and test organism. For instance, at 0.1%, chitosan of 92.1 kDa was treated with a greater growth inhibition of 76.2% on P. italicum than did chitosan of 357.3 kDa (71.4%), while at 0.2%, the antifungal activity was implemented by chitosan of 357.3 kDa was higher than chitosan of 92.1 kDa against P. italicum. Benhamou [81] concluded that 500 and 1000 mg/L chitosan obtained from crab shells was effective in reducing the disease probability caused by F. oxysporum f. sp. radicis-lycopersici. Moreover, El Ghaouth [82] showed that chitosan was effective in inhibiting the mycelial growth of P. aphanidermatum at a concentration of 400 mg/L. These and other studies have demonstrated to be effective in the removal of chitosan plant organisms [33]. In another study conducted on the aquaculture, the antimicrobial activities of the chitosan solutions in the different constants are shown in Table 3.

Chitosan-Based Adsorbents for Wastewater Treatment, Ed. Abu Nasar   Materials Research Forum LLC
Materials Research Foundations **34** (2018) 1-28   doi: http://dx.doi.org/10.21741/9781945291753-1

Table 3 Antimicrobial activity of chitosan against bacteria isolated from oysters [83].

Bacterial Strains	Chitosan Concentration (g/L)			
	0.5	**1.0**	**5.0**	**10.0**
Pseudomonas Sp.	3.8±0.6	6.0±1.0	13.3±0.7	14.0±0.8
Vibrionaceae Sp.	12.3±0.5	14.0±0.6	23.8±0.9	25.0±0.9
Shewanella sp.	0	3.3±0.6	4.5±0.3	5.3±0.5
Alcaligenes sp.	0	2.5±0.7	3.5±0.4	3.3±0.6
Enterobacteriaceae sp.	5.3±0.5	10.3±0.4	13.5±0.6	13.8±0.4
Moraxella sp.	12.8±1.1	24.5±1.4	30.8±0.7	32.3±1.9
Acinetobacter sp.	2.5±0.5	8.0±0.7	13.3±0.5	14.0±0.5
Flavobacterium sp.	4.3±0.4	11.6±0.5	16.0±1.2	20.8±0.6
Corynebacterium sp.	0	4.3±0.4	5.8±0.8	6.1±0.3
Staphylococcus sp.	4.0±0.3	6.8±0.4	12.3±0.4	12.5±0.4
Micrococcus sp.	13.3±0.7	16.5±0.7	20.0±0.5	21.5±0.6
Lactic acid bacteria sp.	14.0	16.0±0.8	22.5±0.7	23.3±1.1
Bacillus sp.	4.3±0.4	11.5±0.3	13.8±0.8	14.0±0.6

Table 4 In vivo applications of chitosan [85, 86].

Chitosan Type	Form	Animal Model	Application	Effect
Chitosan %82 deacetylated 80 kDa from crab shells	Cotton type	Small and large wild or zoo animals	Topical	Wound/infection healing
Methylpyrrolidone chitosan	Freeze dried soft sponge	Rabbit	Topical	Bone healing
Phosphorylated chitosan	With calcium phosphate cement	Rabbit	Implantation	Bone healing
N,N'-dicarboxymethylated chitosan	Soft spongy and hydrophilic material	Sheep	Topical	Bone healing
Chitosan %72 deacetylated from cuttle fish	Powder	Dog	Topical	Bone healing

Chitosan 70 kDa from prawn shells	Solution	Rat	Injection	Tissue regeneration
Chitosan %82 deacetylated 30-80 kDa	Cotton type	Dog	Topical	Wound healing
Chitosan %82 Deacetylated 80 kDa from crab shells	Cotton type	Dog	Implantation	Immuno-potentiatior
Chitosan	Hydrogel	Rat	Topical	Burn replacement of wound/skin healing
Chitosan %100 deacetylated 540 kDa	Non-woven fibers	Rat	Topical	Tissue/bone regeneration
Photocrosslinkable chitosan 800-1000 kDa	Gel	Beagle	Topical	Wound healing
Chitosan crosslinked with glutaraldehyde	Gel	Rat	Topical-Brachytherapy	Implant for brachytherapy

Table 5 Chitosan applications in hair care [3].

Properties	Applications
Electrostatic interaction with hair	Shampoos
Hydrophilic character (antistatic effect)	Hair Tonics, Rinses, Permanent Wave Agents, Hair Colorants, Lacquers, Hair Sprays, Time Release Delivery
Hydrophobic character (removing sebum and oils from hair)	-
Antibacterial and antifungal activity	-
Thickening polymer	-
Surfactant character (stabilize emulsion)	-
Make hair softer, increase the mechanical strength	-
Protect elastic film on hairs	-

3.3   Biomedical and pharmaceutical applications

Besides the biocompatibility, non-toxicity, antimicrobial activity and low immunogenicity, chitosan showing clear potential for development and that makes it propitious material for biomedical applications. It can be used in various conditions in gel, sponge, membrane, bead and scaffold forms in biomedical applications [84]. With protonation of amino groups in its structure, chitosan exhibits polyelectrolyte properties, thus it can be used in various applications by combining many natural and synthetic polymers, metal anions, etc. These applications may be referred to as delivery, drug delivery, bone/wound/tissue repair, artificial skin, pharmaceutical excipients and dietary supplements [85, 86]. The results obtained with different forms of the chitosan-based products used in animal experiments are shown in Table 4.

3.4   Cosmetic applications

Unique features of chitosan provide a polymer available for cosmetic sector applications. At one of these applications, it has been formulated with Inula helenium L. Extract, a perennial plant with antimicrobial properties such chitosan, which has been used in skin lotions and creams [87]. Moreover, it is used as shell material in the microencapsulation of essential oils, an important application in the cosmetics sector [88]. Chitosan can be used as an agent in sun protection products due to its skin compatibility and adhesive properties as well as its ability to absorb UV rays [89].

Except these applications, chitosan has also taken its place in hair care applications (Table 5) [3].

3.5   Adsorbent applications

Chitosan and its derivatives can be used as an adsorbent to remove dyes and heavy metals. The amino and hydroxyl groups present in the chitosan structure act as active sites in this application. At acidic pH values, the amino groups of the chitosan structure become cationic and adsorption of anionic dyes with strong bonds is ensured. In addition, chitosan, which is sensitive to pH, can gel or dissolve according to the pH value [90, 91]. Improvements in adsorbent properties can be achieved by modifying the structure of chitosan. In the grafting reactions, the addition of extra functional groups on the chitosan increases the number of adsorption sites and thus the adsorption capacity. By crosslinking reactions, chitosan resistance is increased, but some functional groups cannot participate in adsorption since they bind to each other [91, 92].

Heavy metal pollution is becoming a serious threat to ecosystem and human health through polluted water, land and air in some parts of the world. Water pollution caused by heavy metals is one of the most serious environmental problems and probably the most difficult to solve. Some metals, iron (Fe), zinc (Zn), copper (Cu), molybdenium (Mo), are important for human health, but others, such as heavy metals, are considered very toxic. These include arsenic (As), nickel (Ni), lead (Pb), which are widely used and frequently found in industrial wastes. In addition to these metals, toxic substances such as chromium (Cr), arsenic (As) or selenium (Se) and also precious metals such as gold (Au), silver (Ag), palladium (Pd) and platinum (Pt) also cause pollution.

Adsorption/biosorption process with chitosan is one of the effective methods for removing heavy metals from water and wastewater (Table 6). Chitosan biosorbate is very effective and very similar to the use of ion exchange resins, which can also perform cleaning work. Commercial ion-exchange resins are sold for 80-100 USD/kg while the cost of new performance-based new chitosan-based biosorbents can be lower than 8-10 USD/kg. Chitosan and chitin were mainly used for removing pollutants through coagulation (4%), precipitation (7%), flocculation (3%), adsorption (28%), flotation (1%), filtration (4%), membranes filtration (53%) [91, 93, 94].

Another cause of environmental pollution is dyed. The synthetic nature of the dyes and the complex molecule structure make them more stable in water and prevent from biodegrading. Dyes can affect photosynthetic activity in water life by reducing light transmission and can be toxic to some water organisms due to aromatics, metals etc. In addition, they can also cause serious damage to people, such as kidney, reproductive system, liver, brain, and central nervous system disfunction [101]. Natural materials or wastes/by-products of synthetic materials can either be used less costly or after a small treatment as an adsorbent. They generally referred to as low-cost adsorbents (LCA). Chitosan, which is one of LCAs, is widely used at adsorbent applications (Table 7) [102].

Table 6   Data on adsorption of metals using chitosan and derivatives

Adsorbent	Metal	Adsorption Capacity (mg/g)	Reference
Chitosan	Cd(II)	5.93	[95, 96]
Chitosan	Hg(II)	815.0	[95, 96]
Chitosan	Cu(II)	222	[95, 96]
Chitosan	Ni(II)	164	[95, 96]
Chitosan	Zn(II)	75	[95, 96]
Chitosan Flakes	Cu^{2+}	1.8-2.2 mmol/g	[95, 96]
Aminated Chitosan Beads	Hg	2.26 mmol/g	[95, 96]
Chitosan	Al(III)	45.45	[95, 96]
Chitosan	VO_4^{3-}	400–450	[95, 96]
Chitosan	Au(III)	30.95	[95, 96]
Chitosan Beads	As(III)	1.83	[95, 96]
Chitosan Beads	As(V)	1.94	[95, 96]
Cross-linked Chitosan	Cr(III)	6.0	[95, 96]
Cross-linked Chitosan	Cr(VI)	215.0	[95, 96]
Cross-linked Chitosan	U(VI)	72.6	[95, 96]
Chitosan	As(III)	2.16	[97]

Cross-linked Chitosan–Fe	As(III)	13.4	[97]
Cross-linked Chitosan–Fe	Cr(VI)	295.0	[97]
Chitosan–Fe(III) Hollow Fiber Membranes	As(V)	0.85	[97]
Chitosan-Polyvinylacetate	Cu(II)	192.57	[97]
Chitosan-Polystyrene	Cu(II)	99.8	[97]
Cross-linked Chitosan	Hg(II)	58.0	[98]
Chitosan-Sargassum	Ni(II)	0.33-0.48 mmol/g	[99]
Magnetic chitosan	Cr(VI)	69.40	[90]
Chitosan-Magnetite	Pb(II)	63.33	[90]
Chitosan-Magnetite	Ni(II)	52.55	[90]
Chitosan-Perlite	Cu(II)	196.07	[90]
Chitosan-Perlite	Cd(II)	178.6	[90]
Chitosan-Silica	Ni(II)	254.3	[90]
Chitosan-Clinoptilolite	Cu(II)	574.49	[90]
Chitosan	Cu(II)	14.75–485.44	[91]
Chitosan	Cd(II)	6.64–449.6	[91]
Chitosan	Hg(II)	51.6–1127.4	[91]
Chitosan Derivatives	Hg(II)	48.1–3170	[91]
Chitosan-Pectin	Pb(II)	30.1	[100]
Chitosan-Magnetite Microparticles	Ni(II)	588.24	[100]
Chitosan-Magnetite Microparticles	Co(II)	833.34	[100]
Chitosan–Silica Hybrid	Co (II)	0.63 mmol/g	[100]

Another cause of environmental pollution is dyed. The synthetic nature of the dyes and the complex molecule structure make them more stable in water and prevent from biodegrading. Dyes can affect photosynthetic activity in water life by reducing light transmission and can be toxic to some water organisms due to aromatics, metals etc. In addition, they can also cause serious damage to people, such as kidney, reproductive system, liver, brain, and central nervous system disfunction [101]. Natural materials or wastes/by-products of synthetic materials can either be used less costly or after a small treatment as an adsorbent. They generally referred to as low-cost adsorbents (LCA). Chitosan, which is one of LCAs, is widely used at adsorbent applications (Table 7) [102].

Table 7   Data on the adsorption of dyes using chitosan and derivatives

Adsorbent	Dye	Adsorption Capacity (mg/g)	Reference
Chitosan/Alumina Composite	Methyl Orange	1–12 g/L	[103]

Chitosan	Acid Orange 10	922.9	[9]
Chitosan	Acid Red 73	728.2	[9]
Chitosan	Acid Red 18	693.2	[9]
Chitosan	Acid Green 25	645.1	[9]
Chitosan-Glutaraldehyde	Cationic Dye (BBR250)	0.8 mmol/g	[104]
Chitosan-Poly(Methyl Methacrylate)	Anionic Dye (RB19)	1498	[104]
Magnetic Chitosan−Graphene Oxide	Methyl Orange	399	[92]
Chitosan−Magnetic Graphitized Multi-Walled Carbon Nanotubes	Cationic Red	263.3	[92]
Chitosan-Polyvinylalcohol	Eosin Yellow	53	[92]
Chitosan−Alunite Composite	Reactive Red 2	569	[92]
Chitosan Flake	Reactive Red 222	380	[9]
Chitosan Flake	Reactive Black 5	1100	[9]
Chitosan Flake	Direct Red 23	155	[9]
Chitosan Flake	Acid Green 25	645.5	[9]
Chitosan Powder	Acid Orange 12	931.85	[9]
Chitosan Powder	Acid blue 9	256.0	[9]
Chitosan-Zeolite A	Bezactive Orange 16	305.8	[92, 104]
Chitosan Grafted with Polypropylene İmine	Reactive Black 5	6250	[92, 104]
Chitosan-Siliceous Mesoporous	Acid Red 18	201.2	[92, 104]
Diatomite-Chitosan−Fe(III) Composite	Direct orange 2GL	1007.5	[105]
Chitosan−Fe(III)−Glutaraldehyde	Reactive Red 120	249.3	[105]
Chitosan-(cetyl) trimethyl ammonium bromide	Weak acid scarlet	91.4	[105]
Glutamine-Chitosan-	Acid Green 25	543.92	[105]

Silica-Fe$_3$O$_4$ Nanoparticles			
Chitosan-Montmorillonite	Congo Red	53.42	[90]
Chitosan-Polyurethane	Acid Violet 48	30.00	[90]
Chitosan-Activated Clay	Methylene Blue	330.0	[90]
Chitosan-Bentonite	Tartrazine	294.1	[90]
Chitosan-Oil Palm	Reactive Blue 19	909.1	[90]
Chitosan–Alkali Lignin Composite	Remazol Brilliant Blue R	111.11	[106]
Magnetic Chitosan Grafted with Graphene Oxide	Methylene Blue	95.2	[107]
Chitosan–Halloysite Nanotubes	Methylene Blue	270.27	[108]
Chitosan-Polyamide Nanofiber	Reactive Black 5	456.9	[109]
Chitosan-Polyamide Nanofiber	Ponceau 4R	502.4	[109]
β-cyclodextrin-Chitosan Magnetic Nanocomposite	Methylene Blue	2788	[92, 104]
Chitosan-Modified Palygorskite	Anionic Dye (RY3RS)	71.38	[92, 104]
Chitosan-Graphite Oxide Composite	Reactive Black 5	277	[92, 104]

4.   Conclusions

This study aims to give information about properties, the production techniques and applications of the chitosan. Chitosan has unique features such as it can be obtained from natural and sustainable sources, it is biodegradability and biocompatibility, it has a functional structure, etc. which provides a wide range of applications. Nowadays, in many industries attentions are turned to natural and sustainable resources, therefore, it is expected that the application areas of chitosan will be further increased.

References

[1]   A. Demir, N. Seventekin, Kitin, kitosan ve genel kullanım alanları. Tekstil Teknolojileri Elektronik Dergisi, 3 (2009) 92-103.

[2]   Ö. İmamoğlu, Biyokontrolde doğal ürünlerin kullanılması; Kitosan. Türk Hijyen ve Deneysel Biyoloji Dergisi, 68 (2011) 215-222.

[3]   M. Rinaudo, Chitin and chitosan: properties and applications. Progress in polymer science, 31 (2006) 603-632. https://doi.org/10.1016/j.progpolymsci.2006.06.001

[4]   K. Ogawa, Effect of heating an aqueous suspension of chitosan on the crystallinity and polymorphs. Agricultural and Biological Chemistry, 55 (1991) 2375-2379.

[5]   N. Cartier, A. Domard, H. Chanzy, Single crystals of chitosan. International journal of biological macromolecules, 12 (1990) 289-294. https://doi.org/10.1016/0141-8130(90)90015-3

[6]   S. Islam, M.R. Bhuiyan, M. Islam, Chitin and chitosan: structure, properties and applications in biomedical engineering. Journal of Polymers and the Environment, 25 (2017) 854-866. https://doi.org/10.1007/s10924-016-0865-5

[7]   M.N.R. Kumar, A review of chitin and chitosan applications. Reactive and functional polymers, 46 (2006) 1-27. https://doi.org/10.1016/S1381-5148(00)00038-9

[8]   H. Tan, C.R. Chu, K.A. Payne and K.G. Marra, Injectable in situ forming biodegradable chitosan–hyaluronic acid based hydrogels for cartilage tissue engineering. Biomaterials, 30 (2009) 2499-2506. https://doi.org/10.1016/j.biomaterials.2008.12.080

[9]   M. Vakili, M. Rafatullaha, B. Salamatinia, A.Z. Abdullah, M.H. Ibrahim, K.B. Tan, Z. Gholami and P. Amouzgar, Application of chitosan and its derivatives as adsorbents for dye removal from water and wastewater: A review. Carbohydrate polymers, 113 (2014) 115-130. https://doi.org/10.1016/j.carbpol.2014.07.007

[10]   M. Miya, R. Iwamoto, S. Yoshikawa and S. Mima, Ir spectroscopic determination of CONH content in highly deacylated chitosan. International journal of biological macromolecules, 2 (1980) 323-324. https://doi.org/10.1016/0141-8130(80)90056-2

[11]   J. Brugnerotto, J. Lizardi, F.M. Goycoolea, W. Argu Èelles-Monal, J. DesbrieÁres and M. Rinaudo, An infrared investigation in relation with chitin and chitosan characterization. Polymer, 42 (2001) 3569-3580. https://doi.org/10.1016/S0032-3861(00)00713-8

[12]   A. Hirai, H. Odani, and A. Nakajima, Determination of degree of deacetylation of chitosan by 1 H NMR spectroscopy. Polymer Bulletin, 26 (1991) 87-94. https://doi.org/10.1007/BF00299352

[13] K.M. Vårum, M.W. Antohonsen, H. Grasdalen and O. Smidsrød, Determination of the degree of N-acetylation and the distribution of N-acetyl groups in partially N-deacetylated chitins (chitosans) by high-field nmr spectroscopy. Carbohydrate Research, 211 (1991) 17-23. https://doi.org/10.1016/0008-6215(91)84142-2

[14] S.C. Tan, E. Khor, T.K. Tan and S.M. Wong, The degree of deacetylation of chitosan: advocating the first derivative UV-spectrophotometry method of determination. Talanta, 45 (1998) 713-719. https://doi.org/10.1016/S0039-9140(97)00288-9

[15] S. Paul, A. Jayan, C.S. Sasikumar and S.M. Cherian, Extraction and purification of chitosan from chitin isolated from sea prawn (Fenneropenaeus indicus). Asian Journal of Pharmaceutical and Clinical Research, 7 (2014).

[16] A.C. Wu, Determination of molecular-weght distribution of chitosan by high-performance liquid chromatography. Journal of Chromatography A, 128 (1976) 87-99. https://doi.org/10.1016/S0021-9673(00)84034-0

[17] A. Domard, M. Rinaudo, Preparation and characterization of fully deacetylated chitosan. International Journal of Biological Macromolecules, 5 (1983) 49-52. https://doi.org/10.1016/0141-8130(83)90078-8

[18] Kasaai, M.R., J. Arul, and G. Charlet, Intrinsic viscosity–molecular weight relationship for chitosan. Journal of Polymer Science Part B: Polymer Physics, 38 (2000) 2591-2598. https://doi.org/10.1002/1099-0488(20001001)38:19<2591::AID-POLB110>3.0.CO;2-6

[19] V. Alexeev, E. A. Kelberg, G. A. Evmenenko, S. V. Bronnikov Improvement of the mechanical properties of chitosan films by the addition of poly (ethylene oxide). Polymer Engineering & Science, 40(2000) 1211-1215. https://doi.org/10.1002/pen.11248

[20] S.F. Wang, L. Shen, W. D. Zhang, and Y. J. Tong, Preparation and mechanical properties of chitosan/carbon nanotubes composites. Biomacromolecules, 6 (2005) 3067-3072. https://doi.org/10.1021/bm050378v

[21] F.S. Kittur, K.R. Kumar, and R.N. Tharanathan, Functional packaging properties of chitosan films. Zeitschrift für Lebensmitteluntersuchung und-Forschung A, 206 (1998) 44-47.

[22] A.T. Paulino, J. I. Simionato, J. C. Garcia and J. Nozaki, Characterization of chitosan and chitin produced from silkworm crysalides. Carbohydrate Polymers, 64 (2006) 98-103. https://doi.org/10.1016/j.carbpol.2005.10.032

[23]   J. Sakamoto, J.Sugiyama, S. Kimura, T. Imai, T. Itoh, T. Watanabe, and S. Kobayashi, Artificial chitin spherulites composed of single crystalline ribbons of α-chitin via enzymatic polymerization. Macromolecules, 33 (2000) 4155-4160. https://doi.org/10.1021/ma000230y

[24]   W. Helbert, J. Sugiyama, High-resolution electron microscopy on cellulose II and α-chitin single crystals. Cellulose, 5 (1998) 113-122. https://doi.org/10.1023/A:1009272814665

[25]   K. Rudall, W. Kenchington, The chitin system. Biological Reviews, 48 (1973) 597-633. https://doi.org/10.1111/j.1469-185X.1973.tb01570.x

[26]   I. Younes, M. Rinaudo, Chitin and chitosan preparation from marine sources. Structure, properties and applications. Marine drugs, 13 (2015) 1133-1174. https://doi.org/10.3390/md13031133

[27]   A. Sahu, P. Goswami, and U. Bora, Microwave mediated rapid synthesis of chitosan. Journal of Materials Science: Materials in Medicine, 20 (2009) 171-175. https://doi.org/10.1007/s10856-008-3549-4

[28]   P. Nahar, U. Bora, Microwave-mediated rapid immobilization of enzymes onto an activated surface through covalent bonding. Analytical biochemistry, 328 (2004) 81-83. https://doi.org/10.1016/j.ab.2003.12.031

[29]   P.K. Dutta, J. Dutta, and V. Tripathi, Chitin and chitosan: Chemistry, properties and applications, 63 (2004) 20-31.

[30]   Chitosan Market Size To Reach USD 17.84 Billion By 2025. 2017, Grand View Research.

[31]   Chitosan Market size worth $7.9bn by 2024. 2016, Global Market Insights, Inc.

[32]   Chitin and Chitosan Derivatives Market Trend. 2016, Global Industry Analysis Inc.

[33]   M.E. Badawy, E.I. Rabea, A biopolymer chitosan and its derivatives as promising antimicrobial agents against plant pathogens and their applications in crop protection. International Journal of Carbohydrate Chemistry, (2011). https://doi.org/10.1155/2011/460381

[34]   E.I. Rabea, M.E.T. Badawy, C. V. Stevens, G. Smagghe, and W. Steurbaut, Chitosan as antimicrobial agent: applications and mode of action. Biomacromolecules, 4 (2003) 1457-1465. https://doi.org/10.1021/bm034130m

[35] L. Orzali, B. Corsi, C. Forni and L. Riccioni, Chitosan in agriculture: a new challenge for managing plant disease, in Biological Activities and Application of Marine Polysaccharides. (2017).

[36] S. Boonlertnirun, C. Boonraung, and R. Suvanasara, Application of chitosan in rice production. Journal of metals, materials and minerals, 18 (2017).

[37] E. G. Lizárraga-Paulín, S. P. M. -Castro, E. M. -Martínez, A. V. L. -Sagahón and I. T. -Pacheco, Maize seed coatings and seedling sprayings with chitosan and hydrogen peroxide: their influence on some phenological and biochemical behaviors. Journal of Zhejiang University Science B, 14 (2013) 87-96. https://doi.org/10.1631/jzus.B1200270

[38] D. Zeng, X. Luo, and R. Tu, Application of bioactive coatings based on chitosan for soybean seed protection. International Journal of Carbohydrate Chemistry, (2012). https://doi.org/10.1155/2012/104565

[39] S. Hirano, C. Itakura, H. Seino, Y.Akiyama, I. Nonaka, N. Kanbara, and T. Kawakami, Chitosan as an ingredient for domestic animal feeds. Journal of agricultural and food chemistry, 38 (1990) 1214-1217. https://doi.org/10.1021/jf00095a012

[40] S. Swiatkiewicz, M. Swiatkiewicz, A. A. -Wlosek and D. Jozefiak, Chitosan and its oligosaccharide derivatives (chito-oligosaccharides) as feed supplements in poultry and swine nutrition. Journal of animal physiology and animal nutrition, 99 (2015) 1-12. https://doi.org/10.1111/jpn.12222

[41] R. Sánchez, G. B. Stringari, J. M. Franco, C. Valencia and C. Gallegos, Use of chitin, chitosan and acylated derivatives as thickener agents of vegetable oils for bio-lubricant applications. Carbohydrate polymers, 85 (2011) 705-714. https://doi.org/10.1016/j.carbpol.2011.03.049

[42] X. Li and W. Xia, Effects of chitosan on the gel properties of salt-soluble meat proteins from silver carp. Carbohydrate Polymers, 82 (2010) 958-964. https://doi.org/10.1016/j.carbpol.2010.06.026

[43] A.A. Tayel, Microbial chitosan as a biopreservative for fish sausages. International journal of biological macromolecules, 2016. 93: p. 41-46. https://doi.org/10.1016/j.ijbiomac.2016.08.061

[44] F. Karadeniz and S.-K. Kim, Antidiabetic activities of chitosan and its derivatives: A mini review, in Advances in food and nutrition research. Elsevier (2014) 33-44. https://doi.org/10.1016/B978-0-12-800268-1.00003-2

[45] M. Sayed, M. T. Islam, M. M. Haque, M. J. H. Shah, R. Ahmed, M. N. Siddiqui and M. A. Hossain, Dietary effects of chitosan and buckwheat (Fagopyrum esculentum) on the performance and serum lipid profile of broiler chicks. South African Journal of Animal Science, 45 (2015) 429-440. https://doi.org/10.4314/sajas.v45i4.9

[46] C. Z. X. Lifang, Moisturizing Property of Oligochitosan and Its Application in Cosmetics. Flavour Fragrance Cosmetics, 6 (2011) 004.

[47] Y.K. Deshmukh and S. Sao, Chitin and Chitosan–Most Convenient Natural Matter That Use In Different Useful Ways. Journal of Industrial Pollution Control, 30 (2014) 213-214.

[48] S. Pokhrel, P.N. Yadav and R. Adhikari, Applications of chitin and chitosan in industry and medical science: a review. Nepal Journal of Science and Technology, 16 (2016) 99-104. https://doi.org/10.3126/njst.v16i1.14363

[49] A. Jimtaisong and N. Saewan, Use of Chitosan and its Derivatives in Cosmetics. Chemistry, 9 (2014) 6.

[50] J. Huang, Z. H. Cheng, H. H. Xie, J. Y. Gong, J. Lou, Q. Ge, Y. J. Wang, Y. F. Wu, S. W. Liu, P. L. Sun and J. W. Mao, Effect of quaternization degree on physiochemical and biological activities of chitosan from squid pens. International journal of biological macromolecules, 70 (2014) 545-550. https://doi.org/10.1016/j.ijbiomac.2014.07.017

[51] K.H. Waibel, B. Haney, M. Moore, B. Whisman and R. Gomez, Safety of chitosan bandages in shellfish allergic patients. Military medicine, 176 (2011) 1153-1156. https://doi.org/10.7205/MILMED-D-11-00150

[52] G.P. De Castro, M. B. Dowling, M. Kilbourne, K. Keledjian, I. R. Driscoll, S. R. Raghavan, J. R. Hess, T. M. Scalea, and G. V. Bochicchio, Determination of efficacy of novel modified chitosan sponge dressing in a lethal arterial injury model in swine. Journal of Trauma and Acute Care Surgery, 72 (2012) 899-907. https://doi.org/10.1097/TA.0b013e318248baa1

[53] P. Sudheesh Kumar, V. K. Lakshmanan, T.V. Anilkumar, C. Ramya, P. Reshmi, A.G. Unnikrishnan, S. V. Nair, and R. Jayakumar, Flexible and microporous chitosan hydrogel/nano ZnO composite bandages for wound dressing: in vitro and in vivo evaluation. ACS applied materials & interfaces, 4 (2012) 2618-2629. https://doi.org/10.1021/am300292v

[54] T. Dai, M. Tanaka, Y. Y. Huang and M. R. Hamblin, Chitosan preparations for
      wounds and burns: antimicrobial and wound-healing effects. Expert review of
      anti-infective therapy, 9 (2011) 857-879. https://doi.org/10.1586/eri.11.59

[55] J. Zhang, W. Xia, P. Liu, Q. Cheng, T. Tahirou, W. Gu and B. Li, Chitosan
      modification and pharmaceutical/biomedical applications. Marine drugs, 8 (2010)
      1962-1987. https://doi.org/10.3390/md8071962

[56] A. Shukla, J. C. Fang, S. Puranam, F. R. Jensen, P. T. Hammond, Hemostatic
      multilayer coatings. Advanced Materials, 24 (2012) 492-496.
      https://doi.org/10.1002/adma.201103794

[57] R. Gu, W. Sun, H. Zhou, Z. Wu, Z. Meng, X. Zhu, Q. Tang, J. Dong and G. Dou,
      The performance of a fly-larva shell-derived chitosan sponge as an absorbable
      surgical hemostatic agent. Biomaterials, 31 (2010) 1270-1277.
      https://doi.org/10.1016/j.biomaterials.2009.10.023

[58] G. Lan, B. Lu, T. Wang, L. Wang, J. Chen, K. Yu, J. Liu, F. Dai and D. Wu,
      Chitosan/gelatin composite sponge is an absorbable surgical hemostatic agent.
      Colloids and surfaces B: Biointerfaces, 136 (2015) 1026-1034.
      https://doi.org/10.1016/j.colsurfb.2015.10.039

[59] X. Huang, Y.Sun, J. Nie, W. Lua, L. Yang, Z. Zhang, H. Yin, Z. Wang and Q. Hu,
      Using absorbable chitosan hemostatic sponges as a promising surgical dressing.
      International journal of biological macromolecules, 75 (2015) 322-329.
      https://doi.org/10.1016/j.ijbiomac.2015.01.049

[60] A. Bernkop-Schnürch and S. Dünnhaupt, Chitosan-based drug delivery systems.
      European Journal of Pharmaceutics and Biopharmaceutics, 81 (2012) 463-469.
      https://doi.org/10.1016/j.ejpb.2012.04.007

[61] X. Hu, H. P. Tan, D. Li and M. Y. Gu, Surface functionalisation of contact lenses
      by CS/HA multilayer film to improve its properties and deliver drugs. Materials
      Technology, 29 (2014) 8-13. https://doi.org/10.1179/1753555713Y.0000000063

[62] N. Bhattarai, J. Gunn, and M. Zhang, Chitosan-based hydrogels for controlled,
      localized drug delivery. Advanced drug delivery reviews, 62 (2010) 83-99.
      https://doi.org/10.1016/j.addr.2009.07.019

[63] R. K. Gautam, A. Mudhoo, G. Lofrano, M. C. Chattopadhyaya, Biomass-derived
      biosorbents for metal ions sequestration: adsorbent modification and activation
      methods and adsorbent regeneration. Journal of environmental chemical
      engineering, 2 (2014) 239-259.

[64] J. Zhao, Y. J. Zhu, J. Wu, J. Q. Zheng, X. Y. Zhao, B. Q. Lu and F. Chen, Chitosan-coated mesoporous microspheres of calcium silicate hydrate: environmentally friendly synthesis and application as a highly efficient adsorbent for heavy metal ions. Journal of colloid and interface science, 418 (2014) 208-215. https://doi.org/10.1016/j.jcis.2013.12.016

[65] Y. Lin, Y. Hong, Q. Song, Z. Zhang, J. Gao and T. Tao, Highly efficient removal of copper ions from water using poly (acrylic acid)-grafted chitosan adsorbent. Colloid and Polymer Science, 295 (2017) 627-635. https://doi.org/10.1007/s00396-017-4042-8

[66] F. Shen, J. Su, X. Zhang, K. Zhang and X. Qi, Chitosan-derived carbonaceous material for highly efficient adsorption of chromium (VI) from aqueous solution. International journal of biological macromolecules, 91 (2016) 443-449. https://doi.org/10.1016/j.ijbiomac.2016.05.103

[67] C. Li, J. Cui, F. Wang, W. Peng and Y. He, Adsorption removal of Congo red by epichlorohydrin-modified cross-linked chitosan adsorbent. Desalination and Water Treatment, 57 (2016) 14060-14066. https://doi.org/10.1080/19443994.2015.1060904

[68] B. Hastuti, A. Masykur and S. Hadi. Modification of chitosan by swelling and crosslinking using epichlorohydrin as heavy metal Cr (VI) adsorbent in batik industry wastes. in IOP Conference Series: Materials Science and Engineering. (2016). IOP Publishing.

[69] H.J.Kumari, P. Krishnamoorthy, T.K. Arumugam, S. Radhakrishnan and D. Vasudevan, An efficient removal of crystal violet dye from waste water by adsorption onto TLAC/Chitosan composite: A novel low cost adsorbent. International journal of biological macromolecules, 96 (2017) 324-333. https://doi.org/10.1016/j.ijbiomac.2016.11.077

[70] L. Zheng, C. Wang, Y. Shu, X. Yan and L. Li, Utilization of diatomite/chitosan– Fe (III) composite for the removal of anionic azo dyes from wastewater: equilibrium, kinetics and thermodynamics. Colloids and Surfaces A: Physicochemical and Engineering Aspects, 468 (2015) 129-139. https://doi.org/10.1016/j.colsurfa.2014.12.015

[71] F. Zhao, E. Repo, M. Sillanpää, Y. Meng, D. Yin, and W. Z. Tang, Green synthesis of magnetic EDTA-and/or DTPA-cross-linked chitosan adsorbents for highly efficient removal of metals. Industrial & Engineering Chemistry Research, 54 (2015) 1271-1281. https://doi.org/10.1021/ie503874x

[72]   A. Li, R. Lin, C. Lin, B. He, T. Zheng, L. Lu and Y. Cao, An environment-friendly and multi-functional absorbent from chitosan for organic pollutants and heavy metal ion. Carbohydrate polymers, 148 (2016) 272-280. https://doi.org/10.1016/j.carbpol.2016.04.070

[73]   E. Vunain, A. Mishra and B. Mamba, Dendrimers, mesoporous silicas and chitosan-based nanosorbents for the removal of heavy-metal ions: a review. International journal of biological macromolecules, 86 (2016) 570-586. https://doi.org/10.1016/j.ijbiomac.2016.02.005

[74]   Y. Wen, J. Ma , J. Chen, C. Shen, H. Li and W. Liu, Carbonaceous sulfur-containing chitosan–Fe (III): a novel adsorbent for efficient removal of copper (II) from water. Chemical engineering journal, 259 (2015) 372-380. https://doi.org/10.1016/j.cej.2014.08.011

[75]   P. Dutta, S. Tripathi, G. K. Mehrotra and J. Dutta, Perspectives for chitosan based antimicrobial films in food applications. Food chemistry, 114 (2009) 1173-1182. https://doi.org/10.1016/j.foodchem.2008.11.047

[76]   J. Dutta, S. Tripathi and P. Dutta, Progress in antimicrobial activities of chitin, chitosan and its oligosaccharides: a systematic study needs for food applications. Revista de Agaroquimica y Tecnologia de Alimentos, 18 (2012) 3-34. https://doi.org/10.1177/1082013211399195

[77]   M. Cruz-Romero, T. Murphy, M. Morris, E. Cummins and J.P. Kerry, Antimicrobial activity of chitosan, organic acids and nano-sized solubilisates for potential use in smart antimicrobially-active packaging for potential food applications. Food Control, 34 (2013) 393-397. https://doi.org/10.1016/j.foodcont.2013.04.042

[78]   S. Tripathi, G. Mehrotra and P. Dutta, Physicochemical and bioactivity of cross-linked chitosan–PVA film for food packaging applications. International Journal of Biological Macromolecules, 45 (2009) 372-376. https://doi.org/10.1016/j.ijbiomac.2009.07.006

[79]   A. Abdel-Mawgoud, A.S. Tantawy, M.A. El-Nemr and Y. Sassine, Growth and yield responses of strawberry plants to chitosan application. European Journal of Scientific Research, 39 (2010) 161.

[80]   P.J. Chien and C.C. Chou, Antifungal activity of chitosan and its application to control post-harvest quality and fungal rotting of Tankan citrus fruit (Citrus tankan Hayata). Journal of the Science of Food and Agriculture, 86 (2006) 1964-1969. https://doi.org/10.1002/jsfa.2570

[81]   N. Benhamou, P. Lafontaine and M. Nicole, Induction of systemic resistance to Fusarium crown and root rot in tomato plants by seed treatment with chitosan. Phytopathology, 84 (1994) 1432-1444. https://doi.org/10.1094/Phyto-84-1432

[82]   A. El Ghaouth, J.Arul, C.Wilson and N.Benhamou, Ultrastructural and cytochemical aspects of the effect of chitosan on decay of bell pepper fruit. Physiological and Molecular Plant Pathology, 44 (1994) 417-432. https://doi.org/10.1016/S0885-5765(05)80098-0

[83]   A. Alishahi and M. Aïder, Applications of chitosan in the seafood industry and aquaculture: a review. Food and Bioprocess Technology, 5 (2012) 817-830. https://doi.org/10.1007/s11947-011-0664-x

[84]   R. Jayakumaret, D. Menon, K. Manzoor, S. V. Nair and H.Tamura, Biomedical applications of chitin and chitosan based nanomaterials—A short review. Carbohydrate Polymers, 82 (2010) 227-232. https://doi.org/10.1016/j.carbpol.2010.04.074

[85]   T. Kean and M. Thanou, Chitin and chitosan: sources, production and medical applications. Renewable resources for functional polymers and biomaterials, (2011) 292-318. https://doi.org/10.1039/9781849733519-00292

[86]   Y. Luo and Q. Wang, Recent development of chitosan-based polyelectrolyte complexes with natural polysaccharides for drug delivery. International journal of biological macromolecules, 64 (2014) 353-367. https://doi.org/10.1016/j.ijbiomac.2013.12.017

[87]   S.B. Seo, C.-S. Ryu, G.-W. Ahn, H.-B. Kim, B.-K. Jo, S.-H. Kim, J.-D. Lee and T. Kajiuchi, Development of a natural preservative system using the mixture of chitosan-Inula helenium L. extract. International journal of cosmetic science, 24 (2002) 195-206. https://doi.org/10.1046/j.1467-2494.2002.00139.x

[88]   I. M. Martins, M. F. Barreiro, M. Coelho, A. E. Rodrigues, Microencapsulation of essential oils with biodegradable polymeric carriers for cosmetic applications. Chemical Engineering Journal, 245 (2014) 191-200. https://doi.org/10.1016/j.cej.2014.02.024

[89]   Y.A. Gomaa, L. K. El-Khordagui, N. A. Boraei and I. A. Darwish, Chitosan microparticles incorporating a hydrophilic sunscreen agent. Carbohydrate Polymers, 81 (2010) 234-242. https://doi.org/10.1016/j.carbpol.2010.02.024

[90]   W.W. Ngah, L. Teong and M. Hanafiah, Adsorption of dyes and heavy metal ions
       by chitosan composites: A review. Carbohydrate polymers, 83 (2011) 1446-1456.
       https://doi.org/10.1016/j.carbpol.2010.11.004

[91]   J. Wang, and C. Chen, Chitosan-based biosorbents: modification and application
       for biosorption of heavy metals and radionuclides. Bioresource technology, 160
       (2014) 129-141. https://doi.org/10.1016/j.biortech.2013.12.110

[92]   G.Z. Kyzas, D.N. Bikiaris and A.C. Mitropoulos, Chitosan adsorbents for dye
       removal: A review. Polymer International, (2017). https://doi.org/10.1002/pi.5467

[93]   G. MR, , Recent Advances in Chitosan Based Biosorbent for Environmental
       Clean-Up. Journal of Bioremediation & Biodegradation, 7 (2016) 173.

[94]   N.G. Kandile, H.M. Mohamed and M.I. Mohamed, New heterocycle modified
       chitosan adsorbent for metal ions (II) removal from aqueous systems. International
       journal of biological macromolecules, 72 (2015) 110-116.
       https://doi.org/10.1016/j.ijbiomac.2014.07.042

[95]   A. Bhatnagar and M. Sillanpää, Applications of chitin-and chitosan-derivatives for
       the detoxification of water and wastewater—a short review. Advances in colloid
       and interface science, 152 (2009) 26-38. https://doi.org/10.1016/j.cis.2009.09.003

[96]   V.K. Thakur and M.K. Thakur, Recent advances in graft copolymerization and
       applications of chitosan: a review. ACS Sustainable Chemistry & Engineering, 2
       (2014) 2637-2652. https://doi.org/10.1021/sc500634p

[97]   L. Zhang, Y. Zeng and Z. Cheng, Removal of heavy metal ions using chitosan and
       modified chitosan: A review. Journal of Molecular Liquids, 214 (2016)175-191.
       https://doi.org/10.1016/j.molliq.2015.12.013

[98]   G.Z. Kyzas and E.A. Deliyanni, Mercury (II) removal with modified magnetic
       chitosan adsorbents. Molecules, 18 (2013) 6193-6214.
       https://doi.org/10.3390/molecules18066193

[99]   J. He and J.P. Chen, A comprehensive review on biosorption of heavy metals by
       algal biomass: materials, performances, chemistry, and modeling simulation tools.
       Bioresource technology, 160 (2014) 67-78.
       https://doi.org/10.1016/j.biortech.2014.01.068

[100] M. Ahmad, S. Ahmed, B. L. Swami and S. Ikram, Adsorption of heavy metal ions:
       role of chitosan and cellulose for water treatment. Langmuir, 79 (2015) 109-155.

[101] M.T. Yagub, T. K. Sen, S. Afroze and H. M. Ang, Dye and its removal from aqueous solution by adsorption: a review. Advances in colloid and interface science, 209 (2014) 172-184. https://doi.org/10.1016/j.cis.2014.04.002

[102] V. Gupta, Application of low-cost adsorbents for dye removal–A review. Journal of environmental management, 90 (2009) 2313-2342. https://doi.org/10.1016/j.jenvman.2008.11.017

[103] J. Zhang, Q. Zhou and L. Ou, Kinetic, isotherm, and thermodynamic studies of the adsorption of methyl orange from aqueous solution by chitosan/alumina composite. Journal of Chemical & Engineering Data, 57 (2011) 412-419. https://doi.org/10.1021/je2009945

[104] G.Z. Kyzas and D.N. Bikiaris, Recent modifications of chitosan for adsorption applications: a critical and systematic review. Marine drugs, 13 (2015) 312-337. https://doi.org/10.3390/md13010312

[105] M. Vakili, S. Deng, L. Shen, D. Shan, D. Liu and G. Yu, Regeneration of Chitosan-Based Adsorbents for Eliminating Dyes from Aqueous Solutions. Separation & Purification Reviews, (2017) 1-13. https://doi.org/10.1080/15422119.2017.1406860

[106] V. Nair, A. Panigrahy and R. Vinu, Development of novel chitosan–lignin composites for adsorption of dyes and metal ions from wastewater. Chemical Engineering Journal, 254 (2014) 491-502. https://doi.org/10.1016/j.cej.2014.05.045

[107] J. Yan, Y. Huang, Y. E. Miao, W. W. Tiju and T. Liu, Polydopamine-coated electrospun poly (vinyl alcohol)/poly (acrylic acid) membranes as efficient dye adsorbent with good recyclability. Journal of hazardous materials, 283 (2015) 730-739. https://doi.org/10.1016/j.jhazmat.2014.10.040

[108] Q. Peng, M. Liu, J. Zheng and C. Zhou, Adsorption of dyes in aqueous solutions by chitosan–halloysite nanotubes composite hydrogel beads. Microporous and Mesoporous Materials, 201 (2015) 190-201. https://doi.org/10.1016/j.micromeso.2014.09.003

[109] G. Dotto, J. M. N. Santos, E. H. Tanabe, D. A. Bertuol, E. L. Foletto, E. C. Lima and F. A. Pavan, Chitosan/polyamide nanofibers prepared by Forcespinning® technology: A new adsorbent to remove anionic dyes from aqueous solutions. Journal of cleaner production, 144 (2017) 120-129. https://doi.org/10.1016/j.jclepro.2017.01.004

Chitosan-Based Adsorbents for Wastewater Treatment, Ed. Abu Nasar   Materials Research Forum LLC
Materials Research Foundations **34** (2018) 29-56   doi: http://dx.doi.org/10.21741/9781945291753-2

Chapter 2

Adsorptive Removal of Mercury from Water and Wastewater by Chitosan and its Derivatives

Fouzia Mashkoor, Abu Nasar*

Department of Applied Chemistry, Z.H College of Engineering & Technology

Aligarh Muslim University, Aligarh – 202002, India

*abunasaramu@gmail.com

Abstract

Mercury is one of the most poisonous heavy metals present in industrial wastewater. There is evidence showing its bioaccumulation in organism and biomagnification in food chains and having long-term toxicity. Among various adsorbents available for the confiscation of mercury from water and wastewater, chitosan has gained considerable attention. Chitosan derivatives produced by grafting, polymerization, crosslinking, N/O substitution have received significant attraction during recent years. There has been an increasing trend to produce variety of derivatized chitosan to improve the adsorption properties. This chapter briefly reviews the current researches on the applications of chitosan-based adsorbents for the elimination Hg(II). It is apparent from the literature that chitosan and its derivatives provide a better opportunity for scientists for the effective elimination of toxic Hg(II) from wastewater.

Keywords

Adsorption, Chitosan, Chitosan Derivatives, Wastewater Treatment, Mercury, Hg(II)

Contents

1.   Introduction

Environmental pollution is one of the utmost issues that the world is facing. With the intensification of industrialization and modernization in our lives, pollution has reached its topmost which is causing grave and irreparable destruction to the natural world and human society [1,2]. Besides other prerequisites, water is a precious natural resource without which life is dreadful. The ever growing demand of water in agricultural, domestic and industrial sectors lead to the generation of high volume of polluted water containing dangerous and virulent contaminants (like heavy metals, dyes, detergents, pesticides, etc.) which causes environmental pollution and their toxicity leads to major problems around the globe [3–8]. Toxic heavy metals such as mercury, zinc, nickel, arsenic, cadmium, lead, chromium, etc., are discharged from the industries such as pharmaceutical, leather, paint, automobiles, chemical, cosmetic, printing, paper, polymer, etc. and finds its way to the water bodies [9–12]. The industrial release of untreated metal contaminated effluent directly into the natural water systems affects the aquatic life and the food chain [13]. Most of the metals can cause serious health ailments and can be lethal to living organism. Among the variety of heavy metals, Hg(II) is highly toxic and severely affect the human health and aqueous system [14]. It remains in the system and causes long-term pollution complications. The major sources of Hg emission are volcanic eruption, electrical and electronic manufacturing industries, burning of fossil fuels, chlor-alkali manufacturing plants, sulfide ore roasting operations, etc. [15–20]. Hg(II) in water may transform chemically or biologically to its organic form known as methylmercury. It is more poisonous than Hg(II) and bio-accumulated in the environment. A distinct nature of mercury is its firm holding to living tissues and slow removal from the living system. When Hg reaches the bloodstream of the human body, it oxidizes into the reactive Hg(II), and easily penetrates into the membranes of the cells and causes damage to the neural and cardiovascular system. It is carcinogenic, teratogenic and mutagenic. The toxicity of Hg(II) ions, even at remarkably low concentrations, can cause several serious human health problems, including blindness, mental illness, involuntary mobilization, unconsciousness, emotional deterioration, etc. The photochemical oxidation of Hg into

Chitosan-Based Adsorbents for Wastewater Treatment, Ed. Abu Nasar   Materials Research Forum LLC
Materials Research Foundations **34** (2018) 29-56   doi: http://dx.doi.org/10.21741/9781945291753-2

the reactive Hg^{2+} lead to the accumulation on soil, rivers, lakes and sea, and bacterial reduction of mercury into the methylated mercury resulting in mercury concentration in fish and humans [21–29]. From the adverse conditions caused by the release of mercury through untreated effluent directly to the environment and for the concern of human health lead to the need of wastewater treatment technologies. The various wastewater treatment methods such as electrodialysis, photocatalysis, electroflotation, electrokinetics, coagulation and flocculation, nanofiltration, reverse osmosis, chemical method, ozonation, chemical precipitation, photochemical, electrochemical, and biological method are used to remove different dangerous and virulent pollutants (dyes, heavy metals, pesticides and other organic matters) from water and wastewater [30–39]. However, among the different available techniques, adsorption method was recognized to be one of the highly effective, simple, sustainable and cost-effective wastewater treatment techniques to reduce hazardous pollutants present in the effluent [40].

2.   Adsorptive treatment of Hg(II) contaminated wastewater

The choice of a suitable adsorbent is one of the most important tasks for the confiscation of the different contaminants from polluted water. An ideal adsorbent should possess high porosity and surface area, greater adsorption capacity, easy availability, mechanical stability, economic feasibility, ease of regeneration, eco-friendly, compatibility, and excellent selectivity. A variety of adsorbents have been employed which are derived from agricultural, industrial, and domestic wastes such as acrylamide [41], alginate [42], almond shell [43], bamboo leaf [44], banana peel [45], cellulose [46,47], citrus limetta peel [48], clays and clay minerals [49,50], cucumis sativus peel [51], cyclodextrin [52], fly ash [53], kaolinite [54], microbes [55], orange peel [56], polyaniline [57], polystyrene [58], polyvinyl alcohol [59], punica granatum [60], rubber [61], starch [62], saw dust [63,64], sunflower biomass [65], tomato waste [66], vermiculite [67], walnut shell [68], zeolites [69], etc. During recent years there has been increasing trends to develop adsorbents from natural polymeric materials. It has been observed that the chitosan-based adsorbents have gained prominent attention for their use as potentially effective adsorbents. The objective of the present study is to review the current researches on the chitosan-based adsorbents employed for the elimination of water contaminants in general with the greater emphasis on mercury.

3.   Chitosan as water decontaminant

3.1   Production and properties

Chitosan is a linear polysaccharide and found from partial N-deacetylation of chitin which is widely distributed in nature. Chitin is the second most plentiful natural biopolymer after cellulose and is made of (1-4)-linked-2-acetamido-2-deoxy-β-D-glucopyranose units [70]. It is biorenewable, biocompatible, eco-friendly, biodegradable and has chelating properties. Chitin is found in natural sources such as squid, fungi, insects, algae and mainly from the shells of mollusks and exoskeleton of crustacean sources (shrimp, crab, lobster, and crayfish) [71]. Shukla et al. [72] described that the crustacean shells were made of calcium carbonate (30-50%), protein (30-40%) and chitin (20-30%). The processes generally adopted for the preparation of chitosan from crustacean source is shown in Fig. 1.

Chitosan is a non-toxic crystalline polymer. The crystalline nature of chitosan ascribed to extensive inter- and intra-molecular hydrogen bonding between the chains and sheets. It is primarily characterized by its degree of deacetylation and molecular weight. Deacetylation comprises the elimination of acetyl groups from the molecular framework of chitin, which gives chitosan. Because of the highly reactive amino group, chitosan has a high degree of deacetylation. Commercially available chitosan is more than 85% deacetylated. The degree of deacetylation is the key factor to distinguish between the chitin and chitosan as it decides the content of free amino groups in the two polysaccharides [73,74].

3.2   Chitosan as an adsorbent

Chitosan and its derivatives have appealed significant consideration from multidisciplinary research groups and found applications in agricultural, cosmetics, environmental remediation, biomedical and pharmaceutical purposes and also contribute as a part of our food supply. However, presently available usages are still less as compared to its potential, as chitosan has a variety of application areas. It is a well-known biosorbent used for the confiscation of heavy metals, dyes, and other contaminants. The high adsorption ability of chitosan and its derivatives is because of the availability of a large number of hydroxyl and amino groups, which resulted in high chemical reactivity, high selectivity, excellent ion exchange pathway, and chelation properties. Many physical and chemical approaches have been employed for the modification of chitosan to increase its adsorption capacity. Chitosan is used as an adsorbent in the form of powder, beads, flakes, sponge, fibers, membranes, gels, etc. for the efficient confiscation of heavy metals from water and wastewater [75,76]. However, the low porosity and small surface area of flakes and powder forms of chitosan make them unsuitable for an adsorption process. Chitosan flakes modified into beads and gels allow the expansion of polymer network and improvement in adsorption sites hence shows better adsorption [77]. The structural modifications of chitosan have been frequently conducted to increase the metal adsorption property and stability in water and acidic medium [78].

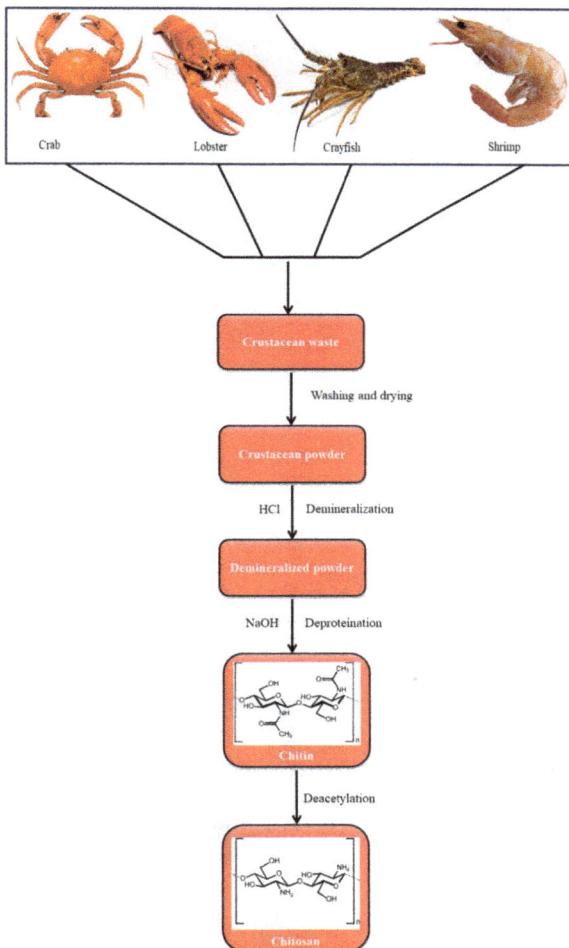

Figure 1 General processes for the chitosan production.

Chitosan-Based Adsorbents for Wastewater Treatment, Ed. Abu Nasar   Materials Research Forum LLC
Materials Research Foundations **34** (2018) 29-56   doi: http://dx.doi.org/10.21741/9781945291753-2

3.3   Derivatization of chitosan

The insolubility of chitosan in many solvents due to its crystalline structure restricts its application in many areas. To overcome this drawback there has been a rising trend in its structural modification by derivatization. Some frequently used methods for the modification of chitosan are illustrated in Fig. 2.

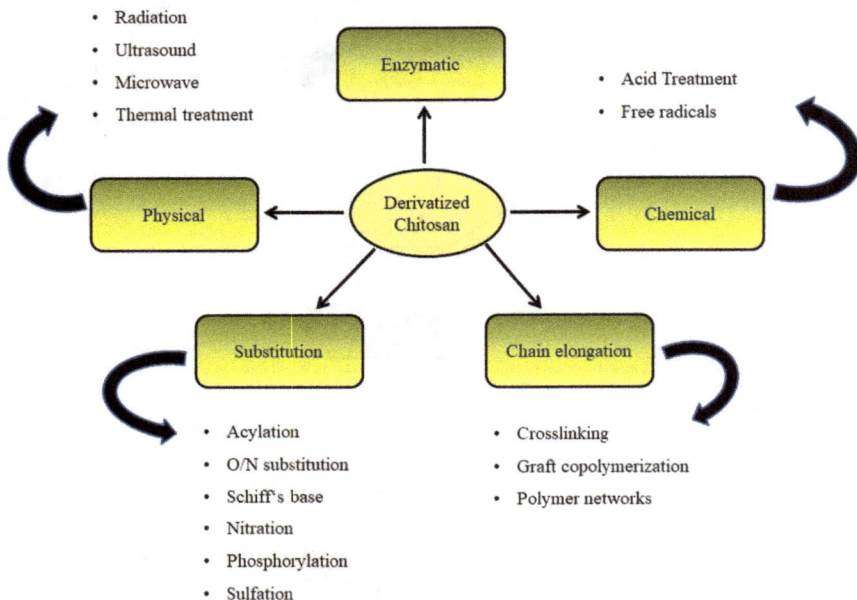

Figure 2.   Multifaceted derivatization potential of chitosan.

In order to improve the adsorption characteristics, a large number of chitosan derivatives are being synthesized by insertion of the small functional group to the structure of chitosan. Among the different available methods, graft copolymerization has been used extensively. The grafting of functional groups onto the chitosan backbone permits the formation of functional derivatives. Chitosan has two types of reactive groups that can be grafted easily. First the free amine groups in the C_2 position of deacetylated units and secondly the hydroxyl groups onto the C_3 and C_6 position of acetylated or deacetylated units. The properties like chelating and adsorption capacity of chitosan can be much-enhanced by grafting. Further, the nature of the resulted graft copolymers is influenced by many factors such as the characteristic of the side chain, molecular structure, chain length, and their numbers [79]. Another modification method of chitosan is the substitution of N/O groups [80,81]. In N-substitution the designed groups react with the amino groups of the chitosan and O-substitution involves the reaction between designed small molecules and hydroxyl groups of chitosan. Chitosan is soluble in acidic medium and to make the chitosan insoluble in acidic medium, cross-linking agents are used. The cross-linkers can be of variable length and comprise the different functional groups than those participate in cross-linking. A cross-linking step is essential to strengthen the chemical stability of the chitosan in an acidic medium. Although cross-linking reduces the removal capacity of chitosan it enhances the resistance to alkali, acid and other chemicals, and can change the crystalline nature of chitosan [82].

4.   Adsorptive removal of Hg(II) from water and wastewater by chitosan

Many investigators have explored the removal efficiency of various forms of chitosan. It has been used by many investigators as an adsorbent for Hg(II) removal from contaminated water. The equilibrium measurement of the sorption of Hg(II) from industrial effluents by chitosan was conducted by Shafaei et al. [83]. The researchers reported that the process obeyed the Langmuir adsorption model with a very high adsorption capacity of 1127.1 mg/g. They also reported that the degree of Hg(II) ions uptake improved with the increase in pH and decrease in the adsorbent particle size. Gamage and Shahidi [84] obtained the chitosan from the waste of crab and studied the adsorption behavior of Hg(II) onto the different molecular weight of chitosan. Three different type of chitosan were obtained by varying the deacetylation time (type I: 20h with 91.30% of the degree of deacetylation, type II: 10 h with 89.30% of the degree of deacetylation, type III: 4h with 86.40% of the degree of deacetylation). They observed that Hg(II) shows the different chelation capacity, viz., 95.8 for type I, 90.7 for type II and 89.8 for type III at pH 5, 6 and 7, respectively. Chitosan flakes with an average length ranging from 0.35 to 0.45 mm, obtained from red lobster shells with 20% degree of deacetylation were utilized for the adsorptive treatment of Hg(II) [85]. It was reported that the adsorption capacity for Hg(II) increases from 361.1 mg/g at pH 2.5 to 461.4 mg/g at pH 4.5. The isotherm model was best explained by Langmuir model. Chitosan membrane prepared by overnight stirring with CH_3COOH was utilized to examine the kinetics associated with the adsorption of Hg(II) ions [86]. It has been reported that the mass transfer of Hg(II) into the chitosan membrane increases with the rise of temperature. Thien et al. [87] prepared deacetylated chitosan for adsorptive removal of Hg(II) ion from contaminated water. Their isotherm results were best obeyed with the Langmuir adsorption model and the adsorption capacity was significantly affected by the change in pH and temperature. Further, their time-dependent results indicated the validity of pseudo-second-order kinetics. A chitosan of a sponge-like structure produced by freeze-drying technique was used for the adsorption of Hg(II) [88]. Equilibrium data were best obeyed by the Langmuir isotherm. Based on the adsorption capacity, it was observed that a chitosan flake was much more efficient than foam. At pH 4, the adsorption capacity onto the flakes was found up to 850 mg/g and only 350 mg/g onto the chitosan foam. A two-region model was demonstrated for examining the adsorption capacity of the foam. Pseudo-second-order kinetic and cylindrical diffusion models were reported to be followed for short and end sorption times, respectively. Some representative studies on the adsorptive removal of Hg(II) by the unmodified chitosan are presented in Table 1.

Table 1 Adsorption of Hg(II) from water and wastewater by unmodified chitosan.

Chitosan source	Adsorption capacity (mg/g)	Physical form	pH	Isotherm	Ref.
Commercial chitosan	1127.1	Particles	6	Langmuir	[83]
Crab discards	-	Particles	5.7	Langmuir	[84]
Red lobster shells	361.1/461.4	Flakes	2.5/4.5	Langmuir	[85]
Shrimp shells	-	membrane	6	Langmuir	[86]
Squid	1016.99	-	6-6.5	Langmuir	[87]
Crustacean shell	850/350	Flakes/cylindrical form	4	Langmuir	[88]
Commercial chitosan	25.3		6	Langmuir	[89]
Chitosan	454		4.5	Langmuir	[90]

5.   Adsorptive removal of Hg(II) from water and wastewater by chitosan derivatives

Chitosan can be cross-linked with ethylene glycol diglycidyl ether, glyoxal, benzoquinone, glutaraldehyde, epichlorohydrin, cyclodextrin etc to increase its chemical stability in acidic solution. Adsorption ability of chitosan and cross-linked chitosan with glutaraldehyde and epichlorohydrin to remove Hg(II) at different experimental conditions such as initial concentration of Hg(II), pH, crosslinking agent has been studied [89]. The adsorption capacity of chitosan (25.3 mg/g) was improved on crosslinking with epichlorohydrin (30.3 mg/g) and glutaraldehyde (75.5 mg/g). The experimental results were best followed by Langmuir adsorption model. A new chelating chitosan microsphere was prepared in which epichlorohydrin cross-linked chitosan was modified by 2-(chloromethyl) benzimidazole and successfully exploited for the elimination of Hg(II) [91]. This adsorbent was reported to have high selectivity toward Hg(II) and the isotherm equilibrium data followed Langmuir model. Jeon and Höll [92], after studying a number of different chemical modifications, reported that the aminated chitosan bead cross-linked with ethylenediamine had an adsorption capacity of 461.36 mg/g for Hg(II) at pH 7. The chitosan/amine and chitosan/azole resins were prepared by performing the chemical modification to glutaraldehyde cross-linked chitosan resin by ethylenediamine and 3-amino-1,2,4-triazole-5-thiol, respectively [93]. The chitosan/amine resin showed 461.4 mg/g uptake capacity and chitosan/azole showed 443.3 mg/g towards Hg(II). The higher uptake of chitosan/amine resin was ascribed to the fact that the amine active sites furnish effective attachment to the Hg(II) compared to the azole moiety. The adsorption performance of barbital-glutaraldehyde cross-linked and glutaraldehyde cross-linked chitosans was studied for the removal of Hg(II), $CH_3Hg(II)$ and $C_6H_5Hg(II)$ [94]. Adsorption of these forms of mercury exhibited similar nature with the exception of thermodynamic behavior as it was endothermic for Hg(II) and exothermic for $CH_3Hg(II)$ and $C_6H_5Hg(II)$. Glutaraldehyde cross-linked chitosan exhibited higher adsorption capacity as compared to barbital-glutaraldehyde cross-linked chitosan. Rocha et al. [95] have developed an easy and efficient chitosan-based film for the confiscation of Hg from aqueous solutions and studied the efficiency of chitosan cross-linked with genipin, and genipin grafted with caffeic acid for water decontamination. It was reported that the chitosan cross-linked with genipin and genipin cross-linked/caffeic acid grafted chitosan were effective for the selective adsorption of Hg(II) coexisting with cadmium and lead. The removal capacity was reported to be much higher for genipin cross-linked/caffeic acid grafted chitosan film in comparison with the ungrafted one.

The complications regarding separation and recovery of powdered chitosan can be easily overcome by cross-linking it with magnetic materials. Kyzas and Deliyanni [96] studied the adsorptive treatment of Hg(II) by chitosan cross-linked with glutaraldehyde and functionalized with magnetic nanoparticles. It was reported that magnetically functionalized cross-linked chitosan showed faster adsorption of Hg(II) than cross-linked chitosan. The adsorption capacity was observed to be increased by enhancing the temperature. A novel chitosan magnetic adsorbent (ethylenediamine modified chitosan magnetic microspheres) was prepared with chemical cross-linking and seed swelling method and successfully utilized for the adsorption of Hg^{2+} and UO_2^{2+} [97]. The selective separation of these ions was achieved at pH < 2.5. They also established that the adsorption capacity increases with the increase of pH. In another study, the magnetic chitosan microsphere was crosslinked with glutaraldehyde and grafting of sulfur were carried out via epichlorohydrin as a crosslinker for the synthesis of thiourea-modified magnetic chitosan microsphere [98]. This adsorbent was exploited for the very efficient elimination of Hg(II) with a very high monolayer adsorption capacity of 625.2 mg/g. In a notable work of Zhou et al. [99], elimination of Hg(II) was demonstrated with ethylenediamine modified magnetic crosslinking chitosan microspheres. The equilibrium data followed Langmuir adsorption model with monolayer adsorption capacity of 539.59 mg/g for Hg(II) at pH 5. Monier and Abdel-Latif [100] synthesized cross-linked magnetic chitosan-phenylthiourea resin for the confiscation of Hg(II) from aqueous solutions. They observed that the modification of the cross-linked chitosan resin by insertion of phenylthiourea moieties significantly increase the Hg(II) uptake. The potential of Hg(II) removal via chitosan–thioglyceraldehyde Schiff's base cross-linked magnetic resin and non-functionalized cross-linked magnetic chitosan was evaluated [101]. The uptake of Hg(II) was higher in case of chitosan–thioglyceraldehyde Schiff's base cross-linked magnetic resin. This was attributed due to the more accessibility of the active functional groups of the inserted thioglyceraldehyde moieties which are capable to chelate the Hg(II). The adsorption capacity of 561.652 for Hg(II) was obtained using the magnetic chitosan resins modified with Schiff's base derived from glutaraldehyde and thiourea [102]. The highest uptake capacity for Hg(II) was reported at pH 5. The higher affinity of Hg(II) to the active sites of resin was suggested to be due to the soft base nature of sulfur atom which is responsible for Hg(II) interaction. A multi-cyanogunidine modified magnetic chitosan produced by employing the functionalized chitosan and cross-linked with cyanoguanidine was reported [103]. The obtained adsorbent was observed to have high selectivity and effective adsorption for Hg(II). On the basis of theoretical modeling, it has been established that adsorption was best followed by Langmuir adsorption and pseudo-second-order kinetics models. An effective and efficient chitosan coated Fe_3O_4 nanoparticles were prepared by the covalent linking of carboxymethyl via carbodiimide activation [104]. The chitosan-bonded magnetic nanoparticles were observed to be the fairly effective removal of Hg(II). The removal efficiency was achieved to 92.4% within 5 min with high adsorption capacities.

Cardenas et al. [90] synthesized chitosan mercaptanes derivatives utilizing mercaptoacetic acid and 1-chloro-2,3-epoxy propane propionic acid. The adsorption capacity was found to be 454, 435, 588, 164 for the adsorption of Hg(II) by chitosan, *N*-(2-hydroxy-3-methyl aminopropyl)chitosan, *N*-(2-hydroxy-3-mercapto propyl)-chitosan, and 6-*O*-(mercaptoacetate-*N*-mercaptoacetyl)-chitosan, respectively at pH 4.5. The comparatively lower removal capacity for Hg(II) onto the 6-*O*-mercaptoacetate-*N*-

mercaptoacetyl)-chitosan was due to the steric hindrance. The adsorption of Hg(II) onto the chitosan mercaptanes derivatives was observed to be best followed by Langmuir adsorption isotherm.

For the sake of higher affinity of Hg(II) towards the beads of chitosan, its porosity and chemical stability was increased by crosslinking with glutaraldehyde and grafting with cysteine [105]. A very high observed adsorption capacity of 1604.7 mg/g at pH 7 was reported. Sulfur-introduced chitosan hydrogel was well prepared by grafting dimethyl 3,3'- dithiodipropionate onto chitosan and then crosslinked with N,N'-methylene diacrylamide [106]. SEM analysis revealed that the sulfur-modified chitosan hydrogel had much more pores and larger specific surface area than chitosan hydrogel. The experimental results indicated that the sulfur-modified chitosan hydrogel exhibited a noticeable enhancement in the adsorption capacity for Hg(II) removal. The selective and higher adsorption capacity for Hg(II) was suggested to be due to its capability of forming covalent bonds with amide groups grafted onto the surface of the chitosan beads. A number of insoluble chitosan derivatives were synthesized by grafting ester- and amino-terminated dendrimer-like polyamidoamine into chitosan and used them for studying their adsorption competencies for Ag^+, Au^{3+}, Cd^{2+}, Cu^{2+}, Hg^{2+}, Ni^{2+}, Pd^{2+}, Pt^{4+}, and Zn^{2+} [107]. It was reported that these chitosan derivatives showed higher adsorption potentials for Au^{3+} and Hg^{2+} than for other metal ions. Further adsorption potential of ester-terminated derivatives was reported to lower than that of amino-terminated products. A thiocarbamoyl derivative was synthesized by the grafting of thiourea on chitosan backbone and demonstrated as a very efficient material for the sorption of Hg(II) in acidic solutions [108]. The researchers observed that the adsorption of Hg(II) in acidic solutions was not influenced by the existence of competitor metals like Cd(II), Cu(II), Ni(II), Pb(II), Zn(II) and nitrate anions in acidic medium. A biodegradable magnetic composite microsphere having glutamine modified chitosan and silica-coated Fe_3O_4 nanoparticles was prepared and utilized for the confiscation of Hg(II) and acid green 25 (amphoteric dye) from water [109]. It has been reported that the glutamine modified adsorbent has higher adsorption capacity for both the contaminants than the unmodified one.

Chitosan poly(vinyl alcohol) hydrogel prepared by cross-linking with glutaraldehyde in association with an alternate freeze-thawed method was developed and utilized for the Hg(II) removal [110,111]. Based on the comparative adsorption studies it has been reported that this product has better adsorption capacity and selectivity for Hg(II). The adsorption capacity for Hg(II) ions was observed to be 585.9 mg/g while the selectivity coefficient of the hydrogel for Hg(II) ions was obtained to be 487.7, 36642.5 and 284298.5 times higher than that for Cu(II) ions, Pb(II) ions and Cd(II) ions, respectively.

In another study, a nanocomposite adsorbent based on chitosan-poly(vinyl alcohol)/bentonite with high adsorption selectivity for Hg(II) ions was also synthesized [112]. It has been observed that bentonite content has a large effect on the microstructure of the nanocomposites and the adsorption capacities for Hg(II) ions having bentonite content of 0%, 10%, 30% and 50% were 460.18, 455.12, 392.19 and 360.73 mg/g, respectively. These values are far greater than that for Cu(II), Cd(II) and Pb(II) ions. This suggests that bentonite is a potential candidate for improving the adsorption selectivity of such nanocomposites for Hg(II). Chang et al. [113] used glutaraldehyde crosslinked chitosan for the removal Hg^{2+} ions from aqueous solution and reported a very high maximum adsorption capacity of 667 mg/g which was over 20 times greater than that of plain alginate bead. An intraparticle diffusion model was observed to be the rate-determining step. Decreasing the bead size of the glutaraldehyde-cross-linked alginate gel containing chitosan resulted in a fast increase in the initial adsorption of Hg(II). Dubey et al. [114] synthesized highly porous novel chitosan-alginate nanoparticles in which tri-polyphosphate and calcium chloride were used as a crosslinker. They suggested that Hg(II) is coordinated with the amine group of the chitosan and bounded electrostatically with the negatively charged chain of alginate.

A composite membrane, Procion Brown MX 5BR immobilized poly(hydroxyethylmethacrylate/chitosan) was used for the elimination of different heavy metals and the system was observed to be significantly selective for Hg(II) ions [115]. Poly(itaconic acid) grafted glutaraldehyde cross-linked chitosan nanoadsorbent was prepared as a nanoadsorbent with an average size of 52.6 nm and utilized for the removal of Hg^{2+} and Pb^{2+} [116]. This nanoadsorbent was characterized by FT-IR, XRD, SEM, TEM, and TGA. The removal process was observed to be best-obeyed to pseudo-second-order kinetics and Langmuir isotherm models. The maximum uptakes of Hg^{2+} and Pb^{2+} based on Langmuir isotherm were 870.1 and 1320 mg/g, respectively. Hg(II) adsorption behavior on carboxymethyl chitosan–hemicellulose resin synthesized by the thermal cross-linking process was reported [117]. The researcher investigated the influence of different batch parameters and observed a value of 28.2 mg/g for the maximum adsorption capacity. Salahi et al. [118] investigated the adsorption of Hg(II) on biocompatible polymeric polypyrrole chitosan nanocomposite. Langmuir adsorption isotherm was observed to be the best fitted model. The maximum monolayer adsorption capacity was 40 mg/g. The possible reactions responsible for effective adsorption of Hg(II) are anion exchange, chemical oxidation, and chelation. A cheap, non-hazardous, biodegradable cerium functionalized polyvinyl(alcohol) chitosan composite nanofiber had been prepared by electrospinning technique and utilized for effective adsorption of Hg(II) in water [119]. Functionalization of polyvinyl(alcohol) chitosan by cerium was found to increase the removal, as small ionic radii, high electric charge and high potential energy of cerium support the adsorption of Hg(II). The maximum adsorption efficiency (31.44 mg/g) for Hg (II) was observed at 3.5% cerium content in polyvinyl(alcohol) chitosan composite nanofibers in pH range of 5.3-6. The adsorption of Hg(II) was observed to follow the pseudo-second-order kinetics. Ge and Hua [120] prepared poly(maleic acid) grafted cross-linked chitosan nanoadsorbent for the elimination of Hg(II). The experimental data showed that the poly(maleic acid) grafted cross-linked chitosan nanomaterial had a higher selectivity for Hg(II) with an adsorption capacity of 1044 mg/g at pH 6. The adsorptive removal followed Langmuir isotherm and pseudo-second-order kinetic models. The increase of temperature favored the adsorption of Hg(II). The chitosan-based granular adsorbent consisting of carboxy and thiourea groups was synthesized by the redox-initiated polymerization in which ascorbic acid/H_2O_2 was used as a redox initiator for the adsorption of Hg(II) [121]. In this study, Langmuir adsorption model was observed to be the best-fitted isotherm with obtained maximum monolayer adsorption capacity of 1106.7 mg/g. The adsorption of Hg(II) was best obeyed by the pseudo-second-order kinetic model.

The adsorption characteristics of Hg(II) onto the different derivatized chitosan have been summarized in Table 2. It is significant to note that the experimental conditions such as adsorbent dose, temperature, the particle size of the adsorbent, the concentration of the adsorbate, ionic strength, the effect of competitive ions etc. greatly influenced the maximum adsorption capacities of the adsorbents.

Table 2 Adsorptive removal of Hg(II) from water and wastewater by modified chitosan.

Chitosan derivatives	Adsorption capacity (mg/g)	Physical forms	pH	Isotherm	References
Epichlorohydrin-crosslinked chitosan membranes	30.3	Membrane	6	Langmuir	[89]
Glutaraldehyde-crosslinked chitosan	75.5	Membrane	6	Langmuir	[89]
2-(chloromethyl) benzimidazole/ epicholorohydrin crosslinked chitosan	257.8	Microsphere	4.5	Langmuir	[91]
Aminated chitosan crosslinked	461.4	Beads	7	Langmuir	[92]

with ethylenediamine					
Chitosan/amine crosslinked with glutaraldehyde	461.4	Resins	5.1	Langmuir	[93]
Chitosan/azole crosslinked with glutaraldehyde	443.3	Resins	5.1	Langmuir	[93]
Chitosan cross-linked with genipin	2.2	Film	6.5	Sips	[95]
Chitosan cross-linked with genipin and grafted with caffeic acid	4	Film	6.5	Sips	[95]
Chitosan cross-linked with glutaraldehyde and functionalized with magnetic nanoparticles	145		5	Langmuir	[96]
Chitosan cross-linked with glutaraldehyde and functionalized with Fe_3O_4 nanoparticles	152		5	Langmuir	[96]
Chitosan crosslinked with ethylenediamine functionalized with magnetic microsphere	455.3	microsphere		Langmuir	[97]
Thiourea-modified magnetic chitosan microsphere	625.2	microsphere	5	Langmuir	[98]
Ethylenediamine modified magnetic crosslinking chitosan microspheres	539.6	Microsphere	5	Langmuir	[99]
Cross-linked magnetic chitosan-phenylthiourea resin	135	Resin	5	Langmuir	[100]
Chitosan–thioglyceraldehyde Schiff's base cross-linked with formaldehyde magnetic resin	98	Resin	5	Langmuir	[101]
Chitosan resins modified with schiff's base derived from thiourea and glutaraldehyde	561.6	Resin	5	Langmuir	[102]
Multi-cyanogunidine modified	285		7	Langmuir	[103]

magnetic chitosan					
Chitosan-coated Fe_3O_4 nanoparticles	10		3	Langmuir	[104]
N-(2-hydroxy-3-methyl aminopropyl)chitosan	435		4.5	Langmuir	[90]
N-(2-hydroxy-3-mercapto propyl)-chitosan	588		4.5	Langmuir	[90]
6-O-(mercaptoacetate-N-mercaptoacetyl)-chitosan	164		4.5	Langmuir	[90]
Chitosan crosslinked with glutaraldehyde and grafted with cysteine	1604.7	Beads	7	Freundlich	[105]
Sulfur-modified chitosan hydrogel	187.5		5		[106]
Thiourea grafted chitosan	450		2	Langmuir	[108]
Glutamine modified chitosan and silica coated Fe_3O_4 nanoparticles	141	microsphere	6	Langmuir	[109]
Chitosan magnetic composite microsphere	73.3	microsphere	6	Langmuir	[109]
Chitosan poly(vinyl alcohol) hydrogel	585.9		5.85		[110]
Chitosan poly(vinyl alcohol)/bentonite (50%) nanocomposites	360.7		5.5		[112]
Glutaraldehyde crosslinked chitosan immobilized with alginate	667	Beads	5	Langmuir	[113]
Chitosan-alginate nanoparticles	217.4		5	Langmuir	[114]
Procion Brown MX 5BR immobilized poly(hydroxyethyl methacrylate) chitosan	68.8	Membrane	5		[115]
Poly(itaconic acid) grafted glutaraldehyde cross-linked	870.1		6	Langmuir	[116]

chitosan nanoadsorbent					
Carboxymethyl chitosan–hemicellulose resin	28.2	Resin	4	Langmuir	[117]
Polypyrrole chitosan nanocomposite	40			Langmuir	[118]
Cerium functionalized polyvinyl(alcohol) chitosan composite nanofiber	31.4		5.3-6	Langmuir	[119]
Poly(maleic acid) grafted crosslinked chitosan nanomaterial	1044		6	Langmuir	[120]
Chitosan based granular adsorbent	1106.7		5	Langmuir	[121]

6.   Conclusion

The pollution of water by heavy metals has become a worldwide environmental issue and therefore treatment of contaminated industrial wastewater is essential before its disposal. The choice of an appropriate adsorbent for the confiscation of contaminants from aqueous media has always been a challenging task. In this context chitosan and its derivatives are becoming potential substitutes to the conventional and expensive adsorbents for the removal of heavy metals from wastewater. Chitosan-based adsorbents have gained noticeable attention by researchers due to their specific features such as abundance, low-cost, effective adsorption ability, biocompatibility, biodegradability, and ease of structural modification. It has been commonly reported that chitosan-based adsorbents have high selectivity for Hg(II) in presence of other heavy metals. In the present work, the adsorptive removal of Hg(II) by using unmodified and modified chitosan-based adsorbents has been briefly reviewed. At present, there is a large scope to develop the different kinds of adsorbents by the derivatization of chitosan and also to examine their suitability for the treatment of polluted wastewater.

References

[1]   S. Ahuti, Industrial Growth and Environmental Degredation, International Education and Research Journal. 1 (2015) 5–7.

[2]   J. Cherniwchan, Economic growth, industrialization, and the environment, Resource and Energy Economics. 34 (2012) 442–467. https://doi.org/10.1016/j.reseneeco.2012.04.004.

[3]   L. Jarup, Hazards of heavy metal contamination, British Medical Bulletin. 68 (2003) 167–182. https://doi.org/10.1093/bmb/ldg032.

[4]   D. Kar, P. Sur, S.K. Mandai, T. Saha, R.K. Kole, Assessment of heavy metal pollution in surface water, International Journal of Environmental Science {&} Technology. 5 (2008) 119–124. https://doi.org/10.1007/BF03326004.

[5]   J.P. Jadhav, D.C. Kalyani, A.A. Telke, S.S. Phugare, S.P. Govindwar, Evaluation of the efficacy of a bacterial consortium for the removal of color, reduction of heavy metals, and toxicity from textile dye effluent, Bioresource Technology. 101 (2010) 165–173. https://doi.org/10.1016/j.biortech.2009.08.027.

[6]   R.P. Schwarzenbach, T. Egli, T.B. Hofstetter, U. von Gunten, B. Wehrli, Global Water Pollution and Human Health, Annual Review of Environment and Resources. 35 (2010) 109–136. https://doi.org/10.1146/annurev-environ-100809-125342.

[7]   Qamruzzaman, A. Nasar, Degradation of acephate by colloidal manganese dioxide in the absence and presence of surfactants, Desalination and Water Treatment. 55 (2015) 2155–2164. https://doi.org/10.1080/19443994.2014.937752.

[8]   A. Nasar, S. Shakoor, Remediation of dyes from industrial wastewater using low-cost adsorbents, in: Inamuddin, A. Al-Ahmed (Eds.), Applications of Adsorption and Ion Exchange Chromatography in Waste Water Treatment, Materials Research Forum LLC, 2017: pp. 1–33. https://doi.org/10.21741/9781945291333-1.

[9]   R.S. Rana, V. Kandari, P. Singh, S. Gupta, Assessment of Heavy Metals in Pharmaceutical Industrial Wastewater of Pharmacity, Selaqui, Dehradun, Uttarakhand, India, Analytical Chemistry Letters. 4 (2014) 29–39. https://doi.org/10.1080/22297928.2014.890772.

[10] M. Junaid, M.Z. Hashmi, Y.-M. Tang, R.N. Malik, D.-S. Pei, Potential health risk of heavy metals in the leather manufacturing industries in Sialkot, Pakistan, Scientific Reports. 7 (2017) 8848. https://doi.org/10.1038/s41598-017-09075-7.

[11] M. Malakootian, J. Nouri, H. Hossaini, Removal of heavy metals from paint industry's wastewater using Leca as an available adsorbent, International Journal of Environmental Science & Technology. 6 (2009) 183–190. https://doi.org/10.1007/BF03327620.

[12] K.G. Akpomie, F.A. Dawodu, Treatment of an automobile effluent from heavy metals contamination by an eco-friendly montmorillonite, Journal of Advanced Research. 6 (2015) 1003–1013. https://doi.org/10.1016/j.jare.2014.12.004.

[13] P. V. Hodson, The effect of metal metabolism on uptake, disposition and toxicity in fish, Aquatic Toxicology. 11 (1988) 3–18. https://doi.org/10.1016/0166-445X(88)90003-3.

[14] K.M. Rice, E.M. Walker, M. Wu, C. Gillette, E.R. Blough, Environmental Mercury and Its Toxic Effects, Journal of Preventive Medicine & Public Health. 47 (2014) 74–83. https://doi.org/10.3961/jpmph.2014.47.2.74.

[15] E.G. Pacyna, J.M. Pacyna, Global Emission of Mercury from Anthropogenic Sources in 1995, Water, Air, and Soil Pollution. 137 (2002) 149–165. https://doi.org/10.1023/A:1015502430561.

[16] O.I. Joensuu, Fossil Fuels as a Source of Mercury Pollution, Science. 172 (1971) 1027–1028. https://doi.org/10.1126/science.172.3987.1027.

[17] J. Nriagu, C. Becker, Volcanic emissions of mercury to the atmosphere: global and regional inventories, Science of The Total Environment. 304 (2003) 3–12. https://doi.org/10.1016/S0048-9697(02)00552-1.

[18] R. Dufault, B. LeBlanc, R. Schnoll, C. Cornett, L. Schweitzer, D. Wallinga, J. Hightower, L. Patrick, W.J. Lukiw, Mercury from chlor-alkali plants: measured concentrations in food product sugar, Environmental Health. 8 (2009) 2. https://doi.org/10.1186/1476-069X-8-2.

[19] L.D. Hylander, R.B. Herbert, Global Emission and Production of Mercury during the Pyrometallurgical Extraction of Nonferrous Sulfide Ores, Environmental Science & Technology. 42 (2008) 5971–5977. https://doi.org/10.1021/es800495g.

[20] J. Wang, X. Feng, C.W.N. Anderson, Y. Xing, L. Shang, Remediation of mercury contaminated sites – A review, Journal of Hazardous Materials. 221–222 (2012) 1–18. https://doi.org/10.1016/j.jhazmat.2012.04.035.

[21] R.A. Bernhoft, Mercury Toxicity and Treatment: A Review of the Literature, Journal of Environmental and Public Health. 2012 (2012) 1–10. https://doi.org/10.1155/2012/460508.

[22] N. Langford, R. Ferner, Toxicity of mercury, Journal of Human Hypertension. 13 (1999) 651–656. https://doi.org/10.1038/sj.jhh.1000896.

[23] M.E. Crespo-López, G.L. Macêdo, S.I.D. Pereira, G.P.F. Arrifano, D.L.W. Picanço-Diniz, J.L.M. do Nascimento, A.M. Herculano, Mercury and human genotoxicity: Critical considerations and possible molecular mechanisms, Pharmacological Research. 60 (2009) 212–220. https://doi.org/10.1016/j.phrs.2009.02.011.

[24] S. De Flora, C. Bennicelli, M. Bagnasco, Genotoxicity of mercury compounds. A review, Mutation Research/Reviews in Genetic Toxicology. 317 (1994) 57–79. https://doi.org/10.1016/0165-1110(94)90012-4.

[25] M.F. Wolfe, S. Schwarzbach, R.A. Sulaiman, Effects of mercury on wildlife: A comprehensive review, Environmental Toxicology and Chemistry. 17 (1998) 146–160. https://doi.org/10.1002/etc.5620170203.

[26] H.E. Ratcliffe, G.M. Swanson, L.J. Fischer, Human Exposure to Mercury: A Critical Assessment of the Evidence of Adverse Health Effects, Journal of Toxicology and Environmental Health. 49 (1996) 221–270. https://doi.org/10.1080/00984108.1996.11667600.

[27] D.W. Boening, Ecological effects, transport, and fate of mercury: a general review, Chemosphere. 40 (2000) 1335–1351. https://doi.org/10.1016/S0045-6535(99)00283-0.

[28] B. Weiss, W. Simon, Quantitative Perspectives on the Long-Term Toxicity of Methylmercury and Similar Poisons, in: Behavioral Toxicology, Springer US, Boston, MA, 1975: pp. 429–437. https://doi.org/10.1007/978-1-4684-2859-9_16.

[29] P. Miretzky, A.F. Cirelli, Hg(II) removal from water by chitosan and chitosan derivatives: A review, Journal of Hazardous Materials. 167 (2009) 10–23. https://doi.org/10.1016/j.jhazmat.2009.01.060.

[30] J. Dhote, S. Ingoleb, A. Chavhana, Review on wastewater treatment technologies, International Journal of Engineering Research & Technology (IJERT). 1 (2012).

[31] C. Zheng, L. Zhao, X. Zhou, Z. Fu, A. Li, Treatment Technologies for Organic Wastewater, in: Water Treatment, InTech, 2013. https://doi.org/10.5772/52665.

[32] E. Forgacs, T. Cserháti, G. Oros, Removal of synthetic dyes from wastewaters: a review, Environment International. 30 (2004) 953–971. https://doi.org/10.1016/j.envint.2004.02.001.

[33] T.A. Kurniawan, G.Y.S. Chan, W.-H. Lo, S. Babel, Physico–chemical treatment techniques for wastewater laden with heavy metals, Chemical Engineering Journal. 118 (2006) 83–98. https://doi.org/10.1016/j.cej.2006.01.015.

[34] F. Fu, Q. Wang, Removal of heavy metal ions from wastewaters: A review, Journal of Environmental Management. 92 (2011) 407–418. https://doi.org/10.1016/j.jenvman.2010.11.011.

[35] C.F. Carolin, P.S. Kumar, A. Saravanan, G.J. Joshiba, M. Naushad, Efficient techniques for the removal of toxic heavy metals from aquatic environment: A review, Journal of Environmental Chemical Engineering. 5 (2017) 2782–2799. https://doi.org/10.1016/j.jece.2017.05.029.

[36]   Qamruzzaman, A. Nasar, Degradation of tricyclazole by colloidal manganese dioxide in the absence and presence of surfactants, Journal of Industrial and Engineering Chemistry. 20 (2014) 897–902. https://doi.org/10.1016/j.jiec.2013.06.020.

[37]   Qamruzzaman, A. Nasar, Kinetics of metribuzin degradation by colloidal manganese dioxide in absence and presence of surfactants, Chemical Papers. 68 (2014) 65–73. https://doi.org/10.2478/s11696-013-0424-7.

[38]   Qamruzzaman, A. Nasar, Treatment of acetamiprid insecticide from artificially contaminated water by colloidal manganese dioxide in the absence and presence of surfactants, RSC Advances. 4 (2014) 62844–62850. https://doi.org/10.1039/c4ra09685a.

[39]   Qamruzzaman, A. Nasar, Degradative treatment of bispyribac sodium herbicide from synthetically contaminated water by colloidal MnO_2 dioxide in the absence and presence of surfactants, Environmental Technology. (2017) 1–7. https://doi.org/10.1080/09593330.2017.1396500.

[40]   D. Lakherwal, Adsorption of Heavy Metals: A Review, International Journal of Environmental Research and Development. 4 (2014) 41–48.

[41]   G. Manju, K. Anoop Krishnan, V. Vinod, T. Anirudhan, An investigation into the sorption of heavy metals from wastewaters by polyacrylamide-grafted iron(III) oxide, Journal of Hazardous Materials. 91 (2002) 221–238. https://doi.org/10.1016/S0304-3894(01)00392-2.

[42]   E.S. Abdel-Halim, S.S. Al-Deyab, Removal of heavy metals from their aqueous solutions through adsorption onto natural polymers, Carbohydrate Polymers. 84 (2011) 454–458. https://doi.org/10.1016/j.carbpol.2010.12.001.

[43]   N. Maaloul, P. Oulego, M. Rendueles, A. Ghorbal, M. Díaz, Novel biosorbents from almond shells: Characterization and adsorption properties modeling for Cu(II) ions from aqueous solutions, Journal of Environmental Chemical Engineering. 5 (2017) 2944–2954. https://doi.org/10.1016/j.jece.2017.05.037.

[44]   D.K. Mondal, B.K. Nandi, M.K. Purkait, Removal of mercury (II) from aqueous solution using bamboo leaf powder: Equilibrium, thermodynamic and kinetic studies, Journal of Environmental Chemical Engineering. 1 (2013) 891–898. https://doi.org/10.1016/j.jece.2013.07.034.

[45]   Y. Li, J. Liu, Q. Yuan, H. Tang, F. Yu, X. Lv, A green adsorbent derived from banana peel for highly effective removal of heavy metal ions from water, RSC Advances. 6 (2016) 45041–45048. https://doi.org/10.1039/C6RA07460J.

[46] D.W. O'Connell, C. Birkinshaw, T.F. O'Dwyer, Heavy metal adsorbents prepared from the modification of cellulose: A review, Bioresource Technology. 99 (2008) 6709–6724. https://doi.org/10.1016/j.biortech.2008.01.036.

[47] D. Zhou, L. Zhang, J. Zhou, S. Guo, Cellulose/chitin beads for adsorption of heavy metals in aqueous solution, Water Research. 38 (2004) 2643–2650. https://doi.org/10.1016/j.watres.2004.03.026.

[48] S. Shakoor, A. Nasar, Removal of methylene blue dye from artificially contaminated water using citrus limetta peel waste as a very low cost adsorbent, Journal of the Taiwan Institute of Chemical Engineers. 66 (2016) 154–163. https://doi.org/10.1016/j.jtice.2016.06.009.

[49] M.K. Uddin, A review on the adsorption of heavy metals by clay minerals, with special focus on the past decade, Chemical Engineering Journal. 308 (2017) 438–462. https://doi.org/10.1016/j.cej.2016.09.029.

[50] D.L. Guerra, R.R. Viana, C. Airoldi, Adsorption of mercury cation on chemically modified clay, Materials Research Bulletin. 44 (2009) 485–491. https://doi.org/10.1016/j.materresbull.2008.08.002.

[51] S. Shakoor, A. Nasar, Adsorptive treatment of hazardous methylene blue dye from artificially contaminated water using cucumis sativus peel waste as a low-cost adsorbent, Groundwater for Sustainable Development. 5 (2017) 152–159. https://doi.org/10.1016/j.gsd.2017.06.005.

[52] A.Z.M. Badruddoza, Z.B.Z. Shawon, W.J.D. Tay, K. Hidajat, M.S. Uddin, Fe_3O_4/cyclodextrin polymer nanocomposites for selective heavy metals removal from industrial wastewater, Carbohydrate Polymers. 91 (2013) 322–332. https://doi.org/10.1016/j.carbpol.2012.08.030.

[53] I.J. Alinnor, Adsorption of heavy metal ions from aqueous solution by fly ash, Fuel. 86 (2007) 853–857. https://doi.org/10.1016/j.fuel.2006.08.019.

[54] K.G. Bhattacharyya, S. Sen Gupta, Adsorption of a few heavy metals on natural and modified kaolinite and montmorillonite: A review, Advances in Colloid and Interface Science. 140 (2008) 114–131. https://doi.org/10.1016/j.cis.2007.12.008.

[55] N. Goyal, S.. Jain, U.. Banerjee, Comparative studies on the microbial adsorption of heavy metals, Advances in Environmental Research. 7 (2003) 311–319. https://doi.org/10.1016/S1093-0191(02)00004-7.

[56] S. Guiza, Biosorption of heavy metal from aqueous solution using cellulosic waste orange peel, Ecological Engineering. 99 (2017) 134–140. https://doi.org/10.1016/j.ecoleng.2016.11.043.

[57]   A. Nasar, Polyaniline (PANI) Based Composites for the Adsorptive Treatment of Polluted Water, in: A. Nasar (Ed.), Smart Polymers and Composites, Materials Research Foundations, 2018: pp. 41–64. https://doi.org/10.21741/9781945291470-2.

[58]   A. Denizli, K. Kesenci, Y. Arica, E. Pişkin, Dithiocarbamate-incorporated monosize polystyrene microspheres for selective removal of mercury ions, Reactive and Functional Polymers. 44 (2000) 235–243. https://doi.org/10.1016/S1381-5148(99)00099-1.

[59]   Z. Abdeen, S.G. Mohammad, M.S. Mahmoud, Adsorption of Mn (II) ion on polyvinyl alcohol/chitosan dry blending from aqueous solution, Environmental Nanotechnology, Monitoring & Management. 3 (2015) 1–9. https://doi.org/10.1016/j.enmm.2014.10.001.

[60]   Ç. Ay, A.S. Özcan, Y. Erdoğan, A. Özcan, Characterization and lead(II) ions removal of modified Punica granatum L. peels, International Journal of Phytoremediation. 19 (2017) 327–339. https://doi.org/10.1080/15226514.2016.1225285.

[61]   E. Manchonvizuete, A. Maciasgarcia, A. Nadalgisbert, C. Fernandezgonzalez, V. Gomezserrano, Adsorption of mercury by carbonaceous adsorbents prepared from rubber of tyre wastes, Journal of Hazardous Materials. 119 (2005) 231–238. https://doi.org/10.1016/j.jhazmat.2004.12.028.

[62]   X. Ma, X. Liu, D.P. Anderson, P.R. Chang, Modification of porous starch for the adsorption of heavy metal ions from aqueous solution, Food Chemistry. 181 (2015) 133–139. https://doi.org/10.1016/j.foodchem.2015.02.089.

[63]   M. Šćiban, B. Radetić, Ž. Kevrešan, M. Klašnja, Adsorption of heavy metals from electroplating wastewater by wood sawdust, Bioresource Technology. 98 (2007) 402–409. https://doi.org/10.1016/j.biortech.2005.12.014.

[64]   S. Shakoor, A. Nasar, Adsorptive decontamination of synthetic wastewater containing crystal violet dye by employing Terminalia arjuna sawdust waste, Groundwater for Sustainable Development. 7 (2018) 30–38. https://doi.org/10.1016/j.gsd.2018.03.004.

[65]   M. Jain, V.K. Garg, K. Kadirvelu, M. Sillanpää, Adsorption of heavy metals from multi-metal aqueous solution by sunflower plant biomass-based carbons, International Journal of Environmental Science and Technology. 13 (2016) 493–500. https://doi.org/10.1007/s13762-015-0855-5.

[66]   A.Ş. Yargıç, R.Z. Yarbay Şahin, N. Özbay, E. Önal, Assessment of toxic copper(II) biosorption from aqueous solution by chemically-treated tomato waste,

Journal of Cleaner Production. 88 (2015) 152–159.
https://doi.org/10.1016/j.jclepro.2014.05.087.

[67] M. Malandrino, O. Abollino, A. Giacomino, M. Aceto, E. Mentasti, Adsorption of
heavy metals on vermiculite: Influence of pH and organic ligands, Journal of
Colloid and Interface Science. 299 (2006) 537–546.
https://doi.org/10.1016/j.jcis.2006.03.011.

[68] M. Zabihi, A. Haghighi Asl, A. Ahmadpour, Studies on adsorption of mercury
from aqueous solution on activated carbons prepared from walnut shell, Journal of
Hazardous Materials. 174 (2010) 251–256.
https://doi.org/10.1016/j.jhazmat.2009.09.044.

[69] E. Erdem, N. Karapinar, R. Donat, The removal of heavy metal cations by natural
zeolites, Journal of Colloid and Interface Science. 280 (2004) 309–314.
https://doi.org/10.1016/j.jcis.2004.08.028.

[70] X. Hu, Y. Du, Y. Tang, Q. Wang, T. Feng, J. Yang, J.F. Kennedy, Solubility and
property of chitin in NaOH/urea aqueous solution, Carbohydrate Polymers. 70
(2007) 451–458. https://doi.org/10.1016/j.carbpol.2007.05.002.

[71] I. Younes, M. Rinaudo, Chitin and Chitosan Preparation from Marine Sources.
Structure, Properties and Applications, Marine Drugs. 13 (2015) 1133–1174.
https://doi.org/10.3390/md13031133.

[72] S.K. Shukla, R. Choubey, A.K. Bajpai, Cationic Nanosorbents Biopolymers:
Versatile Materials for Environmental Cleanup, in: 2018: pp. 75–101.
https://doi.org/10.1007/978-3-319-68708-7_4.

[73] J. Lizardi-Mendoza, W.M. Argüelles Monal, F.M. Goycoolea Valencia, Chemical
Characteristics and Functional Properties of Chitosan, in: Chitosan in the
Preservation of Agricultural Commodities, Elsevier, 2016: pp. 3–31.
https://doi.org/10.1016/B978-0-12-802735-6.00001-X.

[74] V. Zargar, M. Asghari, A. Dashti, A Review on Chitin and Chitosan Polymers:
Structure, Chemistry, Solubility, Derivatives, and Applications, ChemBioEng
Reviews. 2 (2015) 204–226. https://doi.org/10.1002/cben.201400025.

[75] H.K. No, S.P. Meyers, Preparation and Characterization of Chitin and Chitosan—
A Review, Journal of Aquatic Food Product Technology. 4 (1995) 27–52.
https://doi.org/10.1300/J030v04n02_03.

[76] L. Zhang, Y. Zeng, Z. Cheng, Removal of heavy metal ions using chitosan and
modified chitosan: A review, Journal of Molecular Liquids. 214 (2016) 175–191.
https://doi.org/10.1016/j.molliq.2015.12.013.

[77]   M.N.. Ravi Kumar, A review of chitin and chitosan applications, Reactive and
        Functional Polymers. 46 (2000) 1–27. https://doi.org/10.1016/S1381-
        5148(00)00038-9.

[78]   T.R.A. Sobahi, M.Y. Abdelaal, M.S.I. Makki, Chemical modification of Chitosan
        for metal ion removal, Arabian Journal of Chemistry. 7 (2014) 741–746.
        https://doi.org/10.1016/j.arabjc.2010.12.011.

[79]   V.K. Thakur, M.K. Thakur, Recent Advances in Graft Copolymerization and
        Applications of Chitosan: A Review, ACS Sustainable Chemistry & Engineering.
        2 (2014) 2637–2652. https://doi.org/10.1021/sc500634p.

[80]   L. Wang, A. Wang, Adsorption properties of congo red from aqueous solution
        onto N,O-carboxymethyl-chitosan, Bioresource Technology. 99 (2008) 1403–
        1408. https://doi.org/10.1016/j.biortech.2007.01.063.

[81]   V.K. Mourya, N.N. Inamdar, Chitosan-modifications and applications:
        Opportunities galore, Reactive and Functional Polymers. 68 (2008) 1013–1051.
        https://doi.org/10.1016/j.reactfunctpolym.2008.03.002.

[82]   K.V. Harish Prashanth, R.N. Tharanathan, Chitin/chitosan: modifications and their
        unlimited application potential—an overview, Trends in Food Science &
        Technology. 18 (2007) 117–131. https://doi.org/10.1016/j.tifs.2006.10.022.

[83]   A. Shafaei, F.Z. Ashtiani, T. Kaghazchi, Equilibrium studies of the sorption of
        Hg(II) ions onto chitosan, Chemical Engineering Journal. 133 (2007) 311–316.
        https://doi.org/10.1016/j.cej.2007.02.016.

[84]   A. Gamage, F. Shahidi, Use of chitosan for the removal of metal ion contaminants
        and proteins from water, Food Chemistry. 104 (2007) 989–996.
        https://doi.org/10.1016/j.foodchem.2007.01.004.

[85]   E. Taboada, G. Cabrera, G. Cardenas, Retension capacity of chitosan for copper
        and mercury ions, Journal of the Chilean Chemical Society. 48 (2003).
        https://doi.org/10.4067/S0717-97072003000100002.

[86]   E.C.N. Lopes, F.S.C. dos Anjos, E.F.S. Vieira, A.R. Cestari, An alternative
        Avrami equation to evaluate kinetic parameters of the interaction of Hg(II) with
        thin chitosan membranes, Journal of Colloid and Interface Science. 263 (2003)
        542–547. https://doi.org/10.1016/S0021-9797(03)00326-6.

[87]   D.T. Thien, N.T. An, N.T. Hoa, Preparation of Fully Deacetylated Chitosan for
        Adsorption of Hg(II) Ion from Aqueous Solution, Chemical Sciences Journal. 6
        (2015). https://doi.org/10.4172/2150-3494.100095.

[88]   F.-N. Allouche, E. Guibal, N. Mameri, Preparation of a new chitosan-based
        material and its application for mercury sorption, Colloids and Surfaces A:

Physicochemical and Engineering Aspects. 446 (2014) 224–232.
https://doi.org/10.1016/j.colsurfa.2014.01.025.

[89] R.S. Vieira, M.M. Beppu, Interaction of natural and crosslinked chitosan
membranes with Hg(II) ions, Colloids and Surfaces A: Physicochemical and
Engineering Aspects. 279 (2006) 196–207.
https://doi.org/10.1016/j.colsurfa.2006.01.026.

[90] G. Cárdenas, P. Orlando, T. Edelio, Synthesis and applications of chitosan
mercaptanes as heavy metal retention agent, International Journal of Biological
Macromolecules. 28 (2001) 167–174. https://doi.org/10.1016/S0141-
8130(00)00156-2.

[91] C. Xiong, L. Pi, X. Chen, L. Yang, C. Ma, X. Zheng, Adsorption behavior of Hg^{2+}
in aqueous solutions on a novel chelating cross-linked chitosan microsphere,
Carbohydrate Polymers. 98 (2013) 1222–1228.
https://doi.org/10.1016/j.carbpol.2013.07.034.

[92] C. Jeon, W.H. Höll, Chemical modification of chitosan and equilibrium study for
mercury ion removal, Water Research. 37 (2003) 4770–4780.
https://doi.org/10.1016/S0043-1354(03)00431-7.

[93] A.A. Atia, Studies on the interaction of mercury(II) and uranyl(II) with modified
chitosan resins, Hydrometallurgy. 80 (2005) 13–22.
https://doi.org/10.1016/j.hydromet.2005.03.009.

[94] S. Kushwaha, P.P. Sudhakar, Adsorption of mercury(II), methyl mercury(II) and
phenyl mercury(II) on chitosan cross-linked with a barbital derivative,
Carbohydrate Polymers. 86 (2011) 1055–1062.
https://doi.org/10.1016/j.carbpol.2011.06.028.

[95] L.S. Rocha, Â. Almeida, C. Nunes, B. Henriques, M.A. Coimbra, C.B. Lopes,
C.M. Silva, A.C. Duarte, E. Pereira, Simple and effective chitosan based films for
the removal of Hg from waters: Equilibrium, kinetic and ionic competition,
Chemical Engineering Journal. 300 (2016) 217–229.
https://doi.org/10.1016/j.cej.2016.04.054.

[96] G. Kyzas, E. Deliyanni, Mercury(II) Removal with Modified Magnetic Chitosan
Adsorbents, Molecules. 18 (2013) 6193–6214.
https://doi.org/10.3390/molecules18066193.

[97] L.-M. Zhou, Z.-R. Liu, Q.-W. Huang, Adsorption of Hg^{2+} and UO_2^{2+} by
ethylenediamine modified chitosan magnetic microspheres, Journal of Nuclear and
Radiochemistry. 41 (2007) 184–188.

[98]   L. Zhou, Y. Wang, Z. Liu, Q. Huang, Characteristics of equilibrium, kinetics studies for adsorption of Hg(II), Cu(II), and Ni(II) ions by thiourea-modified magnetic chitosan microspheres, Journal of Hazardous Materials. 161 (2009) 995–1002. https://doi.org/10.1016/j.jhazmat.2008.04.078.

[99]   L. Zhou, Z. Liu, J. Liu, Q. Huang, Adsorption of Hg(II) from aqueous solution by ethylenediamine-modified magnetic crosslinking chitosan microspheres, Desalination. 258 (2010) 41–47. https://doi.org/10.1016/j.desal.2010.03.051.

[100] M. Monier, D.A. Abdel-Latif, Preparation of cross-linked magnetic chitosan-phenylthiourea resin for adsorption of Hg(II), Cd(II) and Zn(II) ions from aqueous solutions, Journal of Hazardous Materials. 209–210 (2012) 240–249. https://doi.org/10.1016/j.jhazmat.2012.01.015.

[101] M. Monier, Adsorption of Hg^{2+}, Cu^{2+} and Zn^{2+} ions from aqueous solution using formaldehyde cross-linked modified chitosan–thioglyceraldehyde Schiff's base, International Journal of Biological Macromolecules. 50 (2012) 773–781. https://doi.org/10.1016/j.ijbiomac.2011.11.026.

[102] A.M. Donia, A.A. Atia, K.Z. Elwakeel, Selective separation of mercury(II) using magnetic chitosan resin modified with Schiff's base derived from thiourea and glutaraldehyde, Journal of Hazardous Materials. 151 (2008) 372–379. https://doi.org/10.1016/j.jhazmat.2007.05.083.

[103] Y. Wang, Y. Qi, Y. Li, J. Wu, X. Ma, C. Yu, L. Ji, Preparation and characterization of a novel nano-absorbent based on multi-cyanoguanidine modified magnetic chitosan and its highly effective recovery for Hg(II) in aqueous phase, Journal of Hazardous Materials. 260 (2013) 9–15. https://doi.org/10.1016/j.jhazmat.2013.05.001.

[104] S. Nasirimoghaddam, S. Zeinali, S. Sabbaghi, Chitosan coated magnetic nanoparticles as nano-adsorbent for efficient removal of mercury contents from industrial aqueous and oily samples, Journal of Industrial and Engineering Chemistry. 27 (2015) 79–87. https://doi.org/10.1016/j.jiec.2014.12.020.

[105] J.D. Merrifield, W.G. Davids, J.D. MacRae, A. Amirbahman, Uptake of mercury by thiol-grafted chitosan gel beads, Water Research. 38 (2004) 3132–3138. https://doi.org/10.1016/j.watres.2004.04.008.

[106] X. Zhu, R. Yang, W. Gao, M. Li, Sulfur-modified chitosan hydrogel as an adsorbent for removal of Hg(II) from effluents, Fibers and Polymers. 18 (2017) 1229–1234. https://doi.org/10.1007/s12221-017-7046-6.

[107] R. Qu, C. Sun, C. Ji, C. Wang, H. Chen, Y. Niu, C. Liang, Q. Song, Preparation and metal-binding behaviour of chitosan functionalized by ester- and amino-

terminated hyperbranched polyamidoamine polymers, Carbohydrate Research. 343 (2008) 267–273. https://doi.org/10.1016/j.carres.2007.10.032.

[108] K.C. Gavilan, A.V. Pestov, H.M. Garcia, Y. Yatluk, J. Roussy, E. Guibal, Mercury sorption on a thiocarbamoyl derivative of chitosan, Journal of Hazardous Materials. 165 (2009) 415–426. https://doi.org/10.1016/j.jhazmat.2008.10.005.

[109] X. Tao, K. Li, H. Yan, H. Yang, A. Li, Simultaneous removal of acid green 25 and mercury ions from aqueous solutions using glutamine modified chitosan magnetic composite microspheres, Environmental Pollution. 209 (2016) 21–29. https://doi.org/10.1016/j.envpol.2015.11.020.

[110] X. Wang, W. Deng, Y. Xie, C. Wang, Selective removal of mercury ions using a chitosan–poly(vinyl alcohol) hydrogel adsorbent with three-dimensional network structure, Chemical Engineering Journal. 228 (2013) 232–242. https://doi.org/10.1016/j.cej.2013.04.104.

[111] X. Wang, R. Sun, C. Wang, pH dependence and thermodynamics of Hg(II) adsorption onto chitosan-poly(vinyl alcohol) hydrogel adsorbent, Colloids and Surfaces A: Physicochemical and Engineering Aspects. 441 (2014) 51–58. https://doi.org/10.1016/j.colsurfa.2013.08.068.

[112] X. Wang, L. Yang, J. Zhang, C. Wang, Q. Li, Preparation and characterization of chitosan–poly(vinyl alcohol)/bentonite nanocomposites for adsorption of Hg(II) ions, Chemical Engineering Journal. 251 (2014) 404–412. https://doi.org/10.1016/j.cej.2014.04.089.

[113] Y.-H. Chang, C.-F. Huang, W.-J. Hsu, F.-C. Chang, Removal of Hg^{2+} from aqueous solution using alginate gel containing chitosan, Journal of Applied Polymer Science. 104 (2007) 2896–2905. https://doi.org/10.1002/app.25891.

[114] R. Dubey, J. Bajpai, A.K. Bajpai, Chitosan-alginate nanoparticles (CANPs) as potential nanosorbent for removal of Hg (II) ions, Environmental Nanotechnology, Monitoring & Management. 6 (2016) 32–44. https://doi.org/10.1016/j.enmm.2016.06.008.

[115] Ö. Genç, Ç. Arpa, G. Bayramoğlu, M.. Arıca, S. Bektaş, Selective recovery of mercury by Procion Brown MX 5BR immobilized poly(hydroxyethylmethacrylate/chitosan) composite membranes, Hydrometallurgy. 67 (2002) 53–62. https://doi.org/10.1016/S0304-386X(02)00160-3.

[116] H. Ge, T. Hua, J. Wang, Preparation and characterization of poly (itaconic acid)-grafted crosslinked chitosan nanoadsorbent for high uptake of Hg^{2+} and Pb^{2+}, International Journal of Biological Macromolecules. 95 (2017) 954–961. https://doi.org/10.1016/j.ijbiomac.2016.10.084.

[117] S.-P. Wu, X.-Z. Dai, J.-R. Kan, F.-D. Shilong, M.-Y. Zhu, Fabrication of carboxymethyl chitosan–hemicellulose resin for adsorptive removal of heavy metals from wastewater, Chinese Chemical Letters. 28 (2017) 625–632. https://doi.org/10.1016/j.cclet.2016.11.015.

[118] S. Salahi, M. Parvini, M. Ghorbani, Equilibrium Studies in Adsorption of Hg(II) from Aqueous Solutions using Biocompatible Polymeric Polypyrrole-Chitosan Nanocomposite, Polycyclic Aromatic Compounds. 34 (2014) 225–236. https://doi.org/10.1080/10406638.2014.886077.

[119] R. Sharma, N. Singh, S. Tiwari, S.K. Tiwari, S.R. Dhakate, Cerium functionalized PVA–chitosan composite nanofibers for effective remediation of ultra-low concentrations of Hg(II) in water, RSC Advances. 5 (2015) 16622–16630. https://doi.org/10.1039/C4RA15085F.

[120] H. Ge, T. Hua, Synthesis and characterization of poly(maleic acid)-grafted crosslinked chitosan nanomaterial with high uptake and selectivity for Hg(II) sorption, Carbohydrate Polymers. 153 (2016) 246–252. https://doi.org/10.1016/j.carbpol.2016.07.110.

[121] Y. Zhu, Y. Zheng, W. Wang, A. Wang, Highly efficient adsorption of Hg(II) and Pb(II) onto chitosan-based granular adsorbent containing thiourea groups, Journal of Water Process Engineering. 7 (2015) 218–226. https://doi.org/10.1016/j.jwpe.2015.06.010.

Chitosan-Based Adsorbents for Wastewater Treatment, Ed. Abu Nasar   Materials Research Forum LLC
Materials Research Foundations 34 (2018) 57-80   doi: http://dx.doi.org/10.21741/9781945291753-3

Chapter 3

Novel Chitosan-Based Nanocomposites for Dye Removal Applications

Aysun Savk[1], Betul Sen[1], Buse Demirkan[1], Esra Kuyuldar[1], Aysenur Aygun[1], Mehmet Salih Nas[1,2], Fatih Sen[1,*]

[1]Sen Research Group, Department of Biochemistry, Faculty of Arts and Science, Dumlupınar University, Evliya Çelebi Campus, 43100 Kütahya, Turkey

[2]Department of Environmental Engineering, Faculty of Engineering, Igdir University, 76000 Igdir, Turkey

*fatih.sen@dpu.edu.tr

Abstract

Monodisperse Pd nanoparticles (Pd NPs@CGO) decorated Chitosan-graphene oxide (CGO) are produced to get a nanoadsorbent material to remove methylene blue (MB) from aqueous solutions. X-ray diffraction (XRD), transmission electron microscopy (TEM), high-resolution transmission electron microscopy (HR-TEM) and X-ray photoelectron spectroscopy (XPS) were used to characterize the Pd NPs@CGO. The spectroscopic results showed that Pd NPs@CGO has highly crystalline, monodisperse and colloidal structures. Furthermore, Pd NPs@CGO was highly efficient and stable for methylene blue removal. They provide a high adsorption capacity of 186.42 mg/g and its MB adsorption equilibrium is obtained in ~60 min. Nonetheless, Pd NPs@CGO are reusable and promising nanocomposites for methylene blue removal, keeping 43.05 % of the first efficacy after six adsorption-desorption cycles.

Keywords

Chitosan-Graphene Oxide (CGO), Monodisperse Metal Nanoparticles, Methylene Blue Removal, Recyclable Nanosorbent

Contents

1.   Introduction

In recent years, organic dyes have been used commonly in many industries like paper, coatings, plastics, textile, paint, cosmetics and leather, and hence, those industries produce a significant amount of wastewater containing dyes, which are also quite visible [1-7]. Besides, it has been reported that dyes usually have high toxicity [6, 7]. In previous years, some dye removal methods have been revealed [8-11]. Today, various techniques, such as adsorption, electrochemical degradation, biological treatment, flocculation, ultrafiltration, sonochemical degradation, MnO_2 oxidation, and photocatalytic degradation have been employed for the removal of contaminants from wastewater [7-15]. In some cases, dyes in wastewater cannot be successfully decolorized using the aforementioned techniques, and therefore, inexpensive, simple, and efficient technologies have been explored. Adsorption has been found as the most convenient method to treat wastewaters contaminated with dyes based on the activity, cost and practical use of adsorption [4-8]. Hereof, numerous studies have been done to develop effective and usable adsorbent materials like silica, activated carbon, chitosan, peat, chitin, solid waste and clay materials as shown in Table 1. However, those adsorbents have shown some disadvantageous, such as lack of specificity, expensiveness in production and/or treatment steps, low adsorption efficiency and capacity, problems with reuse, longer process time. For this reason, better adsorbent materials are required for more efficient dye removal.

Generally, nano-structured materials or nanocomposites have been reported as effective materials for many applications [16-30]. For this reason, researchers have modified and prepared different nanocomposites to use them as potential adsorbents for the remediation of wastewaters to remove dyes [1,4,5,9,11,31]. These nanoadsorbents have mechanical flexibility, tunable pore size, chemical stability, good structures, composition developing the ability, large surface area, and hence, allow enhanced contact with dyes and dye removal, which are some of the ideal features of preferred adsorbents. Graphene and graphene oxide, polyurethane foams, carbon nanotubes, polyaniline nanotubes, fullerenes, polypyrrole/TiO_2 composites, PZS nanospheres, iron oxide nanocomposites are some literature examples as they carry promising properties for dye removal as shown in Table 1. Amongst them, graphene, a carbonaceous material, has a single layer of graphite and is a two-dimensional (2-D) nanomaterial consisting of honeycomb crystal structured sp^2 carbon network and its exceptional properties make it very useful for various applications [32-35]. Recently, graphene has been revealed as one of the most researched nanomaterials for its diverse applications by scientists from different disciplines due to its chemical and physical properties [36-43]. However, in practice for dye removal it is not very applicable because of its restricted dispersion in water even though graphene has a great surface area. Based on favorable adsorption properties of chitosan and inherent properties of GO, chitosan-graphene oxide (CGO), which are generated using graphene, becomes a better option as biosorbents, because CGO represents an appropriate structure for many applications owing to different surface functional groups that make CGO highly dispersible in water. These functional groups depend on the reaction type and conditions (i.e. wet or non-wet chemical approaches, and preparation time and temperature). In addition, these functional groups not only enable large surface area but also significant metallic cation sorption capacity. These may make CGO an ideal cationic dye adsorbent like methylene blue (MB), which is used commonly for colorizing cotton, silk, or wood. Hence, the development of economic MB adsorbents from wastewaters has a noticeable environmental importance to reduce its environmental, health and esthetical concerns. The objective of the present study was to prepare Pd NPs@CGO nanocomposites with better adsorption capacity and investigating their properties for efficient methylene blue (MB) removal (Scheme 1). Therefore, in this study, the synthesis of a Pd NPs@CGO nanocomposite through chemical processing method was accomplished. The synthesized nanocomposite was characterized using XRD, TEM, HRTEM, and XPS. To display the practical application of Pd NPs@CGO as a potential adsorbent to remove organic dyes, MB was tested for this purpose. The MB removal efficacy of Pd NPs@CGO nanocomposites was examined using a UV-Visible spectroscopy. Here, the effect of contact time on MB adsorption and the re-usability of Pd NPs@CGO nanocomposite adsorbed per MB unit weight were also investigated. It was shown that this is a novel adsorbent which has ease of operation and low cost for this purpose.

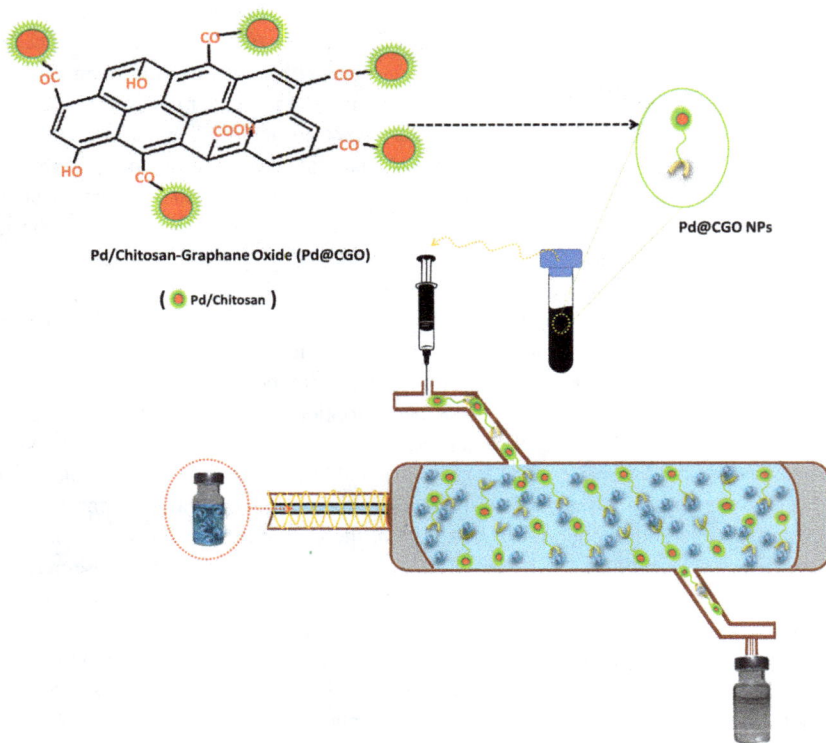

Scheme-1 The schematic view of MB removal with the help of Pd NPs@CGO.

2.   Materials and methods

2.1.   Chemicals and techniques

Palladium (II) chloride (PdCl$_2$ 99%; Alfa), tetrahydrofuran (THF or (CH$_2$)$_3$CH$_2$O; 99.5%; Merck), potassium permanganate (KMnO$_4$; Merck), Chitosan (degree of deacetylation: 90 %, Mw = 4.000-6.000), HAc (Aldrich) sodium nitrate (NaNO$_3$; 99.0%; Merck), hydrogen peroxide (H$_2$O$_2$; 30%; Merck), hydrochloric acid (HCl; 37%; Merck), methylene blue (MB; Merck), ethylene glycol (Aldrich) were used in this study. Water was purified using a Millipore filtration system (18 MΩ). Teflon-coated magnetic stir bars and the glassware were washed with aqua regia and distilled water and then dried.

TEM images of Pd NPs@CGO NPs have been obtained by a JEOL 200 kV TEM instrument. X-ray diffraction (XRD) was performed using a Panalytical Empyrean diffractometer with Ultima + theta-theta high-resolution goniometer, the X-ray generator (Cu K radiation, $\lambda = 1.54056$Å) with operation conditions at 45 kV and 40 mA. A Specs spectrometer was used for X-ray photoelectron spectroscopy (XPS) measurements using K lines of Mg (1253.6 eV, 10 mA) as an X-ray source.

2.2   Preparation of Pd NPs@CGO

By using graphite powder and modifying Hummer's method, graphene oxide (GO) was synthesized [44-52]. Then, graphene oxide powder of 0.20 g was dispersed into 100 mL of ultrapure water and was treated by mild ultrasound for 20 min in a 250 mL beaker, in order to get a homogeneous suspension. Then, 1 ml HAc and 1.5 g chitosan were added into the suspension sequentially under stirring. After 60 min of stirring at room temperature, the CGO mixture solutions were prepared. In order to prepare Pd NPs and mixing them with CGO in order to prepare Pd NPs@CGO, chemical processing method has been performed as described in our previous works [53-55].

2.3   Adsorption experiments

After getting a calibration curve for various concentrations, firstly, 25 mg of Pd NPs@CGO nanocomposite was dispersed in water by using an ultrasonic bath for 2 hours to run the batch adsorption experiments,. Next, the mixture was mixed with 25 mL of MB solution (30 mg/L) and shaken in a water bath (120 rpm) for 24 hours. At the end of 24 h shaking, pH was adjusted to 5.8 using NaOH and HCl solutions. Dye adsorption experiments were done in round bottom flasks at room temperature. After separating the Pd NPs@CGO nanocomposite particles by centrifugation (4000 rpm for 10 min), the supernatant solution was analyzed to measure the absorbance at 664 nm, which is the absorption band of MB in water, by using a UV-Vis spectrophotometer. Using the calibration curve and the absorbance data, the amount of dye adsorbed was calculated using the following equation:

$$q_e = (C_o - C_e) V / m$$

In this equation, q_e, C_o, C_e, V and m represent the concentration of dye adsorbed (mg/g), initial concentration of dye (mg/L), equilibrium concentration of dye (mg/L), mass of the Pd NPs@CGO (g), and volume of solution (L), respectively.

2.4   Testing the reusability of Pd NPs@CGO nanocomposites

For the reusability of the Pd NPs@CGO nanocomposite synthesized in this study for MB removal, 15 mg of the nanocomposite was mixed with 25 mL of MB solution (30 mg/L).

Next, the mixture was sonicated for 30 min at room temperature. After separating the nanocomposite from the mixture by centrifugation, the supernatant was kept for the spectroscopic analyses. Afterward, for desorption, used nanocomposite was washed with 25 mL of ethanol three times at room temperature and then collected by centrifugation. This washed nanocomposite was reused for a next MB adsorption experiment as described above. To figure out the reusability of NPs, experiments were repeated six times.

3.   Results and discussion

After synthesis had successfully been performed, the monodisperse Pd NPs@CGO was firstly described by the help of X-ray diffraction, Transmission electron microscopy, High-resolution transmission electron microscopy and X-ray photoelectron spectroscopy. In Fig. 1a, discrete diffraction patterns of monodisperse Pd NPs@CGO obtained by the XRD analysis were showed. The peak at around 21.8° is ascribed to the CGO. Furthermore, as shown in Figure 1a, diffraction peaks at $2\theta = 40.17°$, 46.47° and 68.17° were observed, which indicates the fcc structure of Pd. The lattice parameter of Pd@CGO was calculated and compared with the nominal Pd values [55-63]. Besides, according to the Debye-Scherrer equation [64-80], it was found that the monodisperse Pd NPs@CGO had a mean value of 4.43 ± 0.51 nm crystallite Pd particle size.

In Fig. 1b, particle size and particle distribution of the prepared nanoparticles, were shown. As shown in this figure, the prepared nanoparticles were uniformly distributed and spherical. Besides, in Fig. 1b, atomic lattice fringes were calculated on the prepared adsorbent [81-85].

In Fig. 1(c-d), the X-ray photoelectron spectroscopy results explain the investigation of nanoadsorbent surface properties and the oxidation state of metal [86-95]. Therefore, the spectrum of Pd 3d region was evaluated. For this purpose, the Gaussian-Lorentzian method and the Shirley-shaped background was used for fitting of XPS [95-103]. In Fig. 1(c-d), the Pd 3d XPS of the Pd NPs@CGO consist of one doublet at about 335.3 and 341.3 eV which shows the metallic Pd indicated as the most intense doublets. The other doublet at about 337.5 and 343.2 eV is most probably due to the oxidized Pd species which is coming from unreduced or PdOx species. Pd(0)/Pd(II) ratio was calculated as 2.85 for the synthesized adsorbent.

After full characterization of Pd NPs@CGO, MB removal with the Pd NPs@CGO was examined, and for this aim, the calibration curve was gathered using five different MB solutions (2.5, 5, 10, 20, and 30 mg/L). The calibration curve can be seen in the Fig. 2. Later, the effect of contact time on dye adsorption was investigated using 13 mg/L of MB solution. Its results were shown in the Fig. 2. When high MB concentrations were used a little aggregation of MB was seen. The findings presented an equilibrium time of almost 60 min for MB adsorption by Pd NPs@CGO nanocomposites. This 60-min-equilibrium-time is relatively short. Therefore, it can be said that Pd NPs@CGO nanocomposites are outstanding adsorbents for MB removal.

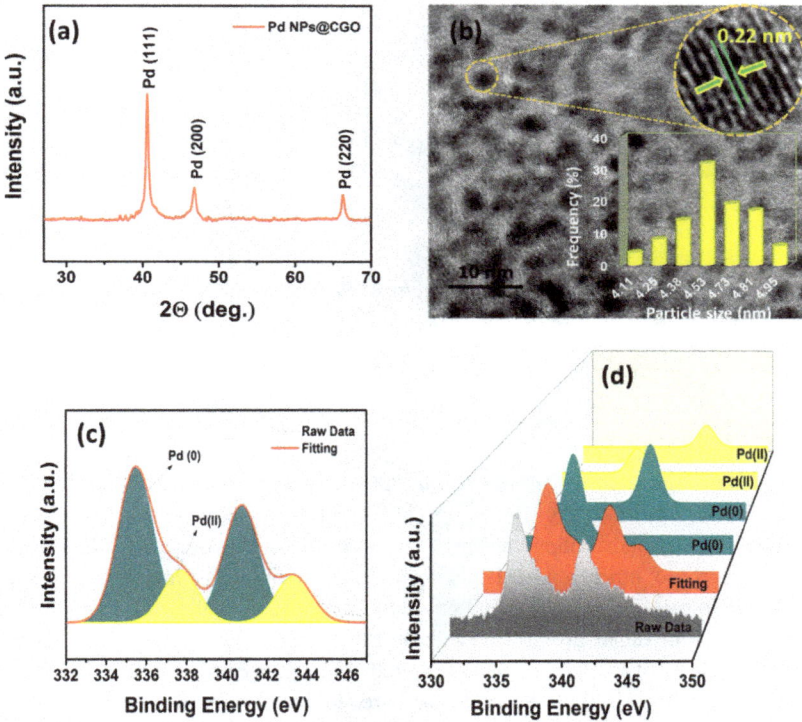

Fig. 1 (a) X-ray diffraction pattern, (b)   Transmission electron microscopy and high-resolution transmission electron microscopy images, and (c-d) X-ray photoelectron spectroscopy image of Pd 3d region of Pd NPs@CGO.

On the other hand, initially, MB removing efficiency of Pd NPs@CGO was rapid, however, it became slower after some time. The possible explanation could be the decrease in methylene blue concentration since methylene blue concentration decreases during the process and the slowing down of the speed of adsorption.

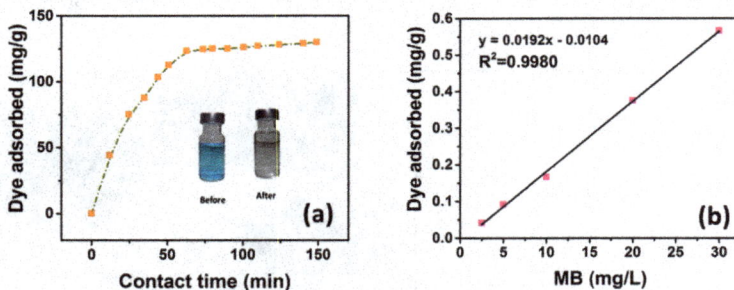

Fig. 2 (a) The change in adsorption capacity of Pd NPs@CGO by the increasing contact time. Here, 13 mg/L methylene blue was used as initial concentration. The two bottles were given as the representative methylene blue solutions for before and after dye adsorption.(b) Calibration curve for the solutions having different concentrations of methylene blue.

The adsorption isotherm (while Pd NPs@CGO nanocomposites exist) was displayed in Fig. 3a. In this figure, q_e, mg/g and C_e, mg/L was shown for the prepared nanomaterials. The maximum dye adsorption capacity was found to be 186.42 mg/g. When this result was compared to other reported adsorbents, this finding has higher dye adsorption capacity (Table-1). This may depend on many factors such as the pore size distribution, surface area, functional groups and polarity of the adsorbent. It can be said that the synthesis method of Pd NPs@CGO supplies the monodispersity, smaller size and larger specific surface area for nanoadsorbent compared to the other prepared ones in Table 1.

It is worth to note that no changes were observed in the Pd NPs@CGO nanocomposites' structure after dye adsorption. This finding underlined the steadiness of the Pd NPs @CGO nanocomposites in aqueous solutions. On the other hand, good adsorbent materials do possess high adsorption capabilities besides perfect desorption features [42]. Therefore, the reusability of Pd NPs@CGO nanocomposites was tested here. For this, 6 successive adsorption-desorption cycles were done, and Fig. 3b was drawn with the results. Although Pd NPs@CGO nanocomposites' adsorption capacity for methylene blue removal showed a decrease, they still had 43.05 % efficiency even after six cycles. According to the results observed in this study, the Pd NPs@CGO nanocomposite is a

Chitosan-Based Adsorbents for Wastewater Treatment, Ed. Abu Nasar   Materials Research Forum LLC
Materials Research Foundations **34** (2018) 57-80   doi: http://dx.doi.org/10.21741/9781945291753-3

reusable efficient material for MB dye removal from aqueous solutions and has a high capacity of adsorption and rate.

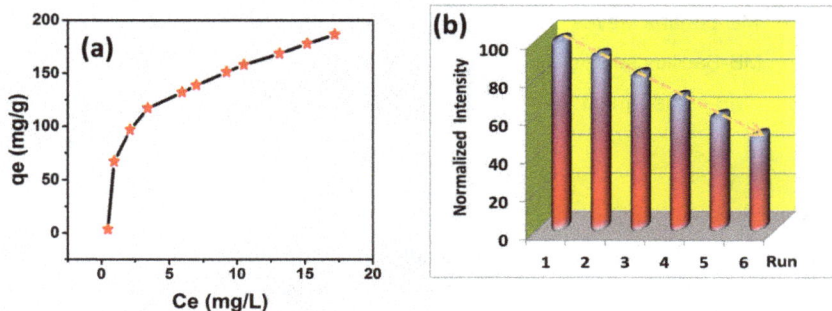

Fig. 3 (a) Isotherm of the MB adsorption of Pd NPs@CGO nanocomposites. (b) The reusability of the Pd NPs@CGO nanocomposites for MB removal which was done in 6 successive cycles. (with 0.25 g/L of Pd NPs@CGO and 13 mg/L MB at 25oC for 30 min contact time).

Table 1 Adsorption capacities of different adsorbents for methylene blue removal.

Adsorbent	Adsorption capacity (mg MB/g)	Reference
Pd NPs@CGO	186.42	This study
MPB-AC	163.3	[104]
PZS nanospheres	20	[105]
GO–Fe_3O_4 hybrids	172.6	[1]
MWCNTs with Fe_2O_3	42.3	[106]
Na-ghassoulite	135	[107]
GO	17.3	[108]
GO-Fe_3O_4	190.14	[109]
Graphene	153.85	[110]
GO-Fe_3O_4-SiO_2	111.1	[111]
MB-wheat straw	274.1	[112]
MB-cotton stalk	147.1	[113]
MB-cucumber peels	111.1	[114]
MB-rice hull ash	17.1	[115]

MB-shaddock peel	309.6	[116]
MB-cottonseed hull	185.2	[117]
MB-banana leaves	109.9	[118]
MB-Bacillus subtilis	169.5	[119]
MB-citrus limetta	227.3	[3]

4.   Conclusions

In summary, an effective, simple, eco-friendly and reusable method was given in this study to successfully synthesize Pd NPs@CGO nanocomposites. Very short process time, easy methodology, outstanding yield, and clean conditions are the most attractive proses of our method. Pd NPs@CGO indicated remarkable catalytic performance together with high dye removal capacity (186.42 mg MB/g nanocomposite) for methylene blue dye in water most probably due to the high specific surface area, monodispersity and metal contents of Pd NPs@CGO. The other reasons for this might be the high electrostatic interactions and π-π interactions between CGO and MB, which induced the MB adsorption on the Pd NPs@CGO nanocomposites. Additionally, considerable steadiness and the potential for reusability were also detected. It was observed that the prepared nanoadsorbents can be used many times for each adsorption-desorption cycles. This just proved that the prepared Pd NPs@CGO has promising nanoadsorbent for methylene blue removal from contaminated waters.

Acknowledgements

This research was financed by the Dumlupinar University Research Fund (grant no. 2014-05, 2015-35 and 2015-50).

References

[1]   K. Meral, Ö. Metin, Graphene oxide-magnetite nanocomposite as an efficient and magnetically separable adsorbent for methylene blue removal from aqueous solution, Turkish J. Chem. 38 (2014) 775-782. https://doi.org/10.3906/kim-1312-28

[2]   S. Shakoor, A. Nasar, Adsorptive treatment of hazardous methylene blue dye from artificially contaminated water using cucumis sativus peel waste as a low-cost

adsorbent, Groundw. Sustain. Dev. 5 (2017) 152–159.
https://doi.org/10.1016/j.gsd.2017.06.005

[3]   S. Shakoor, A. Nasar, Removal of methylene blue dye from artificially
      contaminated water using citrus limetta peel waste as a very low cost adsorbent, J.
      Taiwan Inst. Chem. Eng. 66 (2016) 154–163.
      https://doi.org/10.1016/j.jtice.2016.06.009

[4]   Y. Yıldız, T. Onal Okyay, B. Gezer, Z. Dasdelen, B. Sen, F. Sen, Monodisperse
      Mw-Pt NPs@VC as highly efficient and reusable adsorbents for methylene blue
      removal. Journal of Cluster Science, 27 (2016) 1953–1962.
      https://doi.org/10.1007/s10876-016-1054-3

[5]   Y. Yildiz, T. Onal Okyay, B. Sen, B. Gezer, S. Kuzu, A. Savk, E. Demir, Z.
      Dasdelen and F. Sen, Highly Monodisperse Pt/Rh Nanoparticles Confined in the
      Graphene Oxide for Highly Efficient and Reusable Sorbents for Methylene Blue
      Removal from Aqueous Solutions. Chemistry Select, 2 (2) (2017) 697-70.
      https://doi.org/10.1002/slct.201601608

[6]   F. Liu, H. Zou, J. Hu, H. Liu, J. Peng, Y. Chen, F. Lu, Y. Huo, Fast removal of
      methylene blue from aqueous solution using porous soy protein isolate based
      composite beads, Chemical Engineering Journal. 287 (2016) 410-418.
      https://doi.org/10.1016/j.cej.2015.11.041

[7]   M. A. Khan, S. H. Lee, S. Kang, K. J. Paeng, G. Lee, S. E. Oh, and B. H. Jeon,
      Adsorption Studies for the Removal of Methyl tert-Butyl Ether on Various
      Commercially Available GACs from an Aqueous Medium, Separation Science and
      Tech. 46 (2011) 1121-1130. https://doi.org/10.1080/01496395.2010.551395

[8]   M. S. Chiou, P. Y. Ho, H. Y. Li, Adsorption of Anionic Dyes in Acid Solutions
      Using Chemically Cross- Linked Chitosan Beads, Dyes and Pigments. 60 (2004)
      69-84. https://doi.org/10.1016/S0143-7208(03)00140-2

[9]   L. Bai, Z. Li, Y. Zhang, T. Wang, R. Lu, W. Zhou, H. Gao, and S. Zhang,
      Synthesis of water-dispersible graphene-modified magnetic polypyrrole
      nanocomposite and its ability to efficiently adsorb methylene blue from aqueous
      solution, Chemical Engineering Journal. 279 (2015) 757-766.
      https://doi.org/10.1016/j.cej.2015.05.068

[10] L. Fan, C. Luo, M. Sun, H. Qiu, and X. Li, Synthesis of magnetic-cyclodextrin–
      chitosan/graphene oxide as nanoadsorbent and its application in dye adsorption
      and removal, Colloids and Surfaces B: Biointerfaces. 103 (2013) 601-607.
      https://doi.org/10.1016/j.colsurfb.2012.11.023

[11] F. Liu, S. Chung, G. Oh, T. S. Seo, Three-Dimensional Graphene Oxide Nanostructure for Fast and Efficient Water-Soluble Dye Removal, ACS Applied Materials & Interfaces. 4 (2012) 922-927. https://doi.org/10.1021/am201590z

[12] Qamruzzaman, A. Nasar, Degradation of tricyclazole by colloidal manganese dioxide in the absence and presence of surfactants, J. Ind. Eng. Chem. 20 (2014) 897–902. https://doi.org/10.1016/j.jiec.2013.06.020

[13] Qamruzzaman, A. Nasar, Kinetics of metribuzin degradation by colloidal manganese dioxide in absence and presence of surfactants, Chem. Pap. 68 (2014). doi:10.2478/s11696-013-0424-7.

[14] Qamruzzaman, A. Nasar, Degradation of acephate by colloidal manganese dioxide in the absence and presence of surfactants, Desalin. Water Treat. 55 (2015) 2155–2164. https://doi.org/10.1080/19443994.2014.937752

[15] Qamruzzaman, A. Nasar, Treatment of acetamiprid insecticide from artificially contaminated water by colloidal manganese dioxide in the absence and presence of surfactants, RSC Adv. (2014). https://doi.org/10.1039/C4RA09685A

[16] S. Akocak, B. Şen, N. Lolak, A. Şavk, M. Koca, S. Kuzu, F. Şen, One-pot three-component synthesis of 2-Amino-4H-Chromene derivatives by using monodisperse Pd nanomaterials anchored graphene oxide as highly efficient and recyclable catalyst. Nano-Structures & Nano-Objects, 11 (2017) 25–31. https://doi.org/10.1016/j.nanoso.2017.06.002

[17] F. Sen, A. A. Boghossian, S. Sen, et.al. Application of Nanoparticle Antioxidants to Enable Hyperstable Chloroplasts for Solar Energy Harvesting. Advanced Energy Materials, 3 (7) (2013) 881–893. https://doi.org/10.1002/aenm.201201014

[18] B. Şen, N. Lolak, M. Koca, A. Şavk, S. Akocak, F. Şen, Bimetallic PdRu/graphene oxide-based Catalysts for one-pot three-component synthesis of 2-amino-4H-chromene derivatives. Nano-Structures & Nano-Objects, 12 (2017) 33-40. https://doi.org/10.1016/j.nanoso.2017.08.013

[19] J. P. Giraldo, M. P. Landry, S. M. Faltermeier et.al., A Nanobionic Approach to Augment Plant Photosynthesis and Biochemical Sensing Using Targeted Nanoparticles. Nature Materials, 13 (2014) 400–408. https://doi.org/10.1038/nmat3890

[20] F. Sen, Z. W. Ulissi, X. Gong et.al., Spatiotemporal Intracellular Nitric Oxide Signaling Captured Using Internalized, Near-Infrared Fluorescent Carbon

Nanotube Nanosensors. Nano Letters, 14 (8) (2014) 4887–4894.
https://doi.org/10.1021/nl502338y

[21]   R. Ayranci, G. Baskaya, M. Guzel, S. Bozkurt, M. Ak, A. Savk, F. Sen, Enhanced
        optical and electrical properties of PEDOT via nanostructured carbon materials: A
        Comparative investigation. Nano-Structures and Nano-Objects, 11 (2017) 13–19.
        https://doi.org/10.1016/j.nanoso.2017.05.008

[22]   N. M. Iverson, P. W. Barone, Mia Shandell, et. al. In vivo biosensing via tissue-
        localizable near- infrared-fluorescent single-walled carbon nanotubes, Nature
        Nanotechnology. 8 (11) (2013) 873-880. https://doi.org/10.1038/nnano.2013.222

[23]   B. Şahin, E. Demir, A. Aygün et.al. Investigation of the Effect of Pomegranate
        Extract and Monodisperse Silver Nanoparticle Combination on MCF-7 Cell Line.
        Journal of Biotechnology 260C (2017) 79-83.
        https://doi.org/10.1016/j.jbiotec.2017.09.012

[24]   B. Şen, E. H. Akdere, A. Şavk, E. Gültekin, H. Göksu and F. Şen, A novel
        thiocarbamide functionalized graphene oxide supported bimetallic monodisperse
        Rh-Pt nanoparticles (RhPt/TC@GO NPs) for Knoevenagel condensation of aryl
        aldehydes together with malononitrile, Applied Catalysis B: Environmental. 225 5
        (2018) 148-153. https://doi.org/10.1016/j.apcatb.2017.11.067

[25]   F. Sen, A A. Boghossian, S. Sen, et.al. Observation of Oscillatory Surface
        Reactions of Riboflavin, Trolox, and Singlet Oxygen Using Single Carbon
        Nanotube Fluorescence Spectroscopy, ACS Nano. 6 (12) (2012) 10632-10645.
        https://doi.org/10.1021/nn303716n

[26]   H. Göksu, B. Kilbas and F. Sen, Recent Advances in the Reduction of Nitro
        Compounds by Heterogenous Catalysts, Current Organic Chemistry. 21 (9) (2017)
        794-820. https://doi.org/10.2174/1385272820666160525123907

[27]   B. Şahin, A. Aygün, H. Gündüz, K. Şahin, E. Demir, S. Akocak, F. Şen, Cytotoxic
        Effects of Platinum Nanoparticles Obtained from Pomegranate Extract by the
        Green Synthesis Method on the MCF-7 Cell Line, Colloids and Surfaces B:
        Biointerfaces. 163 (2018) 119–124. https://doi.org/10.1016/j.colsurfb.2017.12.042

[28]   J. Zhang, M. P. Landry, P. W. Barone,et.al. Molecular recognition using corona
        phase complexes made of synthetic polymers adsorbed on carbon nanotubes,
        Nature Nanotechology. 8 (12) (2013) 959-968.
        https://doi.org/10.1038/nnano.2013.236

[29]   J.T. Abrahamson, F. Sen, B. Sempere, et. al. Excess Thermopower and the Theory of Thermopower Waves, ACS Nano.7 (8) (2013) 6533–6544. https://doi.org/10.1021/nn402411k

[30]   S. Sen, F. Sen, A. A. Boghossian, et al. The Effect of Reductive Dithiothreitol and Trolox on Nitric Oxide Quenching of Single Walled Carbon Nanotubes, Journal of Physical Chemistry C. 117 (1) (2013) 593-602. https://doi.org/10.1021/jp307175f

[31]   Y. Yıldız, T. Onal Okyay, B. Sen, B. Gezer, S. Bozkurt, G. Başkaya and F. Sen, Activated Carbon Furnished Monodisperse Pt nanocomposites as a superior adsorbent for methylene blue removal from aqueous solutions, Journal of Nanoscience and Nanotechnology. 17 (2017) 4799–4804. https://doi.org/10.1166/jnn.2017.13776

[32]   S. Eigler, A. Hirsch, Chemistry with Graphene and Graphene Oxide—Challenges for Synthetic Chemists, Angewandte Chemie International Edition. 53 (2014) 7720-7738.

[33]   J. Xu, H. Lv, S. T. Yang and J. Luo, Preparation of graphene adsorbents and their applications in water purification, Reviews in Inorganic Chem. 33 (2013) 139-160. https://doi.org/10.1515/revic-2013-0007

[34]   D. R. Dreyer, S. Park, C. W. Bielawski and R. S. Ruoff, The chemistry of graphene oxide, Chemical Society Reviews. 39 (2010) 228-240. https://doi.org/10.1039/B917103G

[35]   F. Ahmed and D. F. Rodrigues, Investigation of acute effects of graphene oxide on wastewater microbial community: A case study, J. Hazard. Mater. 256–257 (2013) 33-39. https://doi.org/10.1016/j.jhazmat.2013.03.064

[36]   İ. Esirden, E. Erken, M. Kaya and F. Sen, Monodisperse Pt NPs@rGO as highly efficient and reusable heterogeneous catalysts for the synthesis of 5-substituted 1H-tetrazole derivatives. Catal. Sci. Technol., 5 (2015) 4452-4457. https://doi.org/10.1039/C5CY00864F

[37]   H. Pamuk, B. Aday, M. Kaya, Fatih Şen, Pt Nps@GO as Highly Efficient and Reusable Catalyst for One-Pot Synthesis of Acridinedione Derivatives. RSC Advances, 5 (2015) 49295-49300. https://doi.org/10.1039/C5RA06441D

[38]   B. Şen, N. Lolak, Ö. Paralı, M. Koca, A. Şavk, S. Akocak, F. Şen, Bimetallic PdRu/graphene oxide based Catalysts for one-pot three-component synthesis of 2-amino-4H-chromene derivatives, Nano-Structures & Nano-Objects. 12 (2017) 33-40. https://doi.org/10.1016/j.nanoso.2017.08.013

[39] H. Goksu, Y. Yıldız, B. Celik, M. Yazıcı, B. Kılbas and F. Sen, Highly Efficient and Monodisperse Graphene Oxide Furnished Ru/Pd Nanoparticles for the Dehalogenation of Aryl Halides via Ammonia Borane, Chemistry Select. 1 (2016) 953-958. https://doi.org/10.1002/slct.201600207

[40] S. Akocak, B. Şen, N. Lolak, A. Şavk, M. Koca, S. Kuzu, F. Şen, One-pot three-component synthesis of 2-Amino-4H-Chromene derivatives by using monodisperse Pd nanomaterials anchored graphene oxide as highly efficient and recyclable catalyst, Nano-Structures & Nano-Objects. 11 (2017) 25–31. https://doi.org/10.1016/j.nanoso.2017.06.002

[41] Y. Yildiz, E. Erken, H. Pamuk and F. Sen, Monodisperse Pt Nanoparticles Assembled on Reduced Graphene Oxide: Highly Efficient and Reusable Catalyst for Methanol Oxidation and Dehydrocoupling of Dimethylamine-Borane (DMAB) J. Nanosci. Nanotechnol. 16 (2016) 5951-5958. https://doi.org/10.1166/jnn.2016.11710

[42] E. Demir, A. Savk, B. Sen, F. Sen, A novel monodisperse metal nanoparticles anchored graphene oxide as Counter Electrode for Dye-Sensitized Solar Cells. Nano-Structures & Nano-Objects, 12 (2017) 41-45. https://doi.org/10.1016/j.nanoso.2017.08.018

[43] T. Demirci, B. Çelik, Y. Yıldız, S. Eriş, M. Arslan, B. Kilbas and F. Sen, One-Pot Synthesis of Hantzsch Dihydropyridines Using Highly Efficient and Stable PdRuNi@GO Catalyst. RSC Advances, 6 (2016) 76948 – 76956. https://doi.org/10.1039/C6RA13142E

[44] B. Çelik, G. Başkaya, Ö. Karatepe, E. Erken, F. Şen, Monodisperse Pt(0)/DPA@GO nanoparticles as highly active catalysts for alcohol oxidation and dehydrogenation of DMAB. International Journal of Hydrogen Energy, 41 (2016) 5661-5669. https://doi.org/10.1016/j.ijhydene.2016.02.061

[45] H. Goksu, Y. Yıldız, B. Celik, M. Yazıcı, B. Kılbas and F. Sen, Highly Efficient and Monodisperse Graphene Oxide Furnished Ru/Pd Nanoparticles for the Dehalogenation of Aryl Halides via Ammonia Borane. Chemistry Select, 1 (5) (2016) 953-958. https://doi.org/10.1002/slct.201600207

[46] H. Goksu, Y. Yıldız, B. Çelik, M. Yazici, B. Kilbas, and F. Sen, Eco-friendly hydrogenation of aromatic aldehyde compounds by tandem dehydrogenation of dimethylamine-borane in the presence of reduced graphene oxide furnished platinum nanocatalyst. Catalysis Science and Technology, 6 (2016) 2318 – 2324. https://doi.org/10.1039/C5CY01462J

[47]   B. Aday, Y. Yıldız, R. Ulus, S. Eriş, M. Kaya, and F. Sen, One-Pot, Efficient and Green Synthesis of Acridinedione Derivatives using Highly Monodisperse Platinum Nanoparticles Supported with Reduced Graphene Oxide. New Journal of Chemistry, 40 (2016) 748 – 754. https://doi.org/10.1039/C5NJ02098K

[48]   S. Bozkurt, B. Tosun, B. Sen, S. Akocak, A. Savk, M. F. Ebeoğlugil, F. Sen, A hydrogen peroxide sensor based on TNM functionalized reduced graphene oxide grafted with highly monodisperse Pd nanoparticles. Analytica Chimica Acta 989C (2017) 88-94. https://doi.org/10.1016/j.aca.2017.07.051

[49]   Z. Dasdelen, Y. Yıldız, S. Eriş, F. Şen, Enhanced electrocatalytic activity and durability of Pt nanoparticles decorated on GO-PVP hybride material for methanol oxidation reaction. Applied Catalysis B: Environmental 219C (2017) 511-516. https://doi.org/10.1016/j.apcatb.2017.08.014

[50]   R. Ayranci, G. Baskaya, M. Guzel, S. Bozkurt, M. Ak, A. Savk, F. Sen, Carbon based Nanomaterials for High Performance Optoelectrochemical Systems. Chemistry Select, 2 (4) (2017) 1548-1555. https://doi.org/10.1002/slct.201601632

[51]   B. Aday, H. Pamuk, M. Kaya, and F. Sen, Graphene Oxide as Highly Effective and Readily Recyclable Catalyst Using for the One-Pot Synthesis of 1,8-Dioxoacridine Derivatives, J. Nanosci. Nanotechnol. 16 (2016) 6498-6504. https://doi.org/10.1166/jnn.2016.12432

[52]   B. Khodadadi, M. Bordbar, M. Nasrollahzadeh, Facile and green solvothermal synthesis of palladium nanoparticle-nanodiamond-graphene oxide material with improved bifunctional catalytic properties, Journal of the Iranian Chemical Society. 14 (2017) 2503-2512.

[53]   F. Sen, Z. Ozturk, S. Sen, G. Gokagac, The preparation and characterization of nano-sized Pt-Pd alloy catalysts and comparison of their superior catalytic activities for methanol and ethanol oxidation, Journal of Materials Science. 47 (2012) 8134–8144. https://doi.org/10.1007/s10853-012-6709-3

[54]   B. Celik, Y. Yildiz, H. Sert, E. Erken, Y. Koskun, F. Sen, Monodispersed palladium–cobalt alloy nanoparticles assembled on poly(N-vinyl-pyrrolidone) (PVP) as a highly effective catalyst for dimethylamine borane (DMAB) dehydrocoupling, RSC Adv. 6 (2016) 24097 – 24102. https://doi.org/10.1039/C6RA00536E

[55]   B. Celik, S. Kuzu, E. Demir, E. Yıldırır, F. Sen, Highly Efficient Catalytic Dehydrogenation of Dimethly Ammonia Borane via Monodisperse Palladium-

Nickel Alloy Nanoparticles Assembled on PEDOT, Int. J. Hydrogen Energy. 42 (2017) 23307-23314. https://doi.org/10.1016/j.ijhydene.2017.05.115

[56] T. Kim, X. Fu, D. Warther, M.J. Sailor, Green synthesis of Pd nanoparticles at Apricot kernel shell substrate using Salvia hydrangea extract: Catalytic activity for reduction of organic dyes, ACS Nano. 11 (2017) 2773-2784. https://doi.org/10.1021/acsnano.6b07820

[57] M. Bordbar, N. Mortazavimanesh, Green synthesis of Pd/walnut shell nanocomposite using Equisetum arvense L. leaf extract and its application for the reduction of 4-nitrophenol and organic dyes in a very short time, Environmental Science and Pollution Research. 24 (2017) 4093-4104. https://doi.org/10.1007/s11356-016-8183-y

[58] L.L. Carvalho, F. Colmati, A.A. Tanaka, Nickel–palladium electrocatalysts for methanol, ethanol, and glycerol oxidation reactions, International Journal of Hydrogen Energy. 42 (2011) 16118-16116. https://doi.org/10.1016/j.ijhydene.2017.05.124

[59] N.R. Elezovic, P. Zabinski, P. Ercius, U.Č. Lačnjevac, N.V. Krstajic, High surface area Pd nanocatalyst on core-shell tungsten based support as a beneficial catalyst for low temperature fuel cells application, Electrochimica Acta. 247 (2017) 674-684. https://doi.org/10.1016/j.electacta.2017.07.066

[60] J. Fan, K. Qi, L. Zhang, S. Yu, X. Cui, Engineering Pt/Pd Interfacial Electronic Structures for Highly Efficient Hydrogen Evolution and Alcohol Oxidation, ACS Applied Materials and Interfaces. 9 (2017) 18008-18014. https://doi.org/10.1021/acsami.7b05290

[61] P. Qiu, S. Lian, G. Yang, S. Yang, Halide ion-induced formation of single crystalline mesoporous PtPd bimetallic nanoparticles with hollow interiors for electrochemical methanol and ethanol oxidation reaction, Nano Research. 10 (2017) 1064-1077. https://doi.org/10.1007/s12274-016-1367-4

[62] A. Zhang, Y. Xiao, F. Gong, L. Zhang, Y. Zhang, Solid-state synthesis, formation mechanism and enhanced electrocatalytic properties of Pd nanoparticles supported on reduced graphene oxide, ECS Journal of Solid State Science and Technology. 6 (2017) M13-M18. https://doi.org/10.1149/2.0271701jss

[63] J. Zhang, S. Lu, Y. Xiang, J. Liu, S.P. Jiang, Carbon-Nanotubes-Supported Pd Nanoparticles for Alcohol Oxidations in Fuel Cells: Effect of Number of Nanotube Walls on Activity, ChemSusChem. 8 (2015) 2956-2966. https://doi.org/10.1002/cssc.201500107

[64] H. Klug, L. Alexander, X-Ray Diffraction Procedures: For Polycrystalline and Amorphous Materials, 2nd Edition, Wiley, New York, 1954.

[65] E. Erken, İ. Esirden, M. Kaya and F. Sen, A Rapid and Novel Method for the Synthesis of 5-Substituted 1H-tetrazole Catalyzed by Exceptional Reusable Monodisperse Pt NPs@AC under the Microwave Irradiation. RSC Advances, 5 (2015) 68558-68564. https://doi.org/10.1039/C5RA11426H

[66] Ö. Karatepe, Y. Yıldız, H. Pamuk, S. Eriş, Z. Dasdelen and F. Şen, Enhanced electro catalytic activity and durability of highly mono disperse Pt@PPy-PANI nanocomposites as a novel catalyst for electro-oxidation of methanol. RSC Advances, 6 (2016) 50851 – 50857. https://doi.org/10.1039/C6RA06210E

[67] E. Erken, H. Pamuk, Ö. Karatepe, G. Başkaya, H. Sert, O. M. Kalfa, F. Şen, New Pt(0) Nanoparticles as Highly Active and Reusable Catalysts in the C1–C3 Alcohol Oxidation and the Room Temperature Dehydrocoupling of Dimethylamine-Borane (DMAB). Journal of Cluster Science, (2016) 27: 9. https://doi.org/10.1007/s10876-015-0892-8

[68] B. Çelik, Y. Yildiz, E. Erken and Y. Koskun, F. Sen, Monodisperse Palladium-Cobalt Alloy Nanoparticles Assembled on Poly (N-vinyl-pyrrolidone) (PVP) as Highly Effective Catalyst for the Dimethylammine Borane (DMAB) dehydrocoupling. RSC Advances, 6 (2016) 24097 – 24102. https://doi.org/10.1039/C6RA00536E

[69] H. Göksu, B. Çelik, Y. Yıldız, B. Kılbaş and F. Şen, Superior monodisperse CNT-Supported CoPd (CoPd@CNT) nanoparticles for selective reduction of nitro compounds to primary amines with NaBH4 in aqueous medium. Chemistry Select, 1 (10) (2016) 2366-2372. https://doi.org/10.1002/slct.201600509

[70] Y. Yıldız, R. Ulus, S. Eris, B. Aday, M. Kaya and F. Sen, Functionalized multi-walled carbon nanotubes (f-MWCNT) as Highly Efficient and Reusable Heterogeneous Catalysts for the Synthesis of Acridinedione Derivatives. Chemistry Select, 1 (13) (2016) 3861–3865.

[71] Y. Yıldız, İ. Esirden, E. Erken, E. Demir, M. Kaya and F. Şen, Microwave (Mw)-assisted Synthesis of 5-Substituted 1H-Tetrazoles via [3+2] Cycloaddition Catalyzed by Mw-Pd/Co Nanoparticles Decorated on Multi-Walled Carbon Nanotubes. Chemistry Select, 1 (8) (2016) 1695-1701. https://doi.org/10.1002/slct.201600265

[72] Y. Yıldız, H. Pamuk, Ö. Karatepe, Z. Dasdelen and F.Sen, Carbon black hybride material furnished monodisperse Platinum nanoparticles as highly efficient and

reusable electrocatalysts for formic acid electro-oxidation. RSC Advances, 6 (2016) 32858 – 32862. https://doi.org/10.1039/C6RA00232C

[73]   E. Erken, Y. Yildiz, B. Kilbas, and F. Sen, Synthesis and Characterization of Nearly Monodisperse Pt Nanoparticles for C1 to C3 Alcohol Oxidation and Dehydrogenation of Dimethylamine-borane (DMAB). J. Nanosci. Nanotechnol. 16 (2016) 5944-5950. https://doi.org/10.1166/jnn.2016.11683

[74]   B. Çelik, E. Erken, S. Eriş, Y. Yıldız, B. Şahin, H. Pamuk and F. Sen, Highly monodisperse Pt(0)@AC NPs as highly efficient and reusable catalysts: the effect of the surfactant on their catalytic activities in room temperature dehydrocoupling of DMAB. Catalysis Science and Technology, 6 (2016) 1685 – 1692. https://doi.org/10.1039/C5CY01371B

[75]   B. Çelik, S. Kuzu, E. Erken, Y. Koskun, F. Sen, Nearly Monodisperse Carbon Nanotube Furnished Nanocatalysts as Highly Efficient and Reusable Catalyst for Dehydrocoupling of DMAB and C1 to C3 Alcohol Oxidation. International Journal of Hydrogen Energy, 41 (2016) 3093-3101. https://doi.org/10.1016/j.ijhydene.2015.12.138

[76]   G. Baskaya, I. Esirden, E. Erken, F. Sen, and M. Kaya, Synthesis of 5-Substituted-1H-Tetrazole Derivatives Using Monodisperse Carbon Black Decorated Pt Nanoparticles as Heterogeneous Nanocatalysts. J. Nanosci. Nanotechnol. 17 (2017) 1992-1999. https://doi.org/10.1166/jnn.2017.12867

[77]   G. Baskaya, Y. Yıldız, A. Savk, T. Onal Okyay, S. Eris, F. Sen, Rapid, Sensitive, and Reusable Detection of Glucose by Highly Monodisperse Nickel nanoparticles decorated functionalized multi-walled carbon nanotubes. Biosensors and Bioelectronics, 91 (2017) 728–733. https://doi.org/10.1016/j.bios.2017.01.045

[78]   E. Demir, B. Sen, F. Sen, Highly efficient nanoparticles and f-MWCNT nanocomposites based counter electrodes for dye-sensitized solar cells. Nano-Structures & Nano-Objects (Invited), 11 (2017) 39-45. https://doi.org/10.1016/j.nanoso.2017.06.003

[79]   B. Sen, S. Kuzu, E. Demir, E. Yıldırır, F. Sen, Highly Efficient Catalytic Dehydrogenation of Dimethly Ammonia Borane via Monodisperse Palladium-Nickel Alloy Nanoparticles Assembled on PEDOT. International Journal of Hydrogen Energy, 42 (36) 2017 23307-23314. https://doi.org/10.1016/j.ijhydene.2017.05.115

[80]   J.M. Sieben, A.E. Alvarez, V. Comignani, M.M.E. Duarte, Methanol and ethanol oxidation on carbon supported nanostructured Cu core Pt-Pd shell electrocatalysts

synthesized via redox displacement, International Journal of Hydrogen Energy. 39 (2014) 11547-11556. https://doi.org/10.1016/j.ijhydene.2014.05.123

[81]   D.H. Nagaraju, S. Devaraj, P. Balaya, Palladium nanoparticles anchored on graphene nanosheets: Methanol, ethanol oxidation reactions and their kinetic studies, Materials Research Bulletin. 60 (2014) 150-157. https://doi.org/10.1016/j.materresbull.2014.08.027

[82]   Saipanya, S., S. Lapanantnoppakhun, T. Sarakonsri, Electrochemical deposition of platinum and palladium on gold nanoparticles loaded carbon nanotube support for oxidation reactions in fuel cell, Journal of Chemistry. (2014) http://dx.doi.org/10.1155/2014/104514.

[83]   G. Yang, Y. Zhou, H.B. Pan, (...), Zhu, J. J, Y. Lin, Ultrasonic-assisted synthesis of Pd-Pt/carbon nanotubes nanocomposites for enhanced electro-oxidation of ethanol and methanol in alkaline medium, Ultrasonics Sonochemistry. 28 (2016) 192-198. https://doi.org/10.1016/j.ultsonch.2015.07.021

[84]   Y. Gao, F. Wang, Y. Wu, R. Naidu, Z. Chen, Comparison of degradation mechanisms of microcystin-LR using nanoscale zero-valent iron (nZVI) and bimetallic Fe/Ni and Fe/Pd nanoparticles, Chemical Engineering Journal. 285 (2016) 459-466. https://doi.org/10.1016/j.cej.2015.09.078

[85]   K. Mishra, N. Basavegowda, Y.R. Lee, Biosynthesis of Fe, Pd, and Fe-Pd bimetallic nanoparticles and their application as recyclable catalysts for [3 + 2] cycloaddition reaction: A comparative approach, Catalysis Science and Technology. 5 (2015) 2612-2621. https://doi.org/10.1039/C5CY00099H

[86]   F. Sen, and G. Gokagac, Different sized platinum nanoparticles supported on carbon: An XPS study on these methanol oxidation catalysts. Journal of Physical Chemistry C, 111 (2007) 5715-5720. https://doi.org/10.1021/jp068381b

[87]   F. Sen, and G. Gokagac, The activity of carbon supported platinum nanoparticles towards methanol oxidation reaction – role of metal precursor and a new surfactant, tert-octanethiol. Journal of Physical Chemistry C, 111 (2007) 1467-1473. https://doi.org/10.1021/jp065809y

[88]   F. Sen, and G. Gokagac, Improving Catalytic Efficiency in the Methanol Oxidation Reaction by Inserting Ru in Face-Centered Cubic Pt Nanoparticles Prepared by a New Surfactant, tert-Octanethiol. Energy & Fuels, 22 (3) (2008) 1858- 1864. https://doi.org/10.1021/ef700575t

[89] F. Sen, Z. Ozturk, S. Sen, G. Gokagac, The preparation and characterization of nano-sized Pt-Pd alloy catalysts and comparison of their superior catalytic activities for methanol and ethanol oxidation. Journal of Materials Science, 47 (2012) 8134–8144. https://doi.org/10.1007/s10853-012-6709-3

[90] S. Sen, F. Sen, G. Gokagac, Preparation and characterization of nano-sized Pt–Ru/C catalysts and their superior catalytic activities for methanol and ethanol oxidation. Phys. Chem. Chem. Phys., 13 (2011) 6784-6792. https://doi.org/10.1039/c1cp20064j

[91] F. Sen, S. Sen, G. Gokagac, Efficiency enhancement in the methanol/ethanol oxidation reactions on Pt nanoparticles prepared by a new surfactant, 1,1-dimethyl heptanethiol, and surface morphology by AFM. Phys. Chem. Chem. Phys., 13 (2011) 1676-1684. https://doi.org/10.1039/C0CP01212B

[92] F. Sen, S. Ertan, S. Sen, G. Gokagac, Platinum nanocatalysts prepared with different surfactants for C1 to C3 alcohol oxidations and their surface morphologies by AFM. Journal of Nanoparticle Research, 14 (2012) 922-26. https://doi.org/10.1007/s11051-012-0922-5

[93] F. Sen, S. Sen, G. Gokagac, High performance Pt nanoparticles prepared by new surfactants for C1 to C3 alcohol oxidation reactions. Journal of Nanoparticle Research, 15 (2013) 1979. https://doi.org/10.1007/s11051-013-1979-5

[94] F. Sen, and G. Gokagac, Pt Nanoparticles Synthesized with New Surfactans: Improvement in C1-C3 Alcohol Oxidation Catalytic Activity. Journal of Applied Electrochemistry, 44(1) (2014) 199 – 207. https://doi.org/10.1007/s10800-013-0631-5

[95] F. Şen, Y. Karataş, M. Gülcan, M. Zahmakıran, Amylamine stabilized platinum (0) nanoparticles: active and reusable nanocatalyst in the room temperature dehydrogenation of dimethylamine- borane. RSC Advances, 4 (4) (2014) 1526-1531. https://doi.org/10.1039/C3RA43701A

[96] Y. Yıldız, S. Kuzu, B. Sen, A. Savk, S. Akocak, F. Şen, Different ligand based monodispersed metal nanoparticles decorated with rGO as highly active and reusable catalysts for the methanol oxidation. International Journal of Hydrogen Energy, 42 (18) 2017 13061-13069. https://doi.org/10.1016/j.ijhydene.2017.03.230

[97] B. Sen, S. Kuzu, E. Demir, S. Akocak, F. Sen, Monodisperse Palladium-Nickel Alloy Nanoparticles Assembled on Graphene Oxide with the High Catalytic Activity and Reusability in the Dehydrogenation of Dimethylamine-Borane.

International Journal of Hydrogen Energy, 42 (36) (2017) 23276-23283.
https://doi.org/10.1016/j.ijhydene.2017.05.113

[98]   B. Sen, S. Kuzu, E. Demir et.al., Polymer-Graphene hybride decorated Pt
       Nanoparticles as highly eficient and reusable catalyst for the Dehydrogenation of
       Dimethylamine-borane at room temperature. International Journal of Hydrogen
       Energy, 42 (36) (2017) 23284-23291.
       https://doi.org/10.1016/j.ijhydene.2017.05.112

[99]   B. Sen, S. Kuzu, E. Demir, et.al, Highly Monodisperse RuCo Nanoparticles
       decorated on Functionalized Multiwalled Carbon Nanotube with the Highest
       Observed Catalytic Activity in the Dehydrogenation of Dimethylamine Borane.
       International Journal of Hydrogen Energy, 42 (36) (2017) 23292-23298.
       https://doi.org/10.1016/j.ijhydene.2017.06.032

[100] S. Eris, Z. Daşdelen, F. Sen, Investigation of electrocatalytic activity and stability
       of Pt@f-VC catalyst prepared by in-situ synthesis for Methanol electrooxidation,
       International Journal of Hydrogen Energy. 43 (1) (2018) 385-390.
       https://doi.org/10.1016/j.ijhydene.2017.11.063

[101] S. Eris, Z. Daşdelen, F. Sen, Enhanced electrocatalytic activity and stability of
       monodisperse Pt nanocomposites for direct methanol fuel cells, Journal of Colloid
       and Interface Science. 513 (2018) 767–773.
       https://doi.org/10.1016/j.jcis.2017.11.085

[102] B. Sen, S. Kuzu, E. Demir, T. Onal Okyay, F. Sen, Hydrogen liberation from the
       dehydrocoupling of dimethylamine-borane at room temperature by using novel
       and highly monodispersed RuPtNi nanocatalysts decorated with graphene oxide.
       International Journal of Hydrogen Energy, 42 (36) (2017) 23299-23306.
       https://doi.org/10.1016/j.ijhydene.2017.04.213

[103] S. Eris, Z. Daşdelen, Y. Yıldız, F. Sen, Nanostructured Polyaniline-rGO decorated
       platinum catalyst with enhanced activity and durability for Methanol oxidation,
       International Journal of Hydrogen Energy. 43 (3) 2018 1337–1343.
       https://doi.org/10.1016/j.ijhydene.2017.11.051

[104] K. T. Wong, N. C. Eu, S. Ibrahim, H. Kim, Y. Yoon, and M. Jang, Recyclable
       magnetite-loaded palm shell-waste based activated carbon for the effective
       removal of methylene blue from aqueous solution. Journal of Cleaner Production,
       115(2016) 337-342. https://doi.org/10.1016/j.jclepro.2015.12.063

[105] Z. Chen, J. Fu, M. Wang, X. Wang, J. Zhang, Q. Xu, and R. A. Lemons, Self-
       assembly fabrication of microencapsulated n-octadecane with natural silk fibroin

shell for thermal-regulating textiles, Appl. Surf. Sci., 289 (2014) 495–501.
https://doi.org/10.1016/j.apsusc.2013.11.022

[106] S. Qu, F. Huang, S. Yu, G. Chen, and J. Kong, Magnetic removal of dyes from
aqueous solution using multi-walled carbon nanotubes filled with Fe2O3 particles.
Journal of Hazardous Materials, 160 (2008) 643-647.
https://doi.org/10.1016/j.jhazmat.2008.03.037

[107] Y. El Mouzdahir, A. Elmchaouri, R. Mahboub, A. Gil, and S. Korili, Adsorption
of Methylene Blue from Aqueous Solutions on a Moroccan Clay. Journal of
Chemical & Engineering Data, 52 (2007) 1621-1625.
https://doi.org/10.1021/je700008g

[108] GK. Ramesha, AV. Kumara, H.B Muralidhara, S. Sampath, Graphene and
graphene oxide as effective adsorbents toward anionic and cationic dyes. Journal
of Colloid Interface Science, 361 (2011) 270.
https://doi.org/10.1016/j.jcis.2011.05.050

[109] F. He, JT. Fan, D. Ma, L. Zhang, C. Leung, HL. Chan. The attachment of Fe3O4
nanoparticles to graphene oxide by covalent bonding. Carbon, 48 (2010) 3139.
https://doi.org/10.1016/j.carbon.2010.04.052

[110] T. Liu, Y. Li, Q. Du, J. Sun, Y. Jiao, G. Yang, et al. Adsorption of methylene blue
from aqueous solution by graphene. Colloids Surf B, 90 (2012) 197.
https://doi.org/10.1016/j.colsurfb.2011.10.019

[111] Y. Yao, S. Miao, S. Yu, L.P. Ma, H. Sun, S.J. Wang, Fabrication of Fe3O4/SiO2
core/shell nanoparticles attached to graphene oxide and its use as an adsorbent.
Colloid Interface Sci, 379 (2012) 20. https://doi.org/10.1016/j.jcis.2012.04.030

[112] W. Zhang, H. Yan, H. Li, Z. Jiang, L. Dong, X. Kan, et al. Removal of dyes from
aqueous solutions by straw based adsorbents: Batch and column studies. Chem
Eng J, 168 (2011) 1120–7. https://doi.org/10.1016/j.cej.2011.01.094

[113] H. Deng, J. Lu, G. Li, G. Zhang, X. Wang, Adsorption of methylene blue on
adsor- bent materials produced from cotton stalk. Chem Eng J 172 (2011) 326–34.
https://doi.org/10.1016/j.cej.2011.06.013

[114] G. Akkaya, F. Guzel, Application of some domestic wastes as new low-cost
biosorbents for removal of methylene blue: kinetic and equilibrium studies. Chem
Eng Commun 201 (2014) 557–78. https://doi.org/10.1080/00986445.2013.780166

[115] X. Chen, S. Lv, S. Liu, P. Zhang, A. Zhang, J. Sun, et al. Adsorption of methylene blue by rice hull ash. Sep Sci Technol 47 (2012) 147–56. https://doi.org/10.1080/01496395.2011.606865

[116] J. Liang, J .Wu, P. Li, X. Wang, B. Yang, Shaddock peel as a novel low-cost adsor- bent for removal of methylene blue from dye wastewater. Desalination Water Treat 39 (2012) 70–5. https://doi.org/10.1080/19443994.2012.669160

[117] Q. Zhou, W. Gong, C. Xie, X. Yuan, Y. Li, C. Bai, et al. Biosorption of methylene blue from aqueous solution on spent cottonseed hull substrate for pleurotus ostreatus cultivation. Desalination Water Treat 29 (2011) 317–25. https://doi.org/10.5004/dwt.2011.2238

[118] R.R. Krishni, K.Y. Foo, B.H. Hameed, Adsorptive removal of methylene blue using the natural adsorbent-banana leaves. Desalination Water Treat 52 (2014) 6104–12. https://doi.org/10.1080/19443994.2013.815687

[119] A. Ayla, A. Cavus, Y. Bulut, Z. Baysal, C. Aytekin, Removal of methylene blue from aqueous solutions onto Bacillus subtilis : determination of kinetic and equilib- rium parameters. Desalination 51 (2013) 7596–603. https://doi.org/10.1080/19443994.2013.791780

Chitosan-Based Adsorbents for Wastewater Treatment, Ed. Abu Nasar   Materials Research Forum LLC
Materials Research Foundations **34** (2018) 81-98   doi: http://dx.doi.org/10.21741/9781945291753-4

Chapter 4

Effect of Chitosan Modification on its Structure and Specific Surface Area

Krzysztof Barbusiński[1]*, Krzysztof Filipek[2], Szymon Salwiczek[3]

[1]Institute of Water and Wastewater Engineering, The Silesian University of Technology, Konarskiego 18A, 44-100 Gliwice, Poland

[2]FDKF - Advisory Entity – Water and Wastewater Management, Jaworowa 16, 32-590 Libiąż, Poland

[3]Oksydan Ltd., Łużycka 16, 44-100 Gliwice, Poland

*krzysztof.barbusinski@polsl.pl

Abstract

The chapter shows various chitosan modification methods: preparation of hydrogel beads from powdered chitosan, cross-linking of beads using epichlorohydrin or glutaraldehyde and conditioning of beads using $NaHSO_4$. For each modified and non-modified form of chitosan, the surface structure was analyzed using SEM images. Moreover, for chitosan and its modifications the specific surface area, the total pore volume and the total pore surface area were determined using BET nitrogen sorption isotherms.

Keywords

Chitosan, Chitosan Beads, Chitosan Modification, Surface Structure, Specific Surface Area

Contents

1.   Introduction

Activated carbon is the most popular and universal sorbent used to remove contaminants from wastewater. Because of its high costs, several attempts have been performed to find some new, effective and not expensive sorbents to apply in wastewater treatment. The biosorbents, i.e. naturally occurring sorbents as straw, compost, various plant materials, bacteria biomass or some biopolymers could be applied as alternative substances. Their low costs (they are often recognized as waste products), good sorption properties, non-toxic character and sensitivity to biodegradation are their clear advantages. Chitosan is considered as a biosorbent. It is a natural polysaccharide produced from the chitin shells of crustaceans that are fished for food. Chitin and chitosan are formed in the waste products processing to accompany crustaceans (e.g. crabs, shrimps and krill) fishing. The production of chitosan is especially cost-effective because the carotenoids are formed during the process. The seashells and crusts are composed of a high content of astaxanthin – a carotenoid that is very expensive in synthetic production. Astaxanthin is used as a food additive and as a food agent for feeding salmons [1].

Nowadays chitosan is on an industrial scale produced in a reaction of chemical or enzymatic chitin deacetylation and it is applied in many industrial areas. Chemical deacetylation of chitin generally leads to chitosan formation in its pure form. In an initial stage, chitin is purified from mineral salts and lipids in a reaction with acid at a proper temperature and for a period of time. Then, chitin is deacetylated in a solution of NaOH at the proper temperature and for a period of time. The lower reaction time the lower deacetylation degree and the higher molecular weight [2]. Enzymatic deacetylation is always applied when a chitosan of a low molecular weight, low density and a good solubility in water, without a presence of an acid, is required. The enzymatic deacetylation does not affect the polymer structure and is easier to be controlled. Hydrolysis of chitosan is carried out using enzymes as e.g. glucanase, lipase, protease, the most efficient, however, are chitinase and chitosanase – the enzymes that occur in fungi, bacterias, and plants [3,4].

The process conditions of chitosan synthesis determine its crystal structure, molecular weight, polymer acetylation degree and the residues of the other components. The properties of chitin are reflected in the biological and technological usefulness of chitosan (i.e. deacetylation degree, molecular weight) and its chemical purity. Time and temperature are the critical process parameters to affect the required properties of chitosan.

Chitosan shows affinity to many contaminations, e.g. dyes, metal ions, anions, phenols, pesticides, radionuclides and the humic substances [5]. It is composed of a high concentration of amine and hydroxyl groups that have a great ability to bind the contaminations. Chitosan is chemically very stable yet reactive compound and it shows chelating properties what makes it a good sorption material [6]. Chitosan could be applied in a pure form (powdered or in a form of flakes) or it could be changed into membranes or hydrogel beads. The application of chitosan and its modified forms to remove strictly determined wastewater contaminations have been recently examined [7,8]. Thus, chitosan has been modified to get a sorbent of a great sorption capacity, good mechanical strength and very resistant to various process conditions. It is widely used in many industrial applications [9-12].

To increase its sorption properties, the chemical modification of chitosan should meet two basic conditions: sorbent protection against its dissolving in a low pH solution (when such a low pH is required or the most convenient) and improvement of its sorption properties by means of increasing its sorption capacity or a change of sorbent selectivity. Cross-linking, as one of the possible methods of chitosan modification, could be performed using several bifunctional agents, e.g. glutaraldehyde, oxidized β-cyclodextrine, ethylene glicol diglicide ether, hexamethylene diizocyanate or glycerol polyglicide ether. Some monofunctional agents, e.g. epichlorohydrin, could also be applied in cross-linking processes. Their reactivity is based on the incorporation of amine group into ether group, according to the Schiff's reaction, and then a reaction of chloride with the other functional or amine groups. A tri-polyphosphate could also be applied to obtain cross-linked gel chitosan beads in a process of coagulation/neutralization [13]. The cross-linking of chitosan, besides increasing its sorption capacity, could alternatively lead to decreasing the capacity when the amine groups of chitosan are involved in the sorption reaction.

The chitosan modification processes, using the hydrogel beads from a pure chitosan and then their cross-linking modification using epichlorohydrin (ECH) or glutaraldehyde (GLU) and its conditioning with sodium hydrosulphate (NaHSO$_4$), are presented in this chapter. Moreover, the observed changes of surface structure and specific surface area of chitosan and its modifications and the results of these observations, using SEM photos and BET nitrogen sorption isotherms, are also presented.

2. Methodology

Powdered chitosan (purchased from BOC Sciences) of deacetylation degree DD $\geq$ 95%, molecular weight 500000, volume density $\geq$ 0.6 g/cm^3 and water content < 8% was applied in the experiments.

2.1 Formation of chitosan beads

10 g of chitosan was dissolved in 300 cm^3 of 5% CH$_3$COOH to form the chitosan beads. The mixture was adjusted to 1 dm^3 and it was stirred by 12 hours to complete dissolving of chitosan and to get its 1% solution. Then, the solution was added dropwise to a 2M solution of NaOH by means of a 0.8 mm needle. The precipitated chitosan beads of 2-3 mm diameter were formed at the bottom of the beaker. After 12 hours the beads were washed with distilled water to reach a neutral pH. NaOH and CH$_3$COOH solutions were prepared using analytical grade reagents made by POCH (Poland).

2.2 Modification of chitosan beads

- The cross-linking of chitosan using epichlorohydrine (ECH) was performed in a volumetric 2 dm^3 beaker filled up with 500 cm^3 of 1M NaOH and the chitosan beads. Then, ECH 99% solution (Acros Organics) was added in 1:1 relation of ECH/chitosan beads. The reaction time was 6 hours, while slow stirring, at a temperature of 60 °C.The cross-linked beads were washed with distilled water to remove unreacted ECH.

- The cross-linking of chitosan using glutaraldehyde (GLU) was performed in a volumetric 2 dm^3 beaker filled up with a solution of GLU (1 g/1 g of the dry mass of beads) by 24 hours stirring. Then, the beads were separated and washed with distilled water to remove unreacted GLU. The process was carried out using 25% solution of GLU (Acros Organics).

- The conditioning was applied to both the non-cross-linked beads and the ECH cross-linked beads. The reaction was carried out in a volumetric 2 dm^3 beaker with a 500 cm^3 solution of 0.5 mmol/dm^3 NaHSO$_4$ (Acros Organics). The mixture was stirred by 24 hours at 50 °C. Then, the beads were separated and washed with distilled water. The conditioning was done using a 92% granulated NaHSO$_4$.

Chitosan-Based Adsorbents for Wastewater Treatment, Ed. Abu Nasar   Materials Research Forum LLC
Materials Research Foundations 34 (2018) 81-98   doi: http://dx.doi.org/10.21741/9781945291753-4

2.3   SEM photos of chitosan and chitosan beads

To take SEM photos the examined samples were frozen in liquid nitrogen. The frozen samples were shielded from the air and humidity. Then, the samples were heated up to the temperature of -70 °C at a pressure of 10^{-3} Pa in order to evaporate a possible ice-covered surface. After that, the samples were cooled to -130 °C and were observed under a microscope. The high resolution 7600 F (FE-SEM) scanning electron microscope with electron beam gun (produced by JEOL) supplied with X-ray microprobe (EDS) was applied. The microscope was also supplied with a portable "cryo-SEM" to enable in-situ preparation, transfer to a microscope chamber and finally an examination of the frozen samples of water or the other liquids. The environmental samples could also be examined using this equipment.

2.4   Determination of BET isotherms for chitosan and chitosan beads

The ASAP 2010 camera (Micromeritics) was used to measure N_2 adsorption isotherms at liquid nitrogen temperature. First, the samples were filtered through the sieve and were placed in the refrigerator for 24 hours. After that, the samples were moved into a cooling-bath (ice-water-NaCl) at -8 °C. Then, the samples were dried under vacuum for ca 10 hours in the measuring ampoules of ASAP 2010 device and then frozen with a liquid nitrogen and pumped out to the vacuum of 0.001 Tr (1 Tr = 1 mm Hg). Process duration was about 12 hours. The frozen and desorbed samples were weighed and placed into a measurement chamber of ASAP 2010 to determine adsorption isotherms.

3.   Results

3.1   Chitosan and chitosan beads structures

The SEM images of chitosan and chitosan beads - non-modified and modified by means of cross-linking using ECH and GLU and then conditioned with $NaHSO_4$ - are presented and described below.

- **Pure, powdered non-modified chitosan**

The individual molecules of a powdered chitosan, that look like the flakes, are presented in Figure 1A. The substance is not uniform in its structure and appearance and it creates various forms and dimensions. The high magnification image (see Figure 1B) clearly shows the sorbent surface structure. It is flat with a little broken slice, however, it does not show a porous structure and does not have a sponge structure. This is an evidence that chitosan is not a typical porous sorption material, at least at its surface layer.

Figure 1 SEM images of chitosan at a magnification: A) 50x, B) 1700x.

- **Chitosan beads non-modified**

The Figures 2 A, B show the surface structures of non-modified chitosan beads. In comparison with powdered chitosan, the beads look like a porous sponge structure. The sorbent surface porous structure is noticeable. The pores have various shapes and dimensions and they are uniformly situated on the whole beads surface.

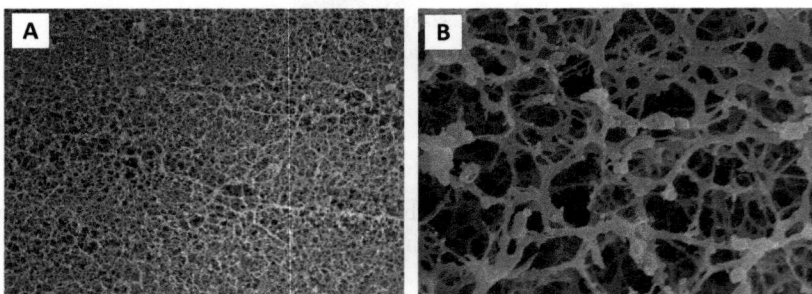

Figure 2 SEM images of non-modified chitosan beads at a magnification:A) 1000x, B) 10000x.

- **Chitosan beads conditioned with NaHSO$_4$**

The Figures 3 A-B present the chitosan beads after conditioning with NaHSO$_4$. Compared to non-modified beads they show apparent changes in the sorbent porous surface structure – it is more compressed and close-packed. A decrease of pores distance

Chitosan-Based Adsorbents for Wastewater Treatment, Ed. Abu Nasar   Materials Research Forum LLC
Materials Research Foundations **34** (2018) 81-98   doi: http://dx.doi.org/10.21741/9781945291753-4

and the dimensions of the pores are easy to notice. A cross-section of the chitosan bead is presented in Figures 3 C-D. One could see a 2 µm thick surface porous structure, whereas a less porous, even sphere is placed underneath.

It clearly shows that some corresponding modifications could be resulted in the partial only surface structure changes, up to a certain distance from the surface. In order to prove it and to determine the influence of the other modifications on the porous layer thickness, in the following experiments it would be required to take some additional photos of the modified beads.

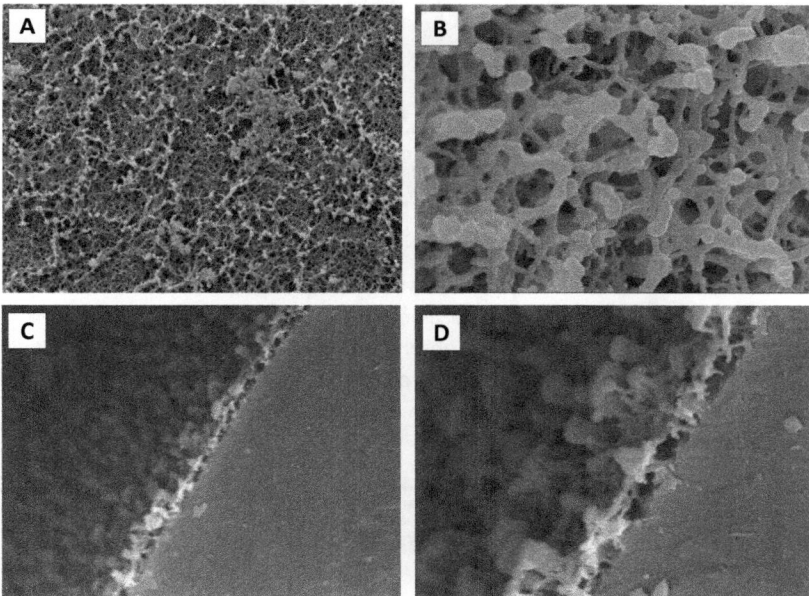

Figure 3 SEM images of chitosan beads conditioned with NaHSO₄ at a magnification: A) 1000x, B) 10000x, C) cross-section 2000x, D) cross-section 6000x

- **GLU cross-linked chitosan beads**

The Figures 4 A-B were taken for the cross-linked beads at the concentration of 1 g GLU/1 g of the dry mass of beads. Compared to non-modified beads and conditioned beads a further increase of surface structure density and compressing of the pores are easily noticed.

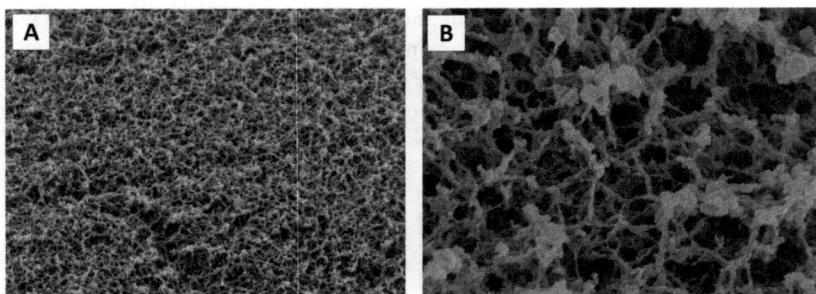

Figure 4   SEM images of chitosan beads cross-linked with GLU at a magnification: A) 1000x, B) 10000x.

- **ECH cross-linked chitosan beads**

The chitosan beads modification by ECH was resulted in the changes of sorbent surface structure (see Figures 5 A-B). The appearance of some ridges, invisible at the previous modifications, was noticed. One could observe a very close internal pores package and a significant decrease of their dimensions, to prove a progressive destruction of the porous surface structure of the beads. These observations should be, however, confirmed in the further experiments.

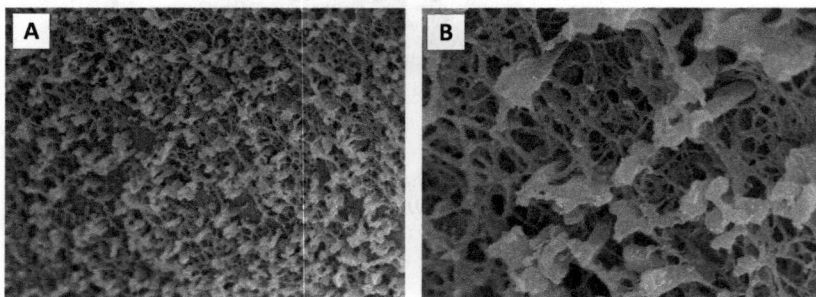

Figure 5   SEM images of chitosan beads cross-linked with ECH at a magnification: A) 2500x, B) 10000x.

Chitosan-Based Adsorbents for Wastewater Treatment, Ed. Abu Nasar   Materials Research Forum LLC
Materials Research Foundations **34** (2018) 81-98   doi: http://dx.doi.org/10.21741/9781945291753-4

- **Chitosan beads cross-linked by ECH and conditioned with NaHSO$_4$**

The final series of the images (see Figures 6 A-B) present the ECH cross-linked beads (1:1 mass relations) and then conditioned with NaHSO$_4$. The clear surface structure changes are noticed. Moreover, one could see a „loosening" of the porous structure and significant disappearance of the pores. The surface structure is less cohesive to prove decay (destruction) of the porous sorbent structure.

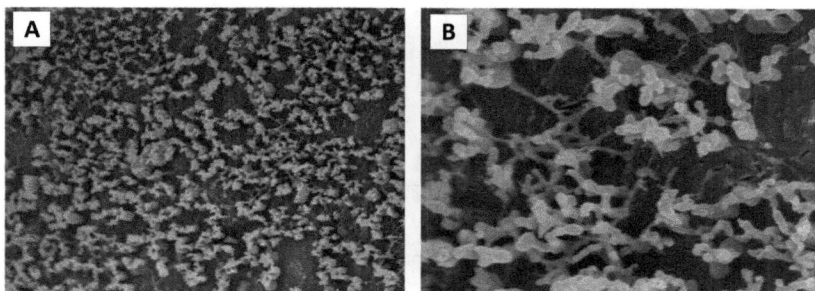

Figure 6   SEM images of chitosan beads cross-linked with ECH and conditioned with NaHSO$_4$ at a magnification: A) 2500x, B) 10000x.

3.2   Analysis of nitrogen sorption isotherms

The experimental BET nitrogen sorption isotherms at a temperature of -196 °C for chitosan and its modifications were analyzed in a form of linear and semi-logarithmic scales of the relative pressure (the pressure was presented in a decimal logarithmic scale). The trajectory was then better observed. A linear scale is most suitable to evaluate the isotherms at the medium and high pressure, whereas a semi-logarithmic one is most suitable at the lower relative pressure. Nitrogen sorption isotherms in the linear and semi-logarithmic scales of a relative pressure for chitosan, non-modified and modified chitosan beads are presented in Figures 7-18.

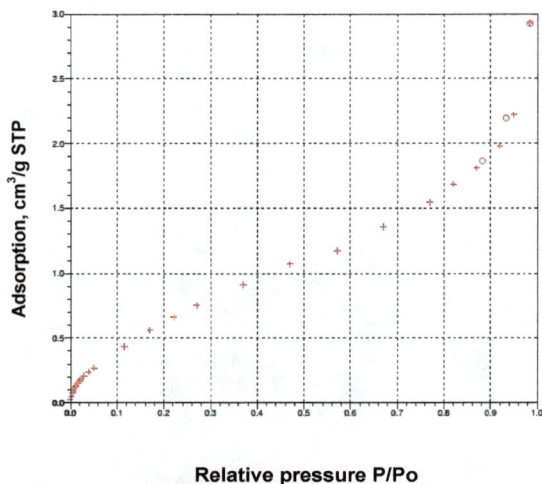

Figure 7   Nitrogen adsorption isotherm for chitosan in the linear scale of a relative pressure.

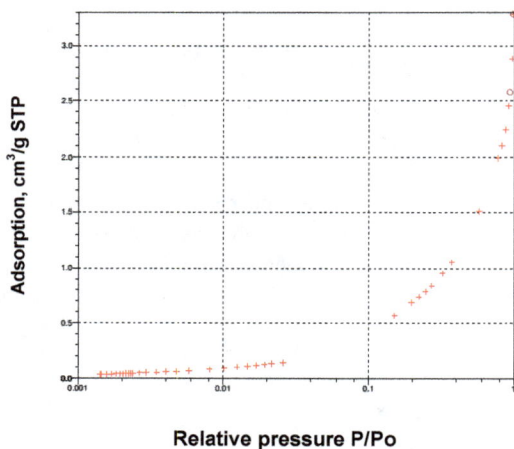

Figure 8   Nitrogen adsorption isotherm for chitosan in the logarithmic scale of a relative pressure.

Chitosan-Based Adsorbents for Wastewater Treatment, Ed. Abu Nasar   Materials Research Forum LLC
Materials Research Foundations **34** (2018) 81-98   doi: http://dx.doi.org/10.21741/9781945291753-4

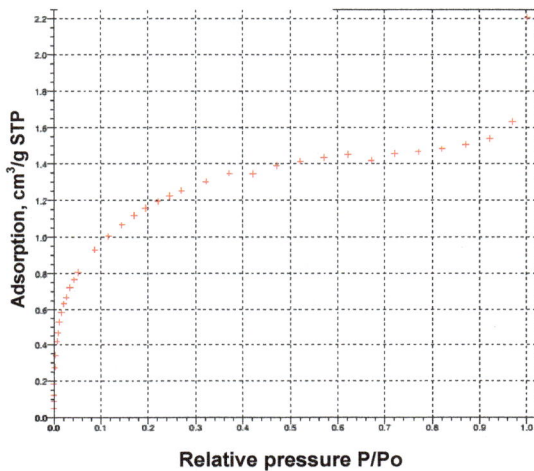

Figure 9 Nitrogen adsorption isotherm for non-modified chitosan beads in the linear scale of a relative pressure.

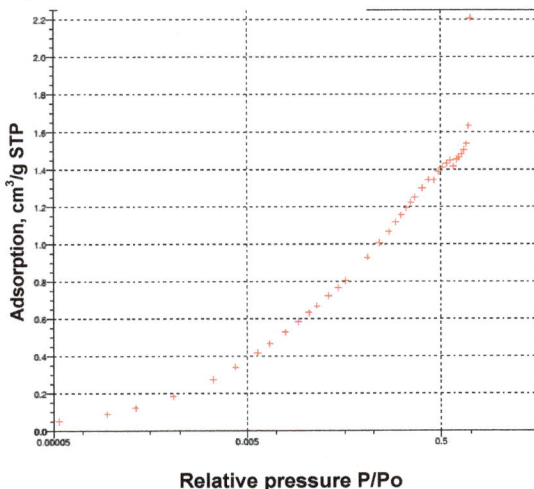

Figure 10   Nitrogen adsorption isotherm for non-modified chitosan beads in the logarithmic scale of a relative pressure.

Chitosan-Based Adsorbents for Wastewater Treatment, Ed. Abu Nasar   Materials Research Forum LLC
Materials Research Foundations **34** (2018) 81-98   doi: http://dx.doi.org/10.21741/9781945291753-4

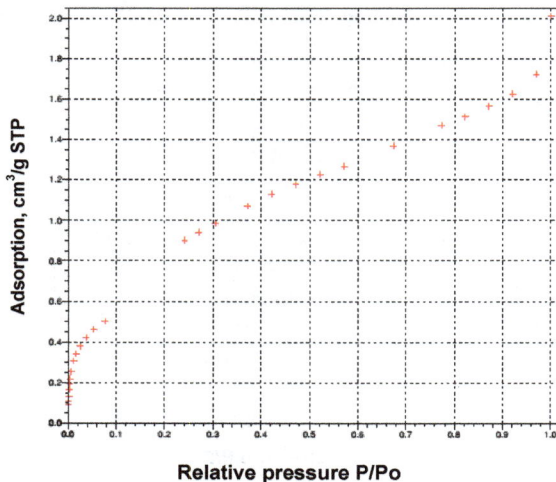

Figure 11   Nitrogen adsorption isotherm for conditioned chitosan beads in the linear scale of a relative pressure.

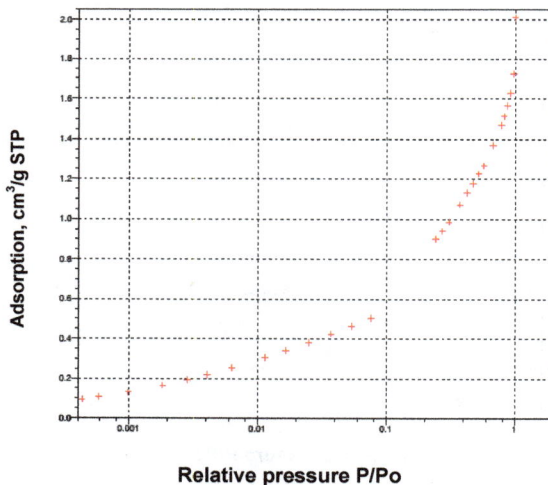

Figure 12   Nitrogen adsorption isotherm for conditioned chitosan beads in the logarithmic scale of a relative pressure.

Chitosan-Based Adsorbents for Wastewater Treatment, Ed. Abu Nasar	Materials Research Forum LLC
Materials Research Foundations **34** (2018) 81-98	doi: http://dx.doi.org/10.21741/9781945291753-4

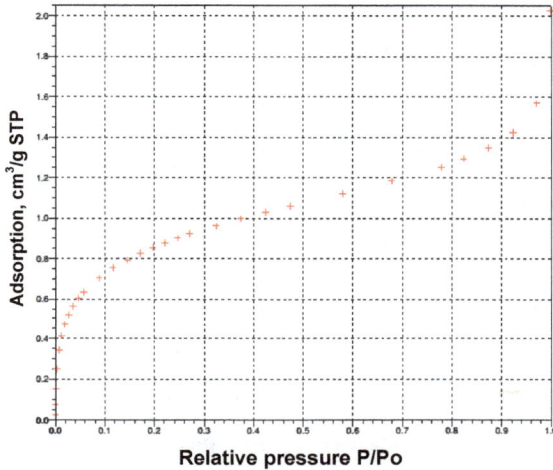

Figure 13 Nitrogen adsorption isotherm for GLU cross-linked chitosan beads in the linear scale of a relative pressure.

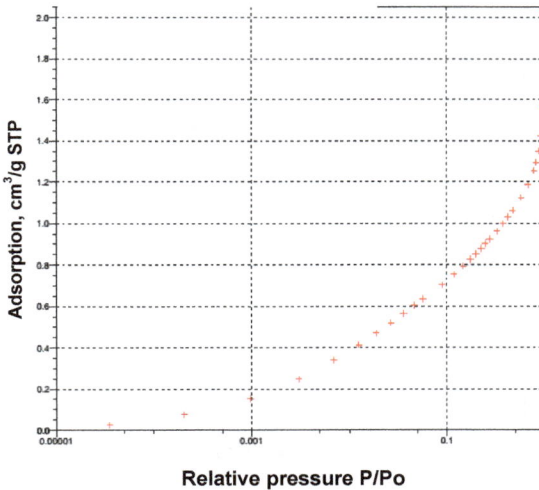

Figure 14 Nitrogen adsorption isotherm for GLU cross-linked chitosan beads in the logarithmic scale of a relative pressure.

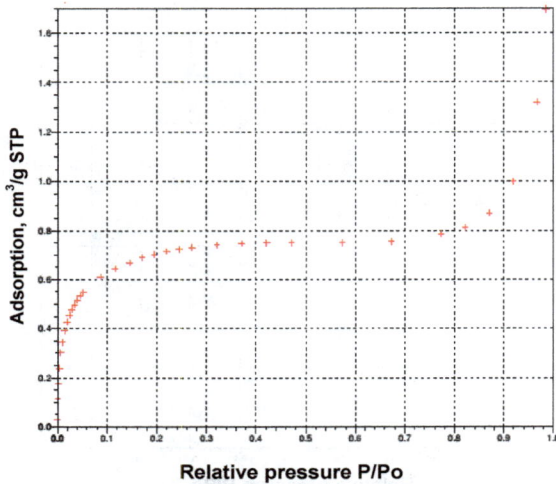

Figure 15   Nitrogen adsorption isotherm for ECH cross-linked chitosan beads in the linear scale of a relative pressure.

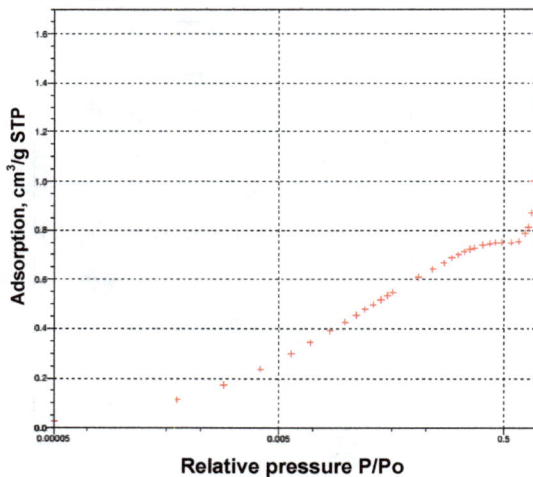

Figure 16   Nitrogen adsorption isotherm for ECH cross-linked chitosan beads in the logarithmic scale of a relative pressure.

Chitosan-Based Adsorbents for Wastewater Treatment, Ed. Abu Nasar   Materials Research Forum LLC
Materials Research Foundations **34** (2018) 81-98   doi: http://dx.doi.org/10.21741/9781945291753-4

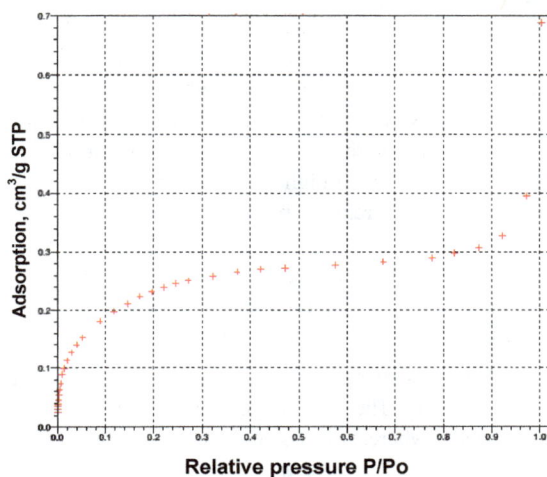

Figure 17 Nitrogen adsorption isotherm for ECH cross-linked and conditioned chitosan beads in the linear scale of a relative pressure.

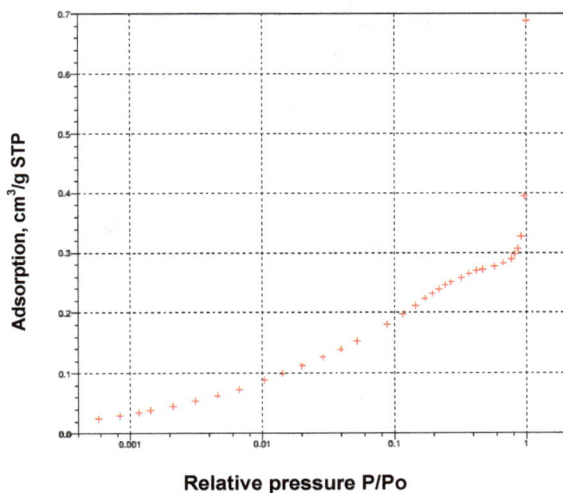

Figure 18 Nitrogen adsorption isotherm for ECH cross-linked and conditioned chitosan beads in the logarithmic scale of a relative pressure

Comparing the determined adsorption isotherms, the GLU conditioned beads and non-modified beads were found to show the highest adsorption at low relative pressure. One could assume that these sorbents should consist of the greatest amount of the slightest pores (micropores) very quickly filled out by liquid nitrogen at low relative pressure. The lowest adsorption properties were shown by ECH cross-linked and conditioned by NaHSO$_4$ beads at low relative pressure. One could assume this modification has none or to have little numbers of micropores. At higher relative pressure no noticeable difference in adsorption properties at medium relative pressure was observed to confirm relatively uniform pore size distribution. The only exception was observed for ECH cross-linked and conditioned beads. The adsorption values were increased at relative pressure scope of 0.9 to 1.0 and it was due to liquid nitrogen condensation inside the largest pores. These pores occupy the majority of the porous sorbent structure.

Estimated nitrogen adsorption isotherms led to determine the standard properties of the examined sorbents (see Table 1). The largest specific surface area was attributed to the non-modified chitosan beads, whereas the lowest area was assigned to the ECH cross-linked and conditioned by NaHSO$_4$ beads, which, in turn, show the lowest volume and pores surface. One could conclude that chitosan modification using hydrogel beads leads to increasing sorbent specific surface area. Similar results could also be observed for GLU cross-linked and conditioned beads. In contrast, decreasing of the specific surface area and diminishing of pores population was observed for ECH cross-linked beads, what is also supported by SEM images (see Figures 5 and 6).

Table 1 Specific surface area, total pore volume, and pore surface area of tested forms of chitosan

Form of chitosan	Specific surface area (S_{BET}) [m^2/g]	Total pore volume (V_p) [cm^3/g]	Total pore surface area (S_p) [m^2/g]
Pure chitosan	2.9483	0.00432	1.242
Non-modified beads	4.2395	0.00225	2.001
beads + NaHSO$_4$	3.4685	0.00246	1.434
GLU modified beads	3.0664	0.00287	1.550
ECH modified beads	2.4183	0.00244	1.256
ECH modified beads + NaHSO$_4$	0.8648	0.00056	0.388

Chitosan-Based Adsorbents for Wastewater Treatment, Ed. Abu Nasar   Materials Research Forum LLC
Materials Research Foundations **34** (2018) 81-98   doi: http://dx.doi.org/10.21741/9781945291753-4

4.   Conclusions

Based on SEM images and BET isotherms one could conclude that a formation of chitosan beads and their modifications make significant changes in their outer structure and the specific surface area in comparison with powdered pure chitosan. Non-modified and modified chitosan beads were characterized by their porous, sponge external structure. Except for ECH cross-linked beads the other modified chitosan beads were observed to show an increase in spreading of their surface structure and pores compressing compared to non-modified chitosan beads. The latter, however, were found to have the largest specific surface area (S_{BET} = 4.2395 m^2/g). The GLU cross-sectioned and conditioned by NaHSO$_4$ chitosan beads were described to show larger specific surface areas (3.0666 and 3.4685 m^2/g, respectively) in comparison with powdered pure chitosan (2.9483 m^2/g). SEM images of these modified structures clearly show considerable spreading of the outer porous structure.

ECH cross-linking and conditioning by NaHSO$_4$ resulted in a relevant decrease in a number of pores in the beads, what was confirmed by the smallest specific surface area (S_{BET} = 0.8648 m^2/g) and the lowest volume and pores area (V_p = 0.00056 cm^3/g and S_p = 0.388 m^2/g, respectively). ECH cross-linking was also found to destruct a chitosan structure, however, on a smaller scale than ECH and NaHSO$_4$ conditioning (S_{BET} = 2.4183 m^2/g). This observation was additionally confirmed by SEM images analysis that shows a clear disappearance of the outer porous structure for the examined modifications.

References

[1]   S. Salwiczek, Ph.D. thesis: The application of chitosan and its modifications for wastewater treatment, (in Polish), Silesian Technical University, Gliwice, 2015 (Poland).

[2]   Shikata Futoshi, Hiroyuki Tokumitsu, Hideki Ichikawa, Yoshinobu Fukumori, In vitro cellular accumulation of gadolinium incorporated into chitosan nanoparticles designed for neutron-capture therapy of cancer, Eur. J. Pharm. Biopharm. 53 (2002) 57-63. https://doi.org/10.1016/S0939-6411(01)00198-9

[3]   A. Wojtasz-Pająk, Ph.D. thesis: Preparation of chitosan with assumed physical and chemical properties from shrimp shells (Crangon cramgon [L.]), (in Polish), Sea Fisheries Institute, Gdynia, 1995 (Poland).

[4]   M. Mucha, Chitosan - A comprehensive polymer from renewable resources, (in Polish), WNT, Warszawa, 2010.

[5] Bhatnagar Amit, Mika Sillanpää, Applications of chitin- and chitosan-derivatives for the detoxification of water and wastewater - A short review, Advances in Colloid and Interface Science, 152 (2009) 26-38. https://doi.org/10.1016/j.cis.2009.09.003

[6] E. Guibal, Interactions of metal ions with chitosan-based sorbents: a review, Sep. Purif. Technol. 38 (2004) 43-74. https://doi.org/10.1016/j.seppur.2003.10.004

[7] P.R. Austin, C.J. Brine, J.E. Castle, J.P. Zikakis, Chitin: New facets of research, Science, 212 (1981) 749-753. https://doi.org/10.1126/science.7221561

[8] K. Barbusiński, S. Salwiczek, A. Paszewska, The use of chitosan for removing selected pollutants from water and wastewater - short review, Architecture Civil Engineering Environment, 2 (2016) 107-115.

[9] P.K. Dutta, M.N.V. Ravikumar, J. Dutta, Chitin and chitosan for versatile applications, Journal of Macromolecular Science Part C Polymer Reviews, C42 (2002) 307-354. https://doi.org/10.1081/MC-120006451

[10] P.K. Dutta, J. Dutta, V.S. Tripathi, Chitin and chitosan: chemistry, properties and applications, Journal of Scientific & Industrial Research, 63 (2004) 20-31.

[11] M.N.V. Ravi Kumar, A review of chitin and chitosan applications, Reactive & Functional Polymers, 46 (2000) 1-27. https://doi.org/10.1016/S1381-5148(00)00038-9

[12] M. Struszczyk, Chitin and chitosan. Part 2. Applications of chitosan, Polimery, 47 (2002) 396-403.

[13] S.T. Lee, F.L. Mi, Y.J. Shen, S.S. Shyu, Equilibrium and kinetic studies of copper (II) ion uptake by chitosan-tripolyphosphate chelating resin, Polymer, 42 (2001) 1879-1892. https://doi.org/10.1016/S0032-3861(00)00402-X

Chitosan-Based Adsorbents for Wastewater Treatment, Ed. Abu Nasar   Materials Research Forum LLC
Materials Research Foundations **34** (2018) 99-122   doi: http://dx.doi.org/10.21741/9781945291753-5

Chapter 5

Applications Chitin and Chitosan-Based Adsorbents for the Removal of Natural Dyes from Wastewater

Sapna Raghav[1], Ritu Painuli[1], Dinesh Kumar[2*]

[1]Department of Chemistry, Banasthali University, Rajasthan–304022, India

[2*]School of Chemical Science, Central University of Gujarat, Gandhinagar–382030, India

*dinesh.kumar@cug.ac.in

Abstract

Adsorption is one of the most advantageous techniques for the removal of pollutants. It is fast, simple, cheap with many opportunities to modify the initial materials after appropriate synthesis routes, etc. Numerous adsorbent materials have been prepared in the last years, having as ultimate scope to remove some pollutants especially from contaminated waters (effluents originated from industries). But the composition of each type of effluents is varying. Dyes are some major components of industrial wastewaters. Chitosan and its derivative have received considerable attention in wastewater treatment for dye removal. This chapter highlights the dye removal affinity of Chitosan and its cross-linked and grafted derivatives.

Keywords

Chitosan, Adsorption, Dyes, Cross-linking, Grafting, Isotherms

Contents

1.   Introduction

Increasing population growth, industrialization, and extensive use of chemicals in agriculture practices lead to environmental deterioration. Anthropogenically produced inorganic and organic are the major concern of water contamination, which intimidates all living organisms [1,2]. Colored substance releases into the water are disastrous for the biological organism and ecology [3]. The foremost resources of industrial wastewater are dying and textile industries [4]. Normally, the rate of releasing wastewater volume from each step of a textile operation is approximately between 40–65 L/kg of the product [5]. Even a very low concentration of dye is clearly visible in water. Even a small concentration of dye can reduce the photosynthetic process in aquatic environments by blocking the penetration of light and oxygen [6]. Dyes caused a severe effect on human health, i.e. allergies, cancer, heart defects, jaundice, mutations, tumors, and skin irritation [8]. Due to their complex structure and synthetic origin, they are non-biodegradable and remain stable in water [7].

Trace amount of pollutant has been removed by utilizing different methods from wastewater, techniques are adsorption, chemical precipitation, electrocoagulation, electrodialysis, ion–exchange, membrane filtration reverse, and osmosis [9–13]. Most of these treatment methods have some restrictions (higher functioning cost, lesser efficiency, and higher waste generation); hence these methods are unsuitable for small-scale use. The adsorption process is dominant over other processes because it is fast, cheap, low initial costs, lesser non–toxic waste generation, and simple design and operation. Adsorption capacity (Q_e) is adsorbent dependent. Organic and inorganic both materials can be utilized for dye banishment. An ideal adsorbent should have following features, cheap, easily available, higher surface area, high Q_e, large pore size and volume, mechanical stability, compatibility, easy regeneration, environmentally friendly, does not require high processing procedures and high selectivity to remove a wide range of dyes. Hence, the recent focus of research is to develop natural polymers–based adsorbent for dye removals such as chitin and Chitosan with enhanced Q_e and cost-effective [14].

Chitosan is easily available low–cost biopolymer attained from natural resources. It is a very specific adsorbent due to its specific features like higher abundance, cationicity, high Q_e, the cheap, and macromolecular structure as compared to others. Chitosan and their amendment derivatives are reported for the effective removal of dyes [15].

The main objective of this study is to provide literature information on the practices of chitin, Chitosan and its derivatives for adsorption of different dye. Different kinds of literature have been studied and critically reviewed to disclose the efficacy of Chitosan and its derivatives in dye removal application. This chapter assembles and presents appropriate information in terms of the use of unmodified and modified Chitosan.

2.   Chitin and Chitosan

Chitin found in the exoskeleton of crustaceans, the cuticles of insects, and the cell walls of fungi, is the most abundant amino polysaccharide in nature. It is composed of β–(1–4)–linked N–acetyl glucosamine and forms a linear homopolymer, structurally like cellulose. While it is an amino polymer, it also contains –$CONH_2$ groups at the C–2 positions in place of the –OH groups. These acetamide groups are responsible for good adsorption performance. Marine crustaceans are the primary source of raw polymer which extracted from marine crustaceans. Slow biodegradation of chitin in crustacean shell waste became a major concern in the seafood processing industry. So, they need to be recycled. Their application in industrial wastewater treatment could be helpful not only for the environment but the economy also. However, chitin has a very poor solubility; hence it cannot be used for large-scale industrial productions. But its derivatives are more beneficial than chitin i.e. Chitosan [16–18].

Chitosan (CS) or poly–(1 $\rightarrow$ 4)–2–amino–2–deoxy–β–d–glucose is linear, nontoxic, and biodegradable polysaccharide and cationic biopolymer with high molecular weight [19]. CS is formed by alkaline deacetylation (DD) of chitin in which acetyl group hydrolyzes and convert into amine groups (Figure. 1) [20,21]. Alkaline DD step determines the degree of deacetylation. The degree of DD affects the Q_e and depends on the concentration of NaOH, temperature, and time used in the DD [22]. The degree of DD is directly proportional Q_e of the Chitosan due to the presence of high amounts of –NH_2 groups, Q_e of CS increases [23]. DD is the unique characteristics of the CS and their other properties such as molecular weight, crystallinity, and distribution of –NH_2 groups physicochemical, biological and the reaction of CS in the solution. From the literature report, molecular weight affects the solubility, bacteriological properties, coagulant–flocculants performance, crystallinity, and tensile strength. The solubility decreases with increasing molecular weight, while, crystallinity affects the Q_e and accessibility of –NH_2 groups. Due to the presence of intermolecular hydrogen bonding, CS is poorly soluble in water, alkaline, and organic solvents, while its solubility is high in acidic medium due protonation of its –NH_2 groups. Due to their above unique properties, CS has high adsorption potential for pollutants, i.e. dyes and heavy metals. The adsorption performance of the CS is inhibited by some drawbacks, i.e. soluble in acid, low mechanical strength, and low surface area. So, it needs some modifications to increase adsorption performance for dye removal and this chapter elaborate recent modification of CS which can be easily characterized as a promising material not only due to its unique physical properties but having applications in many fields i.e. in biotechnology, cosmetics, food, medicine, membranes. The current article is focused on the removal of dyes from effluents with various CS and its derivative adsorbents [24–26].

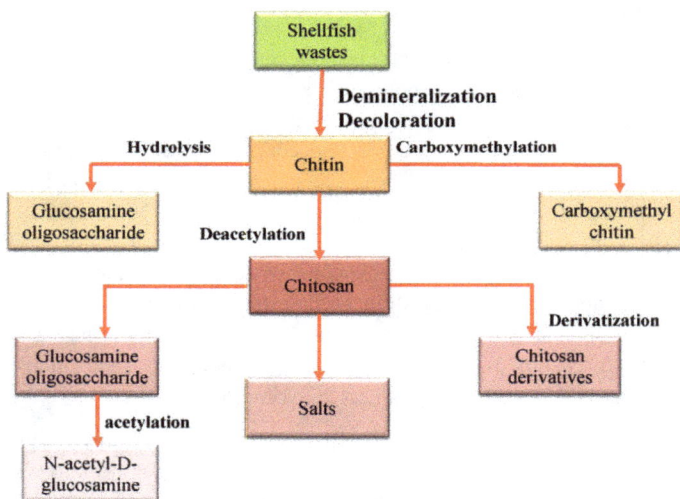

Figure 1 Demonstration of preparation of chitin, Chitosan, and their derivatives.

The major advantage of CS is the presence of amendable positions in its structure which can be modified by grafting (insertion of any functional group) and crosslinking (unite the macromolecular chains each other). These modifications lead to enhancement in the Q_e by the increasing adsorption sites, by the increasing number of functional groups in the polymer chain. In Grafting, the number of functional groups increases in the polymer while in crosslinking some functional groups get bounded with crosslinkers, so with grafting adsorption sites increases while in decrease in adsorption site observed in cross linking, and hence Q_e increases in grafting and decrease in cross-linking.

3.   Adsorption isotherms

To understand the adsorption mechanism process, surface properties, removal capacity, product design of the adsorption systems, and adsorption isotherms (Langmuir, Freundlich, Redlich–Peterson, Dubinin–Radushkevich, Temkin, Toth, Koble Corrigan, Sips, Khan, Hill, Flory–Huggins, and Radke Prausnitz isotherm) are essentially studied, these isotherms explains how the pollutant interact with the adsorbent.

Fundamental thermodynamics intimate about the adsorption mechanism, surface properties and about the extent of the affinity of the adsorbents [27]. Extensive diversity of equilibrium isotherm models had been outlined in terms of three basic approaches. The first approach kinetic consideration and the second one is thermodynamics. An adsorption equilibrium attains when the rate of adsorption and desorption became equal, and this frame many adsorption models which further gives the thermodynamics of the adsorption process. The third approach, i.e. potential theory, helps in the generation of the characteristic curve [28,29] (Table 1).

Table 1 Different Adsorption Isotherms.

Adsorption Isotherm	Formula
Langmuir Isotherm	$Q_e = Q_o b C_e /(1+bC)$
Freundlich Isotherm	$Q_e = K_F (C_e)^{1/n}$
Dubinin–Radushkevich isotherm	$Q_e = (q_s)\, e^{-k\epsilon 2}$
Temkin Isotherm	$Q_e = (RT/b_T)\, Ln\,(A_T C_e)$
Flory Huggins Isotherms	$\theta/C_o = K_{FH}\,(1-\theta)^n$
Hill Isotherm	$Q_e = q_s C_e^n /(K_D + C_e^n)$
Redlich–Peterson isotherm model	$Q_e = K_R C_e^{\,n} /(K_D + a_R C_e^g)$
Sips isotherm	$Q_e = K_S C_e^\beta /(K_D + a_S C_e^\beta)$
Toth isotherm	$Q_e = K_T C_e/(a_T + C_e)^{1/t}$
Koble–Corrigan isotherm	$Q_e = A C_e^n /(1+B C_e^n)$

Table 2 Unmodified chitosan powder utilized for dye removal.

Chitosan powder Particle size (µm)	Dye	Q_e (mg/L)	Temperature (°C)	pH	Ref.
125–710	RB 5	200	–	7	[31]
850	RR 141	68	20	11	[32]
125–710	DB 71	101	–	7	[33]
125–710	AB 1 AG AV 25 AY 25	18 179 152 179	–	7	[33]
355–500	A0 10 A0 12 AR 18 AR 7	696.65 931.85 670.97 695.6	25	4	[34]
70	AB	256.0	25	3	[30]
70	Food yellow 3	352.6	25	3	[30]
125–710	MB 29 M Br 33 MO 10	37 91 157	–	7	[33]

4.   Unmodified Chitosan

The adsorption performance of chitosan was investigated by a number of researchers. The chitosan flakes obtained from chitin is a solid material with high crystallinity and have been used for dye removal from wastewater by researchers [30]. Dotto and Pinto utilized chitosan powder, obtained from shrimp waste also shows good adsorption to remove food dyes i.e. AB 9 and food yellow 3 from aqueous solution. The adsorption capacity influenced by the effect of the pH (with increase pH Q_e decreases), contact time (increases time, Q_e increases) and stirring rate. The parameters for AB 9 and food yellow were pH 3,150 rpm and 60 min and pH 3, 50 rpm and 60 min respectively. The Q_e values for AB 9 and food yellow were 210 and 295 mg/g, respectively. Table 2 summarizes the removal of different dyes from aqueous solutions by unmodified chitosan in the form of flake and powder (Table 2,3). As could be observed in this table, the range of the flake size varied between 125 mm to 1.651 mm and mainly distributed closer to the smaller end of the range. Acidic dyes showed good affinity for the chitosan seen in the Tables 2 and 3.

Table 3 Unmodified chitosan flakes utilized for dye removal.

Chitosan Flakes (μm/mesh)	Dye	Qₑ (mg/L)	Temperature (°C)	pH	Ref.e
177–165.1 μm	Remazole black 13	96	60	6.7	[42]
1000–1410 μm	RR 222 RY 145	339 188			[36]
184–314 μm	RY 84 RR 11 RB 5 RB 8	500 450 650 387	–	7 5 5 5	[37]
125–250 μm	RB 5	477	25	2.3	[38]
125–500 μm	RB5	353	25	2	[39]
228 μm	RB 5	62.92	33	9	[40]
0-125 μm	DB 95	41.84	50	6	[33]
16–30 mesh	RR 222	494	30	–	[35]
100–150 mesh	RR 3	151.5	20	5	[41]

5.   Modified Chitosan

The modification of the raw CS can be achieved by converting it into a conditioned form with the consideration of promising properties and uses of derivatives. They are utilized in various forms such as powder, flakes, composites, and gel (beads, membranes, film, etc.) [43]. In spite of having immense potential, it suffers from various disadvantages for instances, insufficient mechanical properties, low acid stability, low porosity, low thermal resistance, low solubility in acidic solutions, and less surface area [44-47]. Thus, to overcome the limitations of CS, amendment's is an efficient way to formed product with preferred properties. Owing to the presence of the reactive functional groups, its modification is simple in comparison to that of the polysaccharides. [48]. The improvement in the Q_e and the mechanical properties of the chitosan can be achieved via chemical and physical modifications. Transformation of raw CS flakes into beads has been demonstrated as the indispensable approach for enhancing the adsorption capability by increasing the porosity and surface area. The CS membrane and films can be prepared by introducing the CS mixed with alkali into a flat surface for the solvent evaporation. In general, with the physical modification of the chitosan its polymer chains expand, which enhances the access to internal sorption. This, in turn, increases the crystalline state and the diffusion mechanisms of the polymer to be reduced.

The utilization of unmodified CS in industry subjects to disadvantages of reduced mechanical strength and vulnerability in acidic media. Therefore, researchers utilized chemical modification to attain the desirable properties and applications of chitosan in dye removal.

5.1   Chitosan beads, membrane, films, and composites

The foremost factor which enhances the Q_e of chitosan includes particle size, porosity, and crystallinity. The less adsorption capacity limit, enhanced crystallinity, reduced surface area, low porosity are the various parameters which hinder the applications of CS flakes. CS flakes do not perform preferably in columns. This is due to the high crystalline and hydrophobic nature of the CS flakes, which increases their resistance to mass transfer rate, increases pressure drops which results in high operation expenses and column clogging. The Q_e value decrease with a decrease in surface area and also with non-porosity. The $-NH_2$ groups in CS flakes are not available. The H-bonds connected between the monomer units of the same chain or different chains reduce the reactivity of the chitosan flakes. During the dye adsorption of the flakes, large dye molecules are unable to penetrate the porous network with steric hindrance increased with the continuous saturation of the sorbent. As a result, the utilization of CS in its dissolved-state increases the availability of reactive sites. During the polymer dissolving, the breaking of H-bonds between $-NH_2$ groups and between -OH groups, makes the CS accessible for interactions with dyes. This phenomenon may explicate the highly proficient utilization of the amino groups of chitosan utilized in dissolved-state for dye removal.

Moreover, the beads can be merely separated from the treated solution for reuse. It was found from other comparative studies that the Q_e of CS flakes is lesser than that of chitosan beads. The CS flakes can be physically modified into gel beads and CS composite for increasing porosity, surface area, chitosan polymer chains, and decreasing crystallinity, all of which lead to increased Q_e. The removal of various dyes from the aqueous systems by unmodified chitosan flakes and composites are summarized in Tables 4 and 5. The pH effect is clearly observed from the table. Lower pH values tend to cause increased Q_e. The higher Q_e as mentioned earlier could be accredited to the enhanced surface area as the chemical structure does not experience any changes [31,33,35,49].

Table 4 Modified Chitosan utilized for dye removal.

Chitosan beads	Dye	Q_e (mg/L)	Temperature (°C)	pH	Ref.
Beads	Eosin Y	79	30	8	[50]
Beads	AR 37	357.14	27–47	4	[51]
	AB 25	178.57			
Beads	AR 37	130	–	6	[52]
	AB 25	250		4	
Beads	Orange –G	95	–	4	[53]
Beads	Eosin –Y	170.65	25	5	[54]
Beads immobilized with tyrosinate	RR 222	1653	30	–	[36]
	RY 145	885			
Beads	RR 189	1189	25	6	[55]
Beads	RR 189	950	25	6	[56]
Beads	RY 84	690	–	3	[37]
	RR 11	480			
	RB 5	480			
	RB 8	487			
Beads	RR	648	25	2	[57]
	RY	430	25	2	
	RV	398	25	2	
Beads	CR	93	30	6	[58]
Beads	CR	178.32	–	5	[59]
Beads	CR	223.25	30	4	[60]

Beads	CR	200	–	5	[61]
Beads	CR	174.83	25	4	[62]
Beads	Methylene Blue	99.01	30	8	[63]

Table 5 Modified chitosan composites for dye removal.

Chitosan Composite	**Dye**	**Q_e (mg/g)**	**Isotherm**	**Ref.**
CS/PVA	EY	53	L, F, T	[81]
CS/bentonite hybrid composite	OII, RR3BS	362, 497	L	[82]
CS-Starch	DR80	313	F	[83]
CS/r-GO	RR	32	L	[84]
CS-Alunite	AR1, RR2	463, 569	L	[85]
CS/TPP/GLU	DB71	92	L, F	[86]
Ti^{4+} cross-linked CS	OII	1120	L	[87]
Magnetic CS-GO	MO	399	L	[88]
Fly ash-CS/GO	ARG, CRX	39, 65	R-P	[89]
Zr-based CS	OII	926	L	[90]
Modified clay/CS composite	MB	259.8	L, F, R-P	[91]
CS/SBA-15	AR18	201.2	L, F	[92]
CS/PMMA	RB19	1498	L	[93]
Magnetic CS/RR 120	Lysozyme	116.9	L	[94]
CS grafted with polypropylene imine	RB5, MB	6250, 5855	L, F, T	[95]

5.2   Cross-linking and grafting

From a practical perspective, chitosan demonstrates high Q_e to remove dyes from neutral solutions. The variation of pH may be a significant aspect of the dye-binding ability of CS. This occurs because the $-NH_2$ groups in CS are easier to cationize at low pH and they strongly adsorb the dye anions via the electrostatic attraction. In the dying process, acetic acid is frequently used as a stimulator in which the dye solution pH is normally adjusted from 3 to 4. Owing to the dissolution efficiency of CS in the acid effluent, the utilization of CS adsorbent for removal of dye gets severely limited.

Table 6 Cross-linked CS adsorbent for dye removal.

Chitosan	Cross-linking Reagent	Dye	Q_e (mg/g)	Temperature (°C)	pH	Ref.
B (2.3–2.5, 2.5–2.7, 3.5–3.8)	ECH	RR 189	1936, 1686, 1642	30	3	[55]
B	ECH	RR, RR, RR, RY, RB, RB	2252, 2498, 2422, 1911, 722, 2043	30	3	[70]
P	GLA	RB 5	109		4	[33]
B	GLA	RY, RB, RR	9, 10, 7.5	50, 25, 50	2	
B	EDGE	AR1	50	40	5	[71]
	EDGE	AB	150	–	4	
Fiber	Denacol EX841	AO 2	1200–1700	–	4	[72]
B	ECH	DR 81	238	30	4	[73]
P	GLA	DB 71	14	–	4	[33]
Film	GLA	CR	24.18	25	–	[74]

To overcome these activities, modification of the CS can be achieved through cross-linking the polymer chains in order to prevent swelling as well as to improve the mechanical resistance, reinforcement of the chemical stability of CS in acidic solutions. Cross-linking occurs when a reagent (cross-linker) introduces intermolecular bridges among the polysaccharide macromolecules. Cross-linking considerably decreases segment mobility in polymers. The foremost cross-linkers utilized for the modification of CS in the adsorption of dye from the aqueous systems include GLA and tripolyphosphate (TPP) etc. (Table 6). These chemicals are normally called as ionic cross-linkers [18, 49, 55, 60, 64-67].

To improve the dye adsorption tendency of the chitosan, grafting reactions has received a noteworthy interest. The amendments can improve the mechanical and physical properties of the CS polymer. The two types of reactive groups where the grafting occurs in Chitosan include -NH$_2$ and -OH group. Various functional groups have been grafted on the backbone of the chitosan via the covalent bonding. Although the grafting results in modified properties it does not cooperate with the remarkable characteristics of chitosan i.e. much adhesively, biocompatibility and biodegradability. Researchers have studied the graft copolymerization of CS with the focus of increasing CS applications. By grating copolymerization of synthetic polymers onto CS the required properties of the CS get improved and consequently extend the area of the prospective applications of them. APS, PPS, CAN, FAS etc. (Table 7) have been used to investigate grafting copolymerization. The characteristics of the resulting graft copolymers are mostly relying on the properties of the side chains [68,69].

Table 7 Chitosan synthesized by grafting for dye removal.

Chitosan	Modification Reagent	Dye	Q$_e$ (mg/g)	Temperature (°C)	pH	Ref.
Flake	4–formyl–1,3–benzene sodium disulfonate	BB 3	166.5	25	3	[49]
B	Pyromellitic dianhydride	MB NR	935 909	–	5 5	[75]
B	Poly (acrylic acid) Poly(acrylamide)	Remacryl red TGL	510.7 309.8	25	–	[76]
P	Poly (acrylic acid) Poly(acrylamide)	BY BY	363 595	25 25	12 12	[77]
B	Poly(methacrylic acid)	MB MG	1000 523.6	25 25	6 4	[78]

B	Acrylamide	RR	1185	25	2	[79]
		RY	1160			
		RB	1125			
	Polyethyleneimine	RR	1412	25	2	
		RY	1392			
		RB	1329			
B	PMMA	Orange–G	435	–	3–6	[80]
	PEMA	Orange–G	360		3–6	
	PBMA	Orange–G	290		3–6	
	PHMA	Orange–G	265		3–6	

6.   Conclusion

Chitosan and its derivatives are getting noteworthy attention nowadays owing to their efficient adsorption capacity. The present article demonstrated some interesting data which clearly showed the high adsorption capacity of Chitosan. For achieving an enhanced Q_e, the modification of the Chitosan molecule via the grafting and cross-linking reaction results in the formation of the Chitosan materials with enhanced properties. At the same time, CS adsorbents are cheap, eco–friendly, non–toxic, do not require high processing, high Q_e, large surface area, and biodegradability. General, the capability of those materials is very big regarding the decolorization of wastewaters and in the next years, it is expected to be even more used in adsorption applications.

Acknowledgements

We gratefully acknowledge support from the Ministry of Science and Technology and Department of Science and Technology, Government of India under the scheme of Establishment of Women Technology Park, for providing the necessary financial support to carry out this study vide letter No, F. No SEED/WTP/063/2014.

Abbreviations

RR	Reactive Red	MG	Methyl Green
RB	Reactive Blue	MB	Methyl Blue
RY	Reactive Yellow	MO	Mordant Orange
RB.	Reactive Blue	MR	Mordant Red

DS	Direct Scarlet	MBr	Mordant Brown
DB	Direct Blue	MB	Mordant Blue
DR	Direct Red	CR	Congo Red
AB	Acid Black	GLA	Glutaraldehyde
AR	Acid Red	ECH	Epichlorohydrin
AY	Acid yellow	PMMA	Poly(methyl methacrylate)
AV	Acid Violet	PBMA	Poly(butyl methacrylate)
AG	Acid Green	PEMA	Poly(ethyl methacrylate)
AO	Acid Orange	EDGE	Ethylene glycol diglycidyl
B	Beads	P	Powder
L	Langmuir	F	Freundlich
Q_e	Adsorption Capacity	CS	Chitosan
APS	Ammonium persulfate	PPS	potassium persulfate
CAN	ceric ammonium nitrate	FAS	ferrous ammonium sulfate

References

[1]   V. Gupta, Application of low-cost adsorbents for dye removal—A review. J. Environ. Manage. 90 (2009) 2313–2342. https://doi.org/10.1016/j.jenvman.2008.11.017

[2]   O. Ceyhan, D. Baybas, Adsorption of some textile dyes by hexadecyl trimethylammonium bentonite. Turkish J. Chem. 25 (2001) 193–200.

[3]   A. L. Prasad, T. Santhi, Adsorption of hazardous cationic dyes from aqueous solution onto Acacia nilotica leaves as an ecofriendly adsorbent. Sustain. Environ. Res. 22 (2012) 113–122.

[4]   G. Mezohegyi, F. P. van der Zee, J. Font, A. Fortuny, A. Fabregat, Towards advanced aqueous dye removal processes: A short review on the versatile role of activated carbon. J. Environ. Manage. 102 (2012) 148–164. https://doi.org/10.1016/j.jenvman.2012.02.021

[5]   N. Ali, A. Hameed, S. Ahmed, Physicochemical characterization and bioremediation perspective of textile effluent, dyes and metals by indigenous

bacteria. J. Hazard. Mater. 164 (2009) 322–328.
https://doi.org/10.1016/j.jhazmat.2008.08.006

[6]   G. Crini, Non-conventional low-cost adsorbents for dye removal: A review.
        Bioresour. Technol. 97 (2006) 1061–1085.
        https://doi.org/10.1016/j.biortech.2005.05.001

[7]   S. P. Buthelezi, A.O. Olaniran, B. Pillay, Textile dye removal from waste-water
        effluents using bioflocculants produced by indigenous bacterial isolates.
        Molecules, 17 (2012) 14260–14274. https://doi.org/10.3390/molecules171214260

[8]   E. Alver, A. U. Metin, Anionic dye removal from aqueous solutions using
        modified zeolite: Adsorption kinetics and isotherm studies. Chem. Eng. J. 200–
        202 (2012) 59–67. https://doi.org/10.1016/j.cej.2012.06.038

[9]   Y. Xing, X. Chen, D. Wang, Electrically regenerated ion exchange for removal
        and recovery of Cr(VI) from wastewater. Environ. Sci. Technol. 41(2007) 1439–
        1443. https://doi.org/10.1021/es0614991

[10]   M. Chafi, B. Gourich, A. H. Essadki, C. Vial, A. Fabregat, A. Comparison of
        electrocoagulation using iron and aluminum electrodes with chemical coagulation
        for the removal of a highly soluble acid dye. Desalination, 281 (2011) 285–292.
        https://doi.org/10.1016/j.desal.2011.08.004

[11]   T. A. Kurniawan, G. Chan, W. H. Lo, S. Babel, Physicochemical treatment
        techniques for wastewater laden with heavy metals. Chem. Eng. J. 118 (2006) 83–
        98. https://doi.org/10.1016/j.cej.2006.01.015

[12]   M. García-Gabaldón, V. Pérez-Herranz, J. García-Antón, J. Guinon, Electro-
        chemical recovery of tin from the activating solutions of the electroless plating of
        polymers: Galvanostatic operation. Separ. Purif. Technol. 51 (2006) 143–149.
        https://doi.org/10.1016/j.seppur.2005.12.028

[13]   N. A. Mohamed, M. M. Fahmy, Synthesis and antimicrobial activity of some
        novel cross-linked chitosan hydrogels. Int. J. Mol. Sci. 13 (2012) 11194–11209.
        https://doi.org/10.3390/ijms130911194

[14]   R. Menaka, S. Subhashin, Chitosan Schiff base as effective corrosion inhibitor for
        mild steel in acid medium. Polym. Int. 66 (2017) 349-358.
        https://doi.org/10.1002/pi.5245

[15] L. Racine, I. Texier, R. Auzély-Velty, Chitosan-based hydrogels: recent design concepts to tailor properties and functions. Polym. Int. 66 (2017) 981-998. https://doi.org/10.1002/pi.5331

[16] D. B. Lee, D. W. Kim, Y. Shchipunov C. S. Ha, Effects of graphene oxide on the formation, structure and properties of bionanocomposite films made from wheat gluten with chitosan. Polym. Int. 65 (2016) 1039-1045. https://doi.org/10.1002/pi.5148

[17] C. Valencia-Sullca, M. Jiménez, A. Jiménez, L. Atarés, M. Vargas, A. Chiralt, Influence of liposome encapsulated essential oils on properties of chitosan films. Polym. Int. 65 (2016) 979-987. https://doi.org/10.1002/pi.5143

[18] I. Safir, M. Chami, T. Buergi, C. Nardin, Investigation of the thin film crystallization of a DNA copolymer hybrid composed of chitosan. Polym. Int. 65 (2016) 1165-1171. https://doi.org/10.1002/pi.5165

[19] G. Valladares, P. González Audino, M. C. Strumia, Preparation evaluation of alginate/chitosan microspheres containing pheromones for pest control of Megaplatypus mutatus Chapuis (Platypodinae: Platypodidae). Polym. Int. 65 (2016) 216-223. https://doi.org/10.1002/pi.5049

[20] M. P. Di Bello, L. Mergola, S. Scorrano, R. Del Sole, Towards a new strategy of a chitosan-based molecularly imprinted membrane for removal of 4-nitrophenol in real water samples. Polym. Int. 66 (2017) 1055-1063. https://doi.org/10.1002/pi.5360

[21] T. D. A. Senra, J. Desbrières, Using full-factorial design analysis and response surface methodology to better understand the production of cationized chitosan from epoxides. Polym. Int. 65 (2016) 811-819. https://doi.org/10.1002/pi.5137

[22] A. Abdulkarim, M. T. Isa, S. Abdulsalam, A. J. Muhammad, A. O. Ameh, Extraction and characterization of chitin and chitosan from mussel shell. Civil Environ. Res. 3 (2013) 108–114.

[23] J. S. Piccin, M. L. G.Vieira, J. O. Gonc¸ alves, G. L. Dotto, L. A. A. Pinto, Adsorption of FD&C Red No. 40 by chitosan: Isotherms analysis. J. Food Eng. 95 (2009) 16–20. https://doi.org/10.1016/j.jfoodeng.2009.03.017

[24] S. Jana, A. Saha, A. K. Nayak, K. K. Sen, S. K. Basu, Aceclofenac loaded chitosan-tamarind seed polysaccharide interpenetrating polymeric network

microparticles. Colloid. Surfaces B: Biointerfaces 105 (2013) 303–309.
https://doi.org/10.1016/j.colsurfb.2013.01.013

[25] Y. Ren, H. A. Abbood, F. He, H. Peng, K. Huang, Magnetic EDTA-modified chitosan/SiO_2/Fe_3O_4 adsorbent: Preparation, characterization, and application in heavy metal adsorption. Chem. Eng. J. 226 (2013) 300–311.
https://doi.org/10.1016/j.cej.2013.04.059

[26] Y. Peng, D. Chen, J. Ji, Y. Kong, H. Wan, C. Yao, Chitosan-modified paly gorskite Preparation, characterization and reactive dye removal. Appl. Clay Sci. 74 (2013) 81–86. https://doi.org/10.1016/j.clay.2012.10.002

[27] G. Thompson, J. Swain, M. Kay, C. F. Forster, The treatment of pulp and paper mill effluent: a review. Bioresour. Technol. 77 (2001) 275-286.
https://doi.org/10.1016/S0960-8524(00)00060-2

[28] M. I. El-Khaiary, Least-squares regression of adsorption equilibrium data: Comparing the options. J. Hazard. Mater. 158 (2008) 73-87.
https://doi.org/10.1016/j.jhazmat.2008.01.052

[29] E. Bulut, M. Özacar, I. A. Şengil, Adsorption of malachite green onto bentonite: equilibrium and kinetic studies and process design. Micropor. Mesopor. Mater. 115 (2008) 234-246. https://doi.org/10.1016/j.micromeso.2008.01.039

[30] G. L. Dotto, L. A. A. Pinto, Adsorption of food dyes onto chitosan: Optimization process and kinetic. Carbohy. Polym. 84 (2011) 231–238.
https://doi.org/10.1016/j.carbpol.2010.11.028

[31] E. Guibal, E. Touraud, J. Roussy, Chitosan interactions with metal ions and dyes: dissolved-state vs. solid-state application. World J. Microbiol. Biotechnol. 21 (2005) 913–920. https://doi.org/10.1007/s11274-004-6559-5

[32] N. Sakkayawong, P. Thiravetyan, W. Nakbanpote, Adsorption mechanism of synthetic reactive dye wastewater by chitosan. J. Colloid. Interf. Sci. 286 (2005) 36–42. https://doi.org/10.1016/j.jcis.2005.01.020

[33] E. Guibal, P. McCarrick, J. M. Tobin, Comparison of the sorption of anionic dyes on activated carbon and chitosan derivatives from dilute solutions. Sep. Sci. Technol. 38 (2013) 3049–3073. https://doi.org/10.1081/SS-120022586

[34] W. Cheung, Y. Szeto, G. McKay, Intraparticle diffusion processes during acid dye adsorption onto chitosan. Bioresour. Technol. 98 (2007) 2897–2904.
https://doi.org/10.1016/j.biortech.2006.09.045

[35] F. C. Wu, R. L. Tseng, R. S. Juang, Comparative adsorption of metal and dye on flake-and bead-types of chitosans prepared from fishery wastes. J. Hazard. Mater. 73 (2000) 63–75. https://doi.org/10.1016/S0304-3894(99)00168-5

[36] F. C. Wu, R. L. Tseng, R. S. Juang, Enhanced abilities of highly swollen chitosan beads for color removal and tyrosinase immobilization. J. Hazard. Mater. 81 (2000) 167–177. https://doi.org/10.1016/S0304-3894(00)00340-X

[37] U. Filipkowska, Adsorption and desorption of reactive dyes onto chitin and chitosan flakes and beads. Adsorpt. Sci. Technol. 24 (2006) 781–795. https://doi.org/10.1260/026361706781388932

[38] J. Barron Zambrano, A. Szygula, M. Ruiz, A. M. Sastre, E. Guibal, Biosorption of reactive black 5 from aqueous solutions by chitosan: Column studies. J. Environ. Manag. 91 (2010) 2669–2675. https://doi.org/10.1016/j.jenvman.2010.07.033

[39] G. Z. Kyzas, N. K. Lazaridis, Reactive and basic dyes removal by sorption onto chitosan derivatives. J. Colloid. Interf. Sci. 331 (2006) 32–39. https://doi.org/10.1016/j.jcis.2008.11.003

[40] T. K. Saha, N. C. Bhoumik, S. Karmaker, M. G. Ahmed, H. Ichikawa, Y. Fukumori, Y. Adsorption of methyl orange onto chitosan from aqueous solution. J. Water Resource Prot. 2 (2010) 898–906. https://doi.org/10.4236/jwarp.2010.210107

[41] M. E. Ignat, V. Dulman, T. Onofrei, Reactive red 3 and direct brown 95 dyes adsorption onto chitosan. Cellulose Chem. Technol. 46 (2012) 357–367.

[42] G. Annadurai, L. Y. Ling, J. F. Lee, Adsorption of reactive dye from an aqueous solution by chitosan: Isotherm, kinetic and thermodynamic analysis. J. Hazard. Mater. 152 (2008) 337–346. https://doi.org/10.1016/j.jhazmat.2007.07.002

[43] P. Miretzky, A. F. Cirelli, Hg(II) removal from water by chitosan and chitosan derivatives: A review. J. Hazard. Mater. 167 (2009) 10–23. https://doi.org/10.1016/j.jhazmat.2009.01.060

[44] S. Chatterjee, D. S. Lee, M. W. Lee, S. H. Woo, Nitrate removal from aqueous solutions by cross-linked chitosan beads conditioned with sodium bisulfate. J. Hazard. Mater. 166 (2009) 508–513. https://doi.org/10.1016/j.jhazmat.2008.11.045

[45]   A. A. Alhwaige, T. Agag, H. Ishida, S. Qutubuddin, Biobased chitosan hybrid aerogels with superior adsorption: Role of graphene oxide in CO_2 capture. RSC Adv. 3 (2013) 16011–16020. https://doi.org/10.1039/c3ra42022a

[46]   P. Setthamongkol, J. Salaenoi, Adsorption capacity of chitosan beads in toxic solutions. Proceedings of World Academy of Science, Eng. Tech. (2012) 178–183.

[47]   L. Zhou, J. Liu, Z. Liu, Adsorption of platinum (IV) and palladium (II)from aqueous solution by thiourea-modified chitosan microspheres. J. Hazard. Mater. 172 (2009) 439–446. https://doi.org/10.1016/j.jhazmat.2009.07.030

[48]   M. Rinaudo, Chitin and chitosan: properties and applications. Prog. Polym. Sci. 31 (2006) 603–632. https://doi.org/10.1016/j.progpolymsci.2006.06.001

[49]   G. Crini, P. M. Badot, Application of chitosan, a natural amino polysaccharide, for dye removal from aqueous solutions by adsorption processes using batch studies: A review of recent literature. Prog. Polym. Sci. 33 (2008) 399–447. https://doi.org/10.1016/j.progpolymsci.2007.11.001

[50]   S. Chatterjee, S. Chatterjee, B. P. Chatterjee, A. R. Das, A. K. Guha, Adsorption of a model anionic dye, eosin Y, from aqueous solution by chitosan hydro beads. J. Colloid Interface Sci. 288 (2005) 30–35. https://doi.org/10.1016/j.jcis.2005.02.055

[51]   A. Kamari, W. Ngah, M. Chong, M. Cheah, Sorption of acid dyes onto GLA and H_2SO_4cross-linked chitosan beads. Desalination 249 (2009) 1180–1189. https://doi.org/10.1016/j.desal.2009.04.010

[52]   K. Azlan, W. N. Wansaime, L. Lai Ken, Chitosan and chemically modified chitosan beads for acid dyes sorption. J. Environ. Sci. 21 (2009) 296–302. https://doi.org/10.1016/S1001-0742(08)62267-6

[53]   V. K. Konaganti, R. Kota, S. Patil, G. Madras, Adsorption of anionic dyes on chitosan grafted poly(alkyl methacrylate). Chem. Eng. J. 158 (2010) 393–401. https://doi.org/10.1016/j.cej.2010.01.003

[54]   X. Y. Huang, X.-Y., Bin, J.-P. Bu, H.-T. Jiang, G.-B. Jiang, M.-H. Zheng, Removal of anionic dye eosin Y from aqueous solution using ethylene-diamine modified chitosan. Carbohy. Polymer. 84 (2011) 1350–1356. https://doi.org/10.1016/j.carbpol.2011.01.033

[55] M. S. Chiou, H. Y. Li, Equilibrium and kinetic modeling of adsorption of reactive dye on cross-linked chitosan beads. J. Hazard. Mater. 93 (2002) 233–248. https://doi.org/10.1016/S0304-3894(02)00030-4

[56] M. Chiou, H. Li, Adsorption behavior of reactive dye in aqueous solution on chemical cross-linked chitosan beads. Chemosphere 50 (2003) 1095–1105. https://doi.org/10.1016/S0045-6535(02)00636-7

[57] T. T. Kyaw, K. S. Wint, K. M. Naing, Studies on the sorption behavior of dyes on cross-linked chitosan beads in acid medium. In International Conference on Biomedical Engineering and Technology IPCBEE Singapore, (2011) 174–178.

[58] S. Chatterjee, S. Chatterjee, B. P. Chatterjee, A. K. Guha, Adsorptive removal of congo red, a carcinogenic textile dye by chitosan hydro beads: Binding mechanism, equilibrium and kinetics. Colloids Surf. A 299 (2007) 146–152. https://doi.org/10.1016/j.colsurfa.2006.11.036

[59] S. Chatterjee, T. Chatterjee, S. H. Woo, Influence of the polyethylene imine grafting on the adsorption capacity of chitosan beads for reactive black 5 from aqueous solutions. Chem. Eng. J. 166 (2011) 168–175. https://doi.org/10.1016/j.cej.2010.10.047

[60] S. Chatterjee, D. S. Lee, M. W. Lee, S. H. Woo, Congo red adsorption from aqueous solutions by using chitosan hydrogel beads impregnated with non-ionic or anionic surfactant. Bioresour. Technol. 100 (2009) 3862–3868. https://doi.org/10.1016/j.biortech.2009.03.023

[61] S. Chatterjee, T. Chatterjee, S. H. Woo, A new type of chitosan hydrogel sorbent generated by anionic surfactant gelation. Bioresour. Technol. 101 (2010) 3853–3858. https://doi.org/10.1016/j.biortech.2009.12.089

[62] S. Chatterjee, T. Chatterjee, S.-R., Lim, S. H. Woo, Effect of surfactant impregnation into chitosan hydrogel beads formed by sodium dodecyl sulfate gelation for the removal of congo red. Sep. Sci. Technol. 46 (2011) 2022–2031. https://doi.org/10.1080/01496395.2011.592520

[63] S. Chatterjee, T. Chatterjee, S. R. Lim, S. H. Woo, Adsorption of a cationic dye, methylene blue, on to chitosan hydrogel beads generated by anionic surfactant gelation. Environ. Technol. 32 (2011) 1503–1514. https://doi.org/10.1080/09593330.2010.543157

[64] A, Shweta, P. Sonia, Pharamaceutical relevance of cross-linked chitosan in microparticulate drug delivery. Inter. Res. J. Pharm. 4 (2013) 45–51.

[65] Y. Jing, Q. Liu, X. Yu, W. Xia, N. Yin. Adsorptive removal of Pb(II) and Cu(II) ions from aqueous solutions by cross-linked chitosan-polyphosphate-epichlorohydrin beads. Separ. Sci. Technol. 48 (2013) 2132–2139. https://doi.org/10.1080/01496395.2013.794433

[66] K.-J. Hsien, C. M. Futalan, W.-C. Tsai, C.-C. Kan, C.-S. Kung, Y.-H. Shen, M.-W. Wan, Adsorption characteristics of copper(II) onto non-cross-linked and cross-linked chitosan immobilized on sand. Desalin. Water Treat. 51 (2013) 5574–5582. https://doi.org/10.1080/19443994.2013.770191

[67] L. Mengatto, M. G. Ferreyra, A. Rubiolo, I. Rintoul, J. Luna, Hydrophilic and hydrophobic interactions in cross-linked chitosan membranes. Mater. Chem. Phys. 139 (2013) 181–186. https://doi.org/10.1016/j.matchemphys.2013.01.019

[68] N. Alves, J. Mano, Chitosan derivatives obtained by chemical modifications for biomedical and environmental applications. Intern. J. Biol. Macromol. 43 (2008) 401–414. https://doi.org/10.1016/j.ijbiomac.2008.09.007

[69] R. Jayakumar, M. Prabaharan, R. Reis, J. Mano, Graft copolymerized chitosan—Present status and applications. Carbohy. Polym. 62 (2005), 142–158. https://doi.org/10.1016/j.carbpol.2005.07.017

[70] M.-S. Chiou, W.-S. Kuo, H.-Y. Li, Removal of reactive dye from wastewater by adsorption using ECH cross-linked chitosan beads as medium. J. Environ. Sci. Health, A Tox. Hazard. Subst. Environ. Eng. 38 (2003) 2621–2631. https://doi.org/10.1081/ESE-120024451

[71] K. Azlan, W. N. Wansaime, L. Lai Ken, Chitosan and chemically modified chitosan beads for acid dyes sorption. J. Environ. Sci. 21 (2009) 296–302. https://doi.org/10.1016/S1001-0742(08)62267-6

[72] H. Yoshida, A. Okamoto, T. Kataoka, Adsorption of acid dye on cross-linked chitosan fibers: Equilibria. Chem. Eng. Sci. 48 (1993) 2267–2272. https://doi.org/10.1016/0009-2509(93)80242-I

[73] M.-S. Chiou, P.-Y. Ho, H.-Y. Li, Adsorption of anionic dyes in acid solutions using chemically cross-linked chitosan beads. Dyes Pigm. 60 (2004) 69–84. https://doi.org/10.1016/S0143-7208(03)00140-2

[74] T. Feng, F. Zhang, J. Wang, Z. Huang, Adsorption of congo red by cross-linked chitosan film. http://ieeexplore.ieee.org/stamp/stamp.jsp?arnumber=5781167.

[75] Y. Xing, X.-M., Sun, B.-H. Li, Pyromellitic dianhydride-modified chitosan microspheres for enhancement of cationic dyes adsorption. Environ. Eng. Sci. 26 (2009) 551–558. https://doi.org/10.1089/ees.2007.0346

[76] N. K. Lazaridis, G. Z. Kyzas, A. A. Vassiliou, D. N. Bikiaris, Chitosan derivatives as biosorbents for basic dyes. Langmuir, 23 (2007) 7634–7643. https://doi.org/10.1021/la700423j

[77] G. Z.Kyzas, N. K. Lazaridis, Reactive and basic dyes removal by sorption onto chitosan derivatives. J. Colloid Interface Sci. 331 (2009) 32–39. https://doi.org/10.1016/j.jcis.2008.11.003

[78] Y. Xing, X. Sun, B. Li, Poly(methacrylic acid)-modified chitosan for enhancement adsorption of water-soluble cationic dyes. Polym. Eng. Sci. 49 (2009) 272–280. https://doi.org/10.1002/pen.21253

[79] T.T. Kyaw, K. S. Wint, K. M. Naing, Studies on the sorption behavior of dyes on cross-linked chitosan beads in acid medium. https://pdfs.semanticscholar.org/f6dd/de46cbf76a8ad88042562da0b02956f17745.pdf.

[80] V. K. Konaganti, R. Kota, S. Patil, G. Madras, Adsorption of anionic dyes on chitosan grafted poly (alkyl methacrylate)s. Chem. Eng. J. 158 (2010) 393–401. https://doi.org/10.1016/j.cej.2010.01.003

[81] T. Anitha, P. S. Kumar, K. S. Kumar, Synthesis of nano-sized chitosan blended polyvinyl alcohol for the removal of Eosin Yellow dye from aqueous solution. Water Process Eng. 13 (2016) 127-136. https://doi.org/10.1016/j.jwpe.2016.08.003

[82] G. L. Dotto, F. K Rodrigues, E. H. Tanabe, R. Fröhlich, D. A. Bertuol, T. R. Martins, E. L. Foletto, Development of chitosan/bentonite hybrid composite to remove hazardous anionic and cationic dyes from colored effluents. J. Environ. Chem. Eng. 4 (2016) 3230-3239. https://doi.org/10.1016/j.jece.2016.07.004

[83] F. A. Ngwabebhoh, M. Gazi, A. A. Oladipo, Adsorptive removal of multi-azo dye from aqueous phase using a semi-IPN superabsorbent chitosan-starch hydrogel. Eng. Res. Des. 112 (2016) 274-288. https://doi.org/10.1016/j.cherd.2016.06.023

[84] X. Guo, L. Qu, M. Tian, S. Zhu, X. Zhang, X. Tang, K. Sun, Chitosan/Graphene
      Oxide Composite as an Effective Adsorbent for Reactive Red Dye Removal.
      Water Environ. Res. 88 (2016) 579-588.
      https://doi.org/10.2175/106143016X14609975746325

[85] S. T. Akar, E. San, T. Akar, Chitosan–alunite composite: an effective dye remover
      with high sorption, regeneration and application potential. Carbohyd. Polym. 143
      (2016) 318-326. https://doi.org/10.1016/j.carbpol.2016.01.066

[86] R. G. Sánchez-Duarte, J. López-Cervantes, D. I. Sánchez-Machado, M. A. Correa-
      Murrieta, J. A. Núñez-Gastélum, J. R. Rodríguez-Núñez, Chitosan–alunite
      composite: an effective dye remover with high sorption, regeneration and
      application potential. Environ. Eng. Manage. J. 15 (2016) 2469-2478.

[87] J. Gao, L. Zhang, X. Liu W. Zhang, Hierarchically structured, well-dispersed
      Ti^{4+} cross-linked chitosan as an efficient and recyclable sponge-like adsorbent for
      anionic azo-dye removal. RSC Adv. 6 (2016) 106260-106267.
      https://doi.org/10.1039/C6RA24446G

[88] Y. Jiang, J. L. Gong, G. M. Zeng, X. M. Ou, Y. N. Chang, C. H. Deng, J. Zhang,
      H. Y. Liu, S. Y. Huang, Magnetic chitosan–graphene oxide composite for anti-
      microbial and dye removal applications. Int. J. Biol. Macromol. 82 (2016) 702-
      710. https://doi.org/10.1016/j.ijbiomac.2015.11.021

[89] G. Sheng, S. Zhu, S. Wang, Z. Wang, Removal of dyes by a novel fly ash–
      chitosan–graphene oxide composite adsorbent. RSC Adv. 6 (2016) 17987-17994.
      https://doi.org/10.1039/C5RA22091B

[90] L. Zhang, L. Chen, X. Liu, W. Zhang, Effective removal of azo-dye orange II from
      aqueous solution by zirconium-based chitosan micro composite adsorbent. RSC Adv.
      5 (2015) 93840-93849. https://doi.org/10.1039/C5RA12331C

[91] M. Auta, B. H. Hameed, Chitosan–clay composite as highly effective and low-
      cost adsorbent for batch and fixed-bed adsorption of methylene blue. Chem. Eng.
      J. 237 (2014) 352-361. https://doi.org/10.1016/j.cej.2013.09.066

[92] Q. Gao, H. Zhu, W. J. Luo, S. Wang, C. G. Zhou, Preparation, characterization,
      and adsorption evaluation of chitosan-functionalized mesoporous composites.
      Micropor. Mesopor. Mater. 193 (2014) 15-26.
      https://doi.org/10.1016/j.micromeso.2014.02.025

[93] X. Jiang, Y. Sun, L. Liu, S. Wang, X. Tian, Adsorption of C.I. Reactive Blue 19 from aqueous solutions by porous particles of the grafted chitosan. Chem. Eng. J. 235 (2014) 151-157. https://doi.org/10.1016/j.cej.2013.09.001

[94] Z. Li, M. Cao, W. Zhang, L. Liu, J. Wang, W. Ge, Y. Yuan, T. Yue, R. Li, W. W. Yu, Affinity adsorption of lysozyme with Reactive Red 120 modified magnetic chitosan microspheres. Food Chem. 145 (2014) 749-755. https://doi.org/10.1016/j.foodchem.2013.08.104

[95] M. Sadeghi-Kiakhanim, M. Arami M, K. Gharanjig, Dye removal from colored-textile wastewater using chitosan-PPI dendrimer hybrid as a biopolymer: Optimization, kinetic, and isotherm studies. J. Appl. Polym. Sci. 127 (2013) 2607-2619. https://doi.org/10.1002/app.37615

Chitosan-Based Adsorbents for Wastewater Treatment, Ed. Abu Nasar   Materials Research Forum LLC
Materials Research Foundations 34 (2018) 123-132   doi: http://dx.doi.org/10.21741/9781945291753-6

Chapter 6

Adsorptive Treatment of Textile Effluent Using Chemically Modified Chitosan as Adsorbent

Ana Lilia Ramos-Jacques[1], Miriam Estevez[2], Angel Ramón Hernandez-Martinez[2,*]

[1] Independent Researcher. Gorriones, El Canto, Zibatá, 76269 Querétaro, México

[2] Centro de Física Aplicada y Tecnología Avanzada (CFATA), Universidad Nacional Autónoma de México (UNAM), Blvd. Juriquilla 3000, Querétaro, México

*angel.ramon.hernandez@gmail.com

Abstract

The textile industry produces large quantities of wastewater containing several chemicals (used for preparing, purifying, coloring and finishing textile products). Different treatments methods must be used in stages to safely discharge textile effluent. One of the main pollutants of these effluent is a textile dye, which can been removed by adsorption. Chitosan has excellent properties as adsorbent and has been chemically modified to enhance its performance. The best results in modifying Chitosan were achieved using nanomaterials as part of composites. Chitosan could not yet be replaced as an adsorbent by other chemical compounds because it is relatively inexpensive and has high adsorption capacity.

Keywords

Chitosan, Textile Effluents, Treatment, Adsorbent, Decolorization, Dyes

Contents

1.   Introduction

The textile industry is complex, high-demand and a major creator of wastewater. It is significant for the global economy and very important to textile exporters such as China, India, Bangladesh, Pakistan, Malaysia and European Union (the Netherlands, Germany, Spain, and France) [1,2]. As a global industry, it has important environmental challenges: its processing operations of fabrics such as sizing, desizing, bleaching, mercerization, dyeing, printing, and finishing require a huge consumption of water [3] and produce large quantities of wastewater. Textile effluent contains chemicals that cause health and environmental problems, for example, surfactants, dyes, heavy metals, hydrogen peroxide, and acids and alkalis [3,4]. Those characteristics differ greatly among countries due to different processes and regulations. Textile wastewater treatment is the complex and environmental impact of textile effluent is very high if wastewater is not treated properly. Treatment processes include physical, chemical and biological techniques. Among them, decolorization is considered important because of the dyes impact on health and for the negative reaction to colored water of the population. Adsorption has been widely used to capture dye from wastewater from textile effluents. This method is inexpensive and adsorbent material could be recovered. Chitosan has been used as an adsorbent for more than a decade due to its adsorption capacity and in consequence, has been chemically modified using different techniques and as part of composites for wastewater treatment. This chapter analyses current adsorptive treatments of textile effluents using chemically modified chitosan as adsorbent and the perspectives for further research in the area.

2.   Textile effluent characteristics

Generally, textile wastewater has a high temperature and pH, along with approx. 2,000 chemicals that are used in the textile industry [4,6]. Chemicals used could be subdivided into four main groups, textile auxiliaries, process chemicals, colorants and finishes [4]. Nevertheless, textile effluent has different characteristics related to manufacturing processes; some examples of common pollutants in the effluent are shown in Table 1. In

Chitosan-Based Adsorbents for Wastewater Treatment, Ed. Abu Nasar   Materials Research Forum LLC
Materials Research Foundations **34** (2018) 123-132   doi: http://dx.doi.org/10.21741/9781945291753-6

general, the most polluted is the effluent of "wet" processes where the fabric is colored. This effluent has intensive color and the highest pH and salt content (as Total Dissolved Solids, TDS) [3].

Table 1 Common pollutants in textile industry effluent [2,5].

Textile process	Pollutant
Desizing	Starch, ammonia, waxes, lubricants, biocides, anti-static compounds
Scouring	Disinfectants, insecticide residues, solvents, NaOH, detergents oils
Bleaching	H_2O_2, stabilizers (sodium silicate), NaOH, acids
Mercerizing	NaOH
Dyeing	Metals, salt, surfactants, cationic materials, sulfide, spent solvents, dyes
Printing	Urea, solvents, dyes, metals, suspended solids
Finishing	Toxic material, solvents, suspended solids

Table 2 Dyes commonly used in the textile industry. [2,7,8]

Type	Dye	Empirical Formula
Reactive	Remazol Brilliant Blue R	$C_{22}H_{16}N_2Na_2O_{11}S_3$
	Procion Red MX-5B	$C_{19}H_{10}Cl_2N_6Na_2O_7S_2$
	Cibacron F (Yellow, Orange, Red and Blue)	$C_{29}H_{20}ClN_7O_{11}S_3$
Direct	Congo red	$C_{32}H_{22}N_6Na_2O_6S_2$
	Direct yellow 50	$C_{35}H_{24}N_6Na_4O_{13}S_4$
Indigo	Indigo white	$C_{16}H_{10}N_2O_2$
	Tyrian purple	$C_{16}H_8Br_2N_2O_2$
	Indigo carmine	$C_{16}H_8N_2Na_2O_8S_2$
Acid	Blue 5G	$C_{29}H_{20}ClN_7O_{11}S_3$
	Bordeaux B	$C_{20}H_{15}N_2NaO_7S_2$

The two main pollutants in that specific effluent are metal ions and dyes. Trace metals could be present such as As, Zn, Cr, As and mercury from caustic soda. Some commonly used dyes are shown in Table 2. Reactive dyes are used for coloring cellulosic fibers and wool and silk [2] The reactive group of dyes form a chemical bond with the OH-group of the cellulose fiber [6]. Dyes are organic compounds that could interact with other chemicals and make adsorption more difficult.

Dyes could also be classified as natural or synthetic or based on their use or chemical structure. Textile dyes in wastewater could provoke damage to aquatic plants or other aquatic organisms and inhibit microorganisms grown. In human beings, they can cause allergies and respiratory diseases [4]. A very common group of dyes, which are part of acid dyes, are the "azo dyes" because they have a chromophoric azo group (-N=N-) attached to an aromatic and an unsaturated molecule [2]. Several azo dyes could cause damage to DNA that can lead to the genesis of malignant tumors. Also, some of their derivatives could induce cancer in humans and animals [4]. Wastewater treatment focus in dyes adsorption due to that negative impact on health, this process is known as decolorization.

3.   Treatment Processes

Textile wastewater treatment includes several physicochemical procedures to obtain a stream with low environmental impact. However, the objective nowadays is to produce reusable water, remove toxicity, or recover dyes and salt and if at all possible to not produce toxic sludge [1]. Procedures could include flocculation, coagulation, ozonation, biological treatment and dyes and metals removal [2].

3.1   Physical, chemical, biological and hybrid techniques

Common physical methods are filtration, flocculation, adsorption and chemical methods are known as advanced and chemical oxidation (which it is very effective). The biological treatment uses enzymes and microorganisms. As it has been mentioned, wastewater treatment is a function of the content of textile effluent, therefore the selection of the best combination of treatment methods is difficult. It is common to combine one or more oxidation methods along with biological methods considering wastewater components and the cost of the treatment [2]. Figure 1 shows an example of treatment processes, adsorption has been mentioned as a secondary and tertiary treatment, depending on effluent composition and adsorbent material.

Chitosan-Based Adsorbents for Wastewater Treatment, Ed. Abu Nasar   Materials Research Forum LLC
Materials Research Foundations **34** (2018) 123-132   doi: http://dx.doi.org/10.21741/9781945291753-6

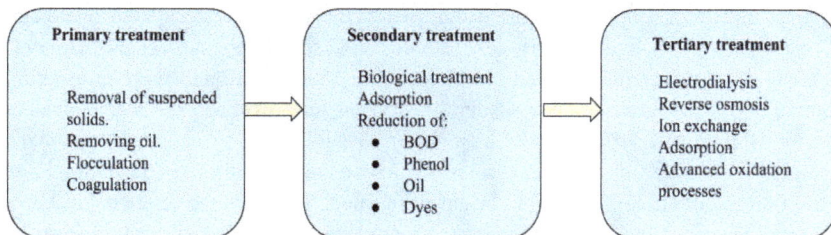

Figure 1 Textile wastewater treatment, commonly used methods [2,5,9].

4.   Adsorption process

Adsorption is a widely used method for wastewater treatment because it is relatively inexpensive. For Textile wastewater adsorption treatment has high decolorization efficiency including for effluent that contains a variety of dyes [2]. In this process, sorbent adsorbs onto its surface the sorbate (water pollutant) as a result of an attractive force [10]. As seen in Figure 2, there are two types of adsorption, physical and chemical; also called physisorption and chemisorption.

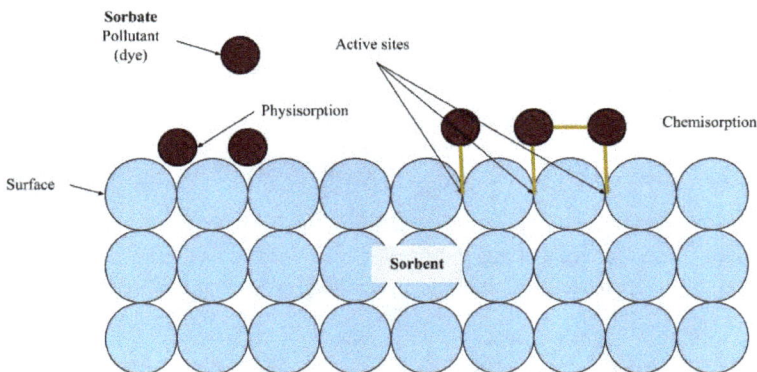

Figure 2 Schematic of physisorption and chemisorption processes [10].

Chitosan-Based Adsorbents for Wastewater Treatment, Ed. Abu Nasar   Materials Research Forum LLC
Materials Research Foundations 34 (2018) 123-132   doi: http://dx.doi.org/10.21741/9781945291753-6

Physisorption include intermolecular forces such as van der Waals and hydrogen bonding, this process does not cause a significant change in the electronic orbital patterns of the species involved, it is less stable and selective than chemisorption. In chemisorption, a reacting molecule forms a definite chemical bond with an unsaturated atom or group of atoms (active center or site) on an adsorbent surface, and electron transfer is involved [10]. Textile wastewater is subject to a secondary treatment of adsorption to reduce dye content, after sufficient amount of time (depending on the adsorption material used), an equilibrium is reached. To study the relationship between pollutants adsorbed quantities compared to the aqueous phase, an adsorption isotherm is obtained. These isotherms can be mathematically modeled using theories such as Langmuir, Freundlich, Halsey, Brunauer-Emmett-Teller (BET), Henderson, Smith, Elovich, and Lagergren [10,11].

4.1   Adsorbents

Adsorbents must have high affinity, adsorption capability, and regeneration ability, as a basis, but shape, dimension, nature (hydrophobic or hydrophilic), size, pH and temperature of the medium must also be considered [1,10]. In selecting a suitable adsorbent, it is possible based on material performance for removing desired pollutants [10]. Activated carbon is an effective adsorbent for a wide range of dyes, but it has a high cost for textile wastewater volumes and its regeneration is difficult, which may represent another increase in cost [3]. Several inorganic adsorbents are available for removing metal ions or dyes, such as Kaolin, Ballclay, Turkish kaolinite, Natural Brazilian bentonite, among others composition of these adsorbents include SiO_2, Al_2O_3, and MgO. The inorganic adsorbent could be oxides, clay minerals, zeolites, fly ash and polymeric resins [11]. They have been used to remove important pollutants but Chitosan is effective removing anionic and cationic dyes as well as heavy metal ions [11].

5.   Chitosan

Chitosan is a polysaccharide that features hydrophilicity, biocompatibility, biodegradability, non-toxicity and high adsorption efficiency. It can remove heavy metals and dyes due to the presence of amino and hydroxyl groups, that can be used as active sites [11].

5.1   Chitosan composites, derivatives and chemical modifications

Chitosan has been used as an adsorbent for more than a decade due to its adsorption capacity and recently it is used as part of diverse composites. However, Chitosan as adsorbent has low acid and thermal stability, inadequate mechanical properties, low

Chitosan-Based Adsorbents for Wastewater Treatment, Ed. Abu Nasar   Materials Research Forum LLC
Materials Research Foundations **34** (2018) 123-132   doi: http://dx.doi.org/10.21741/9781945291753-6

porosity and surface areas, then composites had been prepared using physical (converting Chitosan intro gel or other forms) and chemical (using nanoparticles) methods [11,12,13]. Some of these composites are shown in Table 3. Chemical modifications are intended to enhance Chitosan flexibility, chemical stability, and pH susceptibility; these modifications could include grafting, cross-linking and functionalization (in the amino groups of Chitosan).

Table 3 Chitosan adsorbents.

Year	Adsorbent	Adsorption of	Ref
2004	Chitosan/activated clay	Methylene blue, reactive dye RR222	[12]
2007	Chitosan/montmorillonite	Congo red	[12]
2008	Chitosan/polyurethane	Acid Violet 48	[12]
2008	Chitosan/oil palm	Reactive blue 19	[12]
2010	Chitosan/bentonite	Tartrazine and malachite green	[12]
2010	Chitosan/kaolin/gamma-Fe_2O_3	Methyl orange	[12], [13]
2016	Chitosan/polyurethane/anatase titania porous hybrid composite	Muticomponents(metal ions) mixtures	[14]
2017	Cross-linked magnetic chitosan anthranilic acid glutaraldehyde Schiff's base	As (V) and Cr(VI)	[15]
2017	Chitosan/polyurethane/Inulin composite	Evaluation of inulin replacing Chitosan	[16]

Some adsorbents had been modified due to the need to increase the efficiency of current chitosan adsorption capacity. Nanomaterials incorporation on treatment technologies, specifically in adsorption improves performance and efficiency of those methods [10]. Finally, to select an adequate adsorptive treatment of textile effluents using chemically modified chitosan as an adsorbent, the following proposal should be considered:

1.   Effluents of the specific textile industry must be documented.

2.   Metal ions and dyes present in wastewater must be classified and characterized.

3.   Chitosan adsorbents that remove those pollutants in laboratory conditions must be listed.

4.   Chemically modified chitosan adsorbent with stability on wastewater steam physical properties (pH, temperature, etc.) of effluent would be preferred.

5.   Alternative or combined treatments should be proposed.

6.   Conclusion

Environmental pollution caused by textile effluent poses a worldwide threat to public health. The challenge is obtaining reusable water, non-toxic, and recovering dyes and salt with the lowest impact of textile wastewater on the environment. Treatment techniques include adsorption using Chitosan chemically modified, new composites and derivative have been proposed to enhance adsorption efficiency. Nevertheless, complex textile effluents require selecting proper Chitosan composite considering pH and temperature of adsorption media, adsorbent dosage, co-existing ions.

References

[1]   L, Bilińska L, Gmurek M, Ledakowicz S. Comparison between industrial and simulated textile wastewater treatment by AOPs – Biodegradability, toxicity and cost assessment. Chemical Engineering Journal. 306 (2016) 550-559.

[2]   A.E. Ghaly, R. Ananthashankar, M. Alhattab, V.V. Ramakrishnan, 2014. Production, Characterization and Treatment of Textile Effluents; A Critical Review. Journal of Chemical Engineering & Process Technology. 5:182. https://doi.org/10.4172/2157-7048.1000182

[3]   C. Holkar, A. Jadhav, D. Pinjari, N. Mahamuni, A. Pandit, A critical review on textile wastewater treatments: Possible approaches. Journal of Environmental Management. 182 (2016) 351-366. https://doi.org/10.1016/J.JENVMAN.2016.07.090

[4]   S. Khan, A. Maik, Chapter 4. Environmental and Health Effects of Textile Industry Wastewater, in: Environmental Deterioration and Human Health, Springer, Netherlands, Dordrecht, 2014, pp 55-71. https://doi.org/10.1007/978-94-007-7890-0_4

[5]   G.R. Mettam, L.B. Adams, Nanochemicals and Effluent Treatment in Textile
      Industries, in: Textiles and Clothing Sustainability: Nanotextiles and
      Sustainability, Springer, 2017, pp. 57-96. https://doi.org/10.1007/978-981-10-
      2188-6_2

[6]   Information on http://shop.kremerpigments.com/media/pdf/345110-345170e.pdf

[7]   Information on
      https://pubchem.ncbi.nlm.nih.gov/compound/5491378#section=Top

[8]   Information on https://www.sigmaaldrich.com/mexico.html

[9]   R. Pešoutová, P. Hlavínek, J. Matysíková, Use of advanced oxidation processes
      for textile wastewater treatment: A review, Food and Environment Safety, Journal
      of Faculty of Food Engineering. Universitatii Suceava 10 (2011) 59-65.

[10] R. Das, C.D. Vecitis, A. Schulze, B. Cao, I.A. Fauzi, X, Lu, J. Chen, S.
      Ramakrishna, Recent advances in nanomaterials for water protection and
      monitoring. Chem. Soc. Rev. 46 (2017) 6946-7020.
      https://doi.org/10.1039/c6cs00921b

[11] S.S. Gupta, K.G. Bhattacharyya, Adsorption of metal ions by clays and inorganic
      solids. RSC Adv. 4 (2014) 28537-28586. https://doi.org/10.1039/C4RA03673E

[12] W. Wan Ngah, L. Teong, M. Hanafiah, Adsorption of dyes and heavy metal ions
      by chitosan composites: A review, Carbohydrate Polymers, 83 (2011) 1446-1456.
      https://doi.org/10.1016/J.CARBPOL.2010.11.004

[13] L. Zhang, Y. Zeng, Z. Cheng, Removal of heavy metal ions using chitosan and
      modified chitosan: A review. Journal of Molecular Liquids. 214 (2016) 175-191.
      https://doi.org/10.1016/J.MOLLIQ.2015.12.013

[14] L. Argüello, A.R. Hernandez-Martínez, A. Rodríguez, G.A. Molina, R. Esparza,
      M. Estevez, Novel chitosan/polyurethane/anatase titania porous hybrid composite
      for removal of metal ions waste. Journal of Chemical Technology &
      Biotechnology, 91 (2016) 2185-2197. https://doi.org/10.1002/jctb.4945

[15] Y.G. Abou El-Reash, M. Otto, I.M. Kenawy, A. Ouf, Adsorption of Cr(VI) and
      As(V) ions by modified magnetic chitosan chelating resin. Int. J. Biological
      Macromolecules, 49 (2011) 513-522.
      https://doi.org/10.1016/J.IJBIOMAC.2011.06.001

[16] A.R. Hernández-Martínez, G. Molina, L. Jiménez-Hernández, A. Oskam, G. Fonseca, M. Estevez, Evaluation of Inulin Replacing Chitosan in a Polyurethane/Polysaccharide Material for Pb^{2+} Removal, Molecules, 22 (2017) 2093-. https://doi.org/10.3390/molecules22122093

Chitosan-Based Adsorbents for Wastewater Treatment, Ed. Abu Nasar   Materials Research Forum LLC
Materials Research Foundations **34** (2018) 133-160   doi: http://dx.doi.org/10.21741/9781945291753-7

Chapter 7

Chitosan-Based Adsorbents: Environmental Applications for the Removal of Arsenicals

Lee D. Wilson[1,]*, Brij B. Tewari[2]

[1]Department of Chemistry, University of Saskatchewan, 110 Science Place (Room 165 Thorvaldson Building), Saskatoon S7N 5C9, SK., Canada

[2]Department of Chemistry, University of Guyana (Turkeyan Campus), PO Box 101110, Georgetown, Guyana

*lee.wilson@usask.ca

Abstract

Diverse types of arsenic contaminants occur in aquatic environments due to their multiple oxidation states and ionization properties. To address the removal of arsenic species in water, chitosan has been studied as a platform biopolymer for the design of adsorbent materials. This chapter examines the utility of chitosan and its modified forms as sorbent materials for the removal of arsenic species in aquatic environments. A selected coverage of the literature will provide an overview of various chitosan-based adsorbents and the general utility of chitosan sorbents for the removal of inorganic and organoarsenicals.

Keywords

Chitosan, Chemical Modification, Adsorption, Adsorbent, Contaminants, Inorganic Arsenicals

Contents

1.   Introduction

The presence of arsenic waterborne contaminants has led to concerns over the fate and transport in aquatic environments as a consequence of the build-up and uncontrolled release of inorganic and organic arsenicals in the environment [1-4]. Arsenicals are of great interest due to their global occurrence in locations such as Bangladesh, Vietnam, Argentina, Chile, Mexico, Taiwan, Thailand, USA and Canada [4-6]. Point sources of arsenic include mine tailings, emissions from coal and petrochemicals, agrochemicals, and pigments [4]. Mineral forms of arsenic are co-located in copper, lead, gold, and uranium deposits in various mineral forms ($Pb(AsO_4)_2$, Na_3AsO_3, $Cu_3(AsO_3)_2$, etc.) with an abundance that depends on their mineralogical and geochemistry profiles in ground and surface water environments [1-5]. The occurrence of dissolved arsenic species in ground and surface water environments depend on several factors that include the point source and the solution chemistry conditions; pH, redox potential, mineral species, ionic strength, organic matter content, etc. Dissolution processes of arsenicals at mineral-fluid interfaces are also influenced by the presence of other species (ions, colloids, dissolved solids, suspended solids, etc.) through dissolution, precipitation, ion-exchange and adsorption-desorption processes [7]. In general, groundwater supplies are dominated by As(III) species; whereas, As(V) species are predominant in surface water bodies at greater levels due to the favoured stability of arsenate over arsenite in oxidizing environments. While the toxicity of trivalent arsenic is greater than its pentavalent form, concerns over the levels of As species relate to the exposure and toxicology of arsenic [3-9]. The health risks of arsenic exposure are evidenced by the regulatory safety limits set forth by the WHO to levels below 10 ppb [9]. In general, naturally occurring arsenicals in aquatic environments generally lie below 10 ppb; however, industrial (mining, agriculture, and energy sectors) activities have led to the buildup and uncontrolled transport of arsenicals via agrochemicals, pigments, pesticides, antimicrobials, and pharmaceuticals [10-15]. In the case of organoarsenicals, the fate, distribution, and toxicology of such species have been a source of debate [16,17]. Concerns for human health have been raised that relate to the use of an antimicrobial agent (3-nitro-4-hydroxyphenylarsonic acid; roxarsone) in poultry production, where the breakdown byproducts (mainly the inorganic species) are considered as the toxic components that result from roxarsone upon decomposition or metabolism [18,19].

The occurrence of inorganic and organic arsenic species with variable forms of oxidation and ionization are known in surface and groundwater environments [20]. The chemistry of arsenicals is diverse according to the number reviews and monographs on the subject where the chemical diversity of arsenic and its relevance to environmental science reveal the challenges related to its effective removal [21,22]. In particular, the challenges related to the development of sustainable materials for the remediation waterborne arsenic species provide an inspiration for its coverage in this chapter. This mini-review provides a selected coverage on the use of chitosan and its modified forms as adsorbents materials for the adsorptive removal of arsenic species with a focus on inorganic species. In view of the diverse range of organoarsenicals [23-25], a coverage of the subject will be limited to a model compound (roxarsone) due to its recent widespread use in agriculture [16-19], its suitability for laboratory studies, and the utility of roxarsone as a model system for various As(V) species due to the predominance of this form of arsenic in surface water environments. This chapter will provide a selected coverage of recent studies to highlight the general features and utility of chitosan adsorbent materials for the removal of arsenic-based environmental contaminants.

2.   Arsenic species as environmental contaminants

As mentioned above, arsenic species are chemically diverse according to their existence as inorganic and organic forms with variable oxidation and ionization states due to the environmental conditions. In its elemental form, As^0 is a metalloid among the fifth group of main group elements that share a similar structure and bonding properties to that of phosphorous. In particular, the formation of oxoanion species and general aspects of its solution chemistry are highlighted in Table 1 [7]. In surface water environments, As^V (arsenate) is the found in oxic surface waters, while As^{III} (arsenite) is more commonly found in anoxic groundwater environments. The arsenite form tends to have greater toxicity over the arsenate species, as reported previously [26,27]. A summary of the pK_a values and acid-base speciation at variable pH (low to high) conditions for arsenic acid, arsenous acid, orthophosphoric acid, and roxarsone are listed in Table 1. The accompanying molecular structures for these systems are given in Scheme 1 where it is apparent that arsenic acid (H_3AsO_4) bears structural features and ionization equilibria that compare to phosphoric acid (H_3PO_4) and roxarsone. The role of oxidation state and ionization processes in aquatic environments is a key aspect that contributes to the fate and transport of such oxoanions in water. Thus, the structure and physicochemical properties of arsenic species is an important consideration for the rational design of sorbent materials, where attention will be turned to biopolymers such as chitosan due to its unique biopolymer structure, properties, and synthetic versatility [28-31].

Table 1   *Speciation of arsenic acid (H_3AsO_4), arsenous acid (H_3AsO_3), orthophosphoric acid (H_3PO_4), and roxarsone in water.*

Chemical species	pH (low)	pH (medium)	pH (high)
H_3AsO_4; **As(V)** FW=141.94 g/mol	H_3AsO_4/ $H_2AsO_4^-$ pH = p$K_{a,1}$ = 2.2	$H_2AsO_4^-$/ $HAsO_4^{2-}$ p$K_{a,2}$ =7.08	$HAsO_4^{2-}$/ AsO_4^{3-} p$K_{a,3}$ = 11.5
H_3AsO_3; **As (III)** FW = 125.94 g/mol	H_3AsO_3/ $H_2AsO_3^-$ pH = p$K_{a,1}$ = 9.22	$H_2AsO_3^-$/ $HAsO_3^{2-}$ p$K_{a,2}$ = 12.3	Not Reported
H_3PO_4; **P(V)** FW = 97.99 g/mol	H_3PO_4/ $H_2PO_4^-$ pH = p$K_{a,1}$ (2.17)	$H_2PO_4^-$/ HPO_4^{2-} pH = p$K_{a,2}$ =7.31	HPO_4^{2-}/ PO_4^{3-} pH = p$K_{a,3}$ = 12.36
Roxarsone; **As(V)** FW = 263.04 g/mol	H_3ROX/H_2ROX^- pH = p$K_{a,1}$ (3.49)	H_2ROX^-/$HROX^{2-}$ pH = p$K_{a,2}$ (6.38)	$HROX^{2-}$/ROX^{3-} pH = p$K_{a,3}$ (9.76)

Note: *The use of low, medium, and high pH values relates to the range of pK_a values for each oxospecies. The pK_a values for H_3PO_4 are listed: p$K_{a,1}$ (2.17), p$K_{a,2}$ (7.31), p$K_{a,3}$ (12.36) [4,7]. FW denotes formula weight and ROX denotes the molecular fragment ion $[C_6H_3AsNO_6]^{3-}$.*

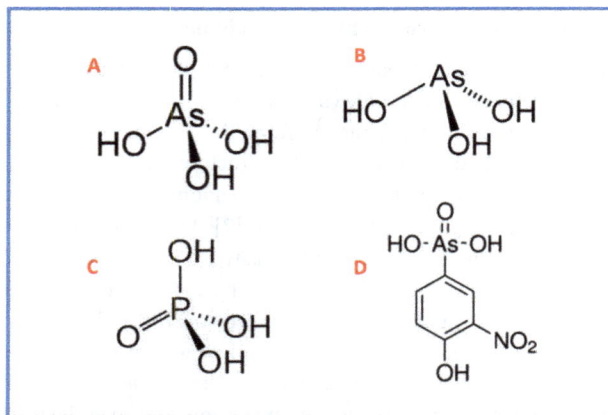

Scheme 1   Molecular structure of inorganic and organic species: A) arsenic acid, B) arsenous acid, C) orthophosphoric acid, and D) roxarsone (3-nitro-4-hydroxyphenyl-arsonic acid).

According to Table 1, arsenous acid exists in its non-ionized form at typical aquatic pH conditions; whereas, the pentavalent species may exist as mono- or di-valent anions. Thus, the water solubility of As^{III} and As^V differ considerably when pH < 9 which can be exploited for the controlled separation of trivalent and pentavalent arsenic by use of ion exchange, precipitation, coagulation-flocculation, and adsorption [18,19,21,22,30-32] processes. The role of chitosan structure and modification contribute importantly to the design of adsorbent materials for adsorption-based processes described below.

3.   Chitosan structure and properties

The removal of waterborne arsenic contaminants includes various approaches such as coagulation, photolysis, bioremediation, oxidation, precipitation, membrane separation, and adsorption [21,22,32,33]. Among the various methods of removal, adsorption is a physical approach with relatively low cost and infrastructure requirements, facile operation, and minimal chemical or energy inputs [30,34,35]. Adsorption-based methods offer a sustainable and scalable technology for the removal of contaminants that is limited largely by the nature of the adsorbent material. To address the need of structural diversity of adsorbents, increasing attention has been directed toward the development of sustainable biopolymer materials with improved properties derived from biomass. Research activity on the utilization of abundant biopolymer materials such as cellulose (poly-β(1-4)-D-glucose) and chitosan (poly-β(1-4)-2-amino-2-deoxy-D-glucose) as adsorbents for the removal of environmental contaminants has gained momentum due to the relative availability and low cost of these materials. Cellulose and chitosan are structurally related polysaccharides but their unmodified (native) forms have limited adsorption properties which can be overcome through synthetic modification [21,30]. While cellulose is an abundant biopolymer obtained from plant biomass, chitosan is a marine polysaccharide derived from chitin (poly-β-(1-4)-2-acetamido-2-deoxy-D-glucose) biomass derived from of crustaceans. Chitosan is a copolymer obtained through the partial deacetylation of chitin under hydrolysis conditions [28], where the degree of deacetylation (DDA) depends on the extent of the hydrolysis. Many commercial forms of chitosan have DDA values ca. 75-85% where the resulting pK_a ($\approx$ 5.5-6.5) is found to depend on the molecular weight and DDA of chitosan [28,36]. Scheme 2 depicts a generalized molecular structure of chitosan where the monomer units differ at C-2 according to the DDA, where the R-substituent is an amine or an acetyl group.

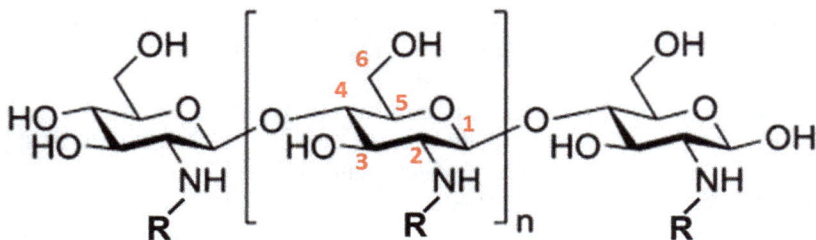

Scheme 2   Molecular structure of chitosan where R represents -H or –COCH₃ (acetyl) substituents at C-2 that depend on the degree of deacetylation (DDA) of the biopolymer chain. The numbering of the carbon atoms (C-1 to C-6) of the D-glucopyranose units are denoted in red where n represents the degree of polymerization.

The molecular structure of chitosan and chitin closely resemble cellulose except for the substituent at the C-2 position where the –NHR group is exchanged for –OH in the case of cellulose. Chitosan is a semi-crystalline biopolymer due to the formation of stable hydrogen bonds between adjacent biopolymer strands that contribute to its relatively low

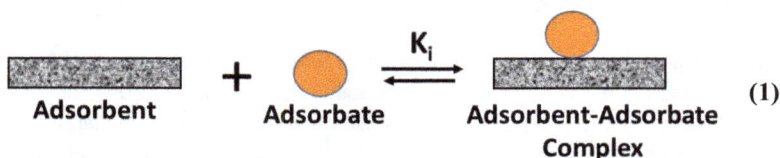

$$\text{Adsorbent} + \text{Adsorbate} \underset{}{\overset{K_i}{\rightleftharpoons}} \text{Adsorbent-Adsorbate Complex} \tag{1}$$

surface area (ca. $<10^1$ m^2/g) and limited pore structure characteristics in the solid state [36,37]. The semi-crystalline nature of chitosan contributes to its low solubility in many organic solvents and water at ambient pH conditions above its pK_a value. In the presence of inorganic/organic acids such as HCl, HNO$_3$, and CH$_3$CO$_2$H, chitosan has favourable water solubility due to ionization of the amine groups, especially when the solution pH lies below the pK_a of chitosan. Chitosan is also soluble in DMSO and various ionic liquids; however, an enhancement of the solubility profile in water and other solvents can be achieved by selective functionalization of the –OH or –NH groups (C-2, -3, and -6) of the biopolymer. For example, carboxymethylation (CM; -CH$_2$CO$_2$H) at the -OH sites of chitosan increases the water solubility due to the ionization of the CM group [38]. Synthetic modification of –OH groups of chitosan with other types of polar and ionic groups (e.g., phosphate, sulfate, quaternary ammonium, etc.) serve to alter the crystalline

Chitosan-Based Adsorbents for Wastewater Treatment, Ed. Abu Nasar   Materials Research Forum LLC
Materials Research Foundations **34** (2018) 133-160   doi: http://dx.doi.org/10.21741/9781945291753-7

nature and hydrophile-lipophile balance (HLB) of chitosan [30-35,39]. Along with changes in the solubility profile of chitosan, the physical and chemical modification can be used to enhance the adsorption properties of this unique biopolymer material. The context of adsorption herein relates to heterogeneous solid-liquid processes at equilibrium conditions, according to Eq. (1). The equilibrium constant (K_i) for the adsorption process in Eq. (1) is a key thermodynamic constant obtained through adsorption isotherm modeling studies that is described in greater detail elsewhere [40]. For example, adsorbent-adsorbate systems with variable adsorption affinity are anticipated by variation of the relative Lewis-acid or -base character of the adsorbent surface sites to yield favourable adsorptive interactions with the adsorbate. As such, the physicochemical properties of the adsorbate system must be considered in the case of heterogeneous adsorption processes in aqueous media. The structure and properties of chitosan can be modified by functionalization of the –OH and/or -NH groups (Scheme 2), as described in the next section.

4.   Chitosan adsorbent design

In general, the native form of chitosan (Scheme 2) displays relatively low to moderate adsorption properties due to its relatively low accessible surface area that relate to its semi-crystalline structure [28,36]. Greater accessibility of the polar functional groups of chitosan can be achieved by increasing the surface area (SA) and pore volume of the biopolymer via physical or chemical treatment methods. Physical treatment may alter the morphology of chitosan by conversion to various forms such as granules, powders, films, beads, and fibers. Physical treatment by thermal or mechanical means generally results in changes to the *surface-to-volume ratio*, along with alteration of the pore structure properties of chitosan, as evidenced by the adsorption properties of micro- versus nano-chitosan [37]. Chemical methods for the structural modification of chitosan may involve a range of methods [34-36]: *Type 1*) functionalization of the hydroxyl and amine groups to alter the HLB profile, *Type 2*) cross-linking of the biopolymer with multifunctional cross-linker units (e.g., glutaraldehyde, epichlorohydrin, diacid chlorides, dicarboxylic acids, etc.), and *Type 3*) composite formation of chitosan with additional components. These three types of synthetic modification have two general outcomes on the structure of chitosan; i) alteration of the biopolymer textural properties (SA and pore structure properties) and ii) modification of the surface chemical properties. The above types of synthetic modification will be outlined in greater detail to highlight the related structural modification with the adsorption properties of chitosan materials with selected inorganic and organic arsenicals.

4.1   Structural modification of Chitosan

Although there exist a number of recent reviews [28-35] related to the topic of synthetic modification of chitosan, an overview of three general types of modification is presented below due to their relevance to adsorbent design for the removal of arsenicals from aqueous media.

4.1.1   Type 1 (Chitosan functionalization)

Chitosan is known as a suitable sorbent for apolar species such as fatty acids due to its lipophilic nature [41]. As well, chitosan is a known sorbent for metal cation species [28,42-47] due to the presence of functional groups that provide potential binding sites along the biopolymer chain (Scheme 2). The presence of abundant amino and hydroxyl groups of chitosan have led to the use of substitution and addition reactions at C-2, -3, and -6 positions of the glucopyranose units of chitosan. The replacement of –OH and – NHR with substituents of variable polarity, size, and electrostatic charge result in changes to the solubility and the HLB profile of chitosan. The introduction of ionic groups such as phosphate, sulfonate, and carboxylate groups often lead to greater water solubility and higher binding affinity with metal cations. The grafting of apolar silane units may result in variable surface properties of chitosan with unique hydration and adsorption properties, according to the enhanced uptake of arsenate [48]. Although the introduction of functional groups with Lewis base character may not favor the sorption of arsenic anion species, the use of acidic pH conditions to protonate chitosan alters the zeta-potential of the biopolymer [37-39]. pH effects offer a facile and useful strategy for increasing the adsorption capacity *via* ion-exchange with anion species. The formation of arsenate anion species is favored at typical pH conditions (pH 6-8) in aquatic environments based on the pK_a values for As(V) species listed in Table 1. As well, protonation of chitosan at pH < pK_a (below pH 5.5 for native chitosan) enhances its water solubility and adsorption properties with oxoanion species (Table 1). [28,37]. In a review by Da Sacco and Masotti [49], the adsorptive interactions between chitosan and arsenic oxide (As_2O_3) were reported, where H-bonding and ion-exchange were concluded as the prominent interactions for chitosan and As(III/V) species at variable pH (*cf.* Fig. 1 in [49]). Depending on the ionization state of the adsorbent-adsorbate system, the type of adsorption may range in modality from a *physisorption* to *chemisorption-like* processes, where ion-exchange contributes favorably to the adsorptive interactions. Dipolar and dispersion interactions are favored when negligible ionization occurs and weaker noncovalent interactions persist. Pontoni and Fabbricino [34] reviewed various types of modified chitosan for the removal of arsenite and arsenate species. Table 2 lists selected examples of equilibrium uptake values for native and modified chitosan adsorbents with inorganic As (III/V) species [38-47].

Table 2 Chitosan and modified materials for the uptake of inorganic arsenate species from aqueous solution.

Chitosan Adsorbent	As (V) Uptake and pH Conditions	Reference
Unmodified Powder	2.5- 4 mg/g (pH 3.5-4.5)	50
Unmodified flakes/beads	22.5/2.24 mg/g (pH 6)	51
	69.3 mg/g (pH 5)	52
Unmodified Powder	42 mg/g (pH 8.0)	53
Unmodified Beads	50 mg/g (pH 8.0)	54
Cross-linked	58 (pH 4.0)	55
Cross-linked	61.1	56
Cross-linked & NDMG grafted	60.9	
Iron doped flakes	16.2*/22.5 (pH 7)	57
Cross-linked	58 mg/g (pH 4)	58
Unmodified Powder	27 mg/g (pH 8.5)*	59
Cross-linked glutaraldehyde	227 mg/g (pH 8.5)	
Cross-linked Glutaraldehyde	270 mg/g	60
Acryloylated grafted	551 mg/g	61
Surface functionalized	70 mg/g (pH 2-3)	62
	230 mg/g (pH 2-3)	
Iron Cross-linked	13.4 mg/g*	63
Tin Cross-linked	17.10 mg/g*	64

Denotes As (III) Uptake as H_3AsO_3

In general, native chitosan generally displays lower uptake toward As(III/V) species, especially above its pK_a, where ion-exchange processes constitute a minor contribution to adsorption. Variation of the pH of the medium results in alteration of the uptake of arsenate over arsenite species, especially when the pH conditions (Table 1) favor ion-exchange processes [37,49]. By contrast, modified forms of chitosan generally reveal greater uptake due to favorable changes to the surface functional groups and/or textural properties. In the case of modification of chitosan via cross-linking, this strategy involves the formation of a cross-linked network that alters the surface functional groups and the textural properties of the biopolymer. The effect is understood according to the creation of defects within the semi-crystalline lattice of chitosan, in agreement with the onset of features related to amorphization of the semi-crystalline network [36]. Cross-linking is a specific category of the Type 1 method where the formation of a cross-linked network alters the SA and pore structure properties of the biopolymer, as outlined in the next section. As well, modification of chitosan affords unique materials with variable uptake capacity as evidenced by the variable monolayer adsorption capacity (Q_m) for arsenate in Table 2, where Q_m varies from 1.84 to 228 mg/g for As(V) species. Estimates of Q_m are based on the Langmuir model or related mathematical variants of this widely used adsorption isotherm model [40].

4.1.2   Type 2 (Chitosan cross-linking)

As mentioned above, native chitosan has been studied as a sorbent for metal cation species but fewer studies report its adsorbent applications for anion species [65-70]. The relatively low affinity of non-ionized chitosan with anions is related to the presence of numerous polar functional groups of the biopolymer with Lewis base character. *Type 2* synthesis uses cross-linkers to alter the Lewis acid-base properties of chitosan by changing the surface chemical and textural properties. The textural properties of cross-linked chitosan are altered due to *"pillaring effects"* between adjacent biopolymer chains [71] by judicious choice of the cross-linker unit (size, flexibility, and electronic character). By contrast, the hydrogen bonding and electrostatic interactions for native chitosan (Scheme 2) are altered upon cross-linking due to the reduction of the crystallinity upon cross-linking. Cross-linking often results in alteration of the relative SA, textural properties, and adsorption site accessibility when compared against the properties of native chitosan in Table 2. The formation of cross-links within the biopolymer network affect the hydration properties and zeta-potential of the surface sites of chitosan due to alteration of the surface functional groups that are known to affect the adsorbate binding affinity (K_i, see eq. 1) [72]. The formation of cross-linked chitosan frameworks share features that resemble that of zeolites where the formation of pre-organized (inclusion) sites may favor cooperative binding interactions that vary with the level of cross-linking of chitosan. Scheme 3 provides a general illustration of cross-linking at variable levels of cross-linker, along with doping of an arbitrary metal ion is shown (b) at Lewis base (-OH or –NHR) sites [73]. The consequence of *Type 2* modification illustrated in Scheme 3 leads to changes in the surface and textural properties of the biopolymer framework.

The use of glutaraldehyde as a cross-linker results in the formation of imine linkages that may impart Lewis acid character that depends on the pH of the medium. Evidence of greater anion binding was reported in the case of phosphate anion species by Mahaninia and Wilson (*cf.* Table IV in ref. 72), upon cross-linking of chitosan with glutaraldehyde, according to the enhanced uptake of phosphate anions. These results are in parallel agreement with the greater uptake of arsenate anions observed for chitosan cross-linked with glutaraldehyde at variable mole ratios (see Table 2 above) and reported elsewhere

[7]. It can be inferred that greater levels of cross-linking result in greater anion uptake due to the presence of the imine cross-link sites. The resulting Lewis acid character of such cross-linked frameworks can be further modified by chelation of Lewis acid species, as shown by incorporation of Cu(II) in Scheme 3b [73]. Chitosan-glutaraldehyde systems along with the chelation of divalent metal ions reveal increased sorption capacity toward urea [73] and orthophosphate [74], in accordance with the greater uptake of arsenic species in Table 2, where Lewis acids are employed [63,64].

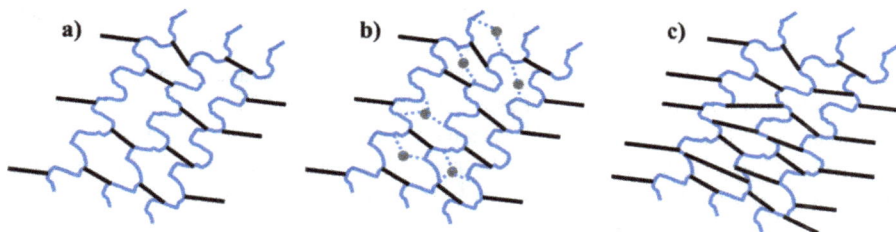

Scheme 3 Structural effects due to cross-linking of chitosan: a) low level cross-linking, b) low level of cross-linking with metal chelation, and c) higher level cross-linking. The cross-linker is shown as a black line segment and the metal ion species are shown as grey spheres. The dashed line indicates metal chelate interaction with an arbitrary functional group, where the counterions and solvent are not shown for clarity.

The use of epichlorohydrin (EPH) as a cross-linker involves reaction at the –OH groups of chitosan, where the adsorption properties are observed to increase upon cross-linking in accordance with the "*pillaring effect*" [72]. As well, a decrease in the Lewis base character of chitosan occurs upon formation of glycerol cross-link sites, as evidenced by increased anion arsenical sorption capacity [75].

4.1.3 Type 3 (Composite formation)

Various approaches have been reported for the design of composites that vary from solid phase synthesis of multi-component systems to self-assembly using a solution-based approach [76-97]. A key feature of composite materials is their distinctive physicochemical properties that differ markedly in a nonlinear manner relative to chitosan and the respective component precursors. A versatile solution-based method for the preparation chitosan composite materials involves the use of phase-inversion synthesis, where the method relies on the acid-base properties of chitosan [72]. In brief, an aqueous chitosan solution is prepared at pH conditions below its pK_a, where additive components (metal oxides, secondary polymers, organics, nanoparticles, etc.) are mixed with chitosan as a slurry, gel, or a solution. The mixture of prepared components is added to an alkaline solution under variable conditions to convert the mixed phase to a solid-phase with variable morphology (beads, capsules, powders, etc.), followed by subsequent isolation and purification steps. In some cases, the composites are further stabilized by cross-linking to improve the mechanical stability and other structural or physicochemical properties that relate to adsorption. The composite structure may vary between a physical blend of a unique chemical combination that depends on the additive components and the synthesis conditions. For example, the composite formation may vary from a two-step preparation where the metal (hydr)oxides are prepared separately with subsequent blending to form multi-component composites, as described above. Alternatively, a single-step method may use chitosan as a biopolymer support where metal (hydr)oxides are prepared *in situ*. The *in situ* one-step method is a useful method for tuning the surface chemistry of an adsorbent, as reported by Kwon *et al.* where various iron (hydr)oxides were supported onto the surface of activated carbon (AC) [98].

Various studies have been reported [76-97] where enhanced uptake of arsenite and arsenate species have been reported for chitosan composites that contain various additives (metal ions, metal oxides, clays, zeolite, and zero-valent iron, etc.), in addition to those illustrated by the examples in Table 3 [21,99-108]. The formation of composite materials with chitosan parallels the outcome for the Type 1 and 2 synthesis according to the variation in structure and surface functionality described above. Although the structure of such multi-component systems (composites) is considerably more complex than cross-linked binary systems, the role of additive components modifies the surface chemical and textural properties. In the case of metal oxides such as iron (hydr)oxides, these additives are of increasing interest due to their variable structural forms (nm- to micron-scale size) and facile preparation using green chemical methods. A general outcome of composite formation is the goal of improving adsorption properties toward arsenicals by structural and surface chemical modification of the adsorption sites.

A key challenge related to the design of composite materials involves their structural characterization and stability. In some cases, there is a need to stabilize the mechanical properties of composites since the synthetic preparation often relies on the physical blending of components to avoid exfoliation or leaching of components during multiple adsorption-desorption cycles. There are challenges that relate to the ability of controlling phase separation during synthesis and the ability to characterize the structure of multi-component systems, especially for amorphous systems. However, a key advantage concerning the use of composites relates to the ability to incorporate additive components (Lewis acids or bases) in a step-wise and facile manner by tuning the composition of additives. The formation of a composite with suitable composition often yields products with improved adsorption properties [109,110].

Table 3 Selected examples of uptake capacity for inorganic arsenic species by various chitosan composite adsorbents in aqueous solution.

Composite Adsorbent	[1]As (III) Uptake and pH Conditions	[2]As (V) Uptake and pH Conditions	Reference
Vermiculite Chitosan/vermiculite	34.9 mg/g (pH 5.0) 72.2 mg/g (pH 5.0	NR	99
Chitosan/biosorbent	56.5 mg/g (pH 4.0)	NR	100
Chitosan/PVA/cerium nanofibres	18.0 mg/g	NR	101
Molybdate impregnated Chitosan beads (MICB)	1.98 mg/g	2.00 mg/g	102
Magnetic/chitosan/biochar	NR	17.9 (pH = 5.0)	103
Fe-Mn chitosan	54.2 (pH 7.0)	39.1 (pH=7.0)	104
Fe-Mn composite	251 mg/g (pH 5.0)	117 mg/g (pH5.0)	21
MnO_2	18 mg/g	13 mg/g	21
Goethite	-	67 mg/g (pH 3–3.3)	21
$Al_2O_3/Fe(OH)_3$	1.7 mg/g (pH 6.6)	62 mg/g (pH 7.2)	21
Fe(III)-loaded sponge	34 mg/g (pH 9.0)	230 mg/g (pH 4.5)	21
Fe-Mn-mineral	23 mg/g (pH 5.5)	11 mg/g (pH 5.5)	21
TiO_2	61 mg/g (pH 7.0)	69 mg/g (pH 7.0)	21
Nanocrystalline TiO_2	14 mg/g (pH < 8)	6.3 mg/g (pH > 7.5)	105
Nanocrystalline β-FeOOH (Akaganeite)	-	120 mg/g (pH 7.5)	106
Nanoscale zero-valent Iron	3.5 mg/g; pH 7	-	107
Fe(II) treated zeolite	-	22.5 mg/g (pH 5.5)	108

[1]*Calculated based on the uptake of H_3AsO_4 (Molar Mass = 141.94 g/mol)*
[2]*Calculated based on the uptake of H_3AsO_3 (Molar Mass = 125.94 g/mol)*

4.2   Uptake of roxarsone and selected organoarsenicals

The widespread use of organoarsenicals as agrochemicals, pigments, pharmaceuticals, and anti-microbials is reflected by their diverse chemical structure (e.g., Paris green, cacodylic acid, Salvarsan, etc.) [3, 8,16,23,25]. In the case of antimicrobials, roxarsone has been used extensively in poultry production to control coccidiosis, a parasitic disease in chickens. In 2012, roxarsone was withdrawn from the USA market and was subsequently replaced by nitarsone (4-nitrophenyl)arsonic acid). A key controversy surrounding the use of these antimicrobials relate to their potential metabolism or breakdown into inorganic arsenic species, where the inorganic species are considered as

145

the toxic components [3,8,9]. Other types of organoarsenicals result from the reduction of H_3AsO_4 and H_3AsO_3 to yield methylated derivatives (methyl arsonic acid and dimethyl arsines) that have variable fate and transport in the environment [4,111]. Roxarsone (ROX) is considered as a suitable model organoarsenical due to its relevance to inorganic arsenate species due to its structure and physicochemical properties (*cf.* Table 1) [112]. In contrast to inorganic arsenate, ROX has a lower water solubility at ambient pH, where the variable hydration behavior of such As-species relate to differences in their adsorption profile with chitosan and mineral surfaces. Selected examples of studies on the removal of roxarsone are reported with some other organoarsenicals in Table 4.

Table 4   Uptake capacities for organic arsenicals for various chitosan and modified adsorbent materials.

Adsorbent	[1]As (III) Uptake & pH Conditions	[2]As (V) Uptake & pH Conditions	Refeference
Unmodified Chitosan Powder Activated Carbon	NR	**ROX**: 109 mg/g (pH 7.0); 10 mM KH₂PO4 **ROX**: 623` mg/g (pH 7.0); 10 mM KH₂PO4	98
Chitosan Cross-linked at low, medium, & high levels	NR	**ROX**: 211-473 mg/g (pH 7.0); 10 mM KH₂PO4	19
Fe-Mn Chitosan Composite Fe(OH)₃	NR	**PANA**: 280 mg/g (pH 7.5) **PANA**: 80.3 mg/g (pH 7.5)	112
Chitosan Modified with N-methylglucamine	16.2 (pH 7)	**MMA**: 15.4 mg/g (pH 3.4) **DMA**: 7.1 mg/g (pH 5)	113

Note: ROX (Roxarsone); PANA (para-arsanilic acid); MMA (monomethylarsonic acid); DMA (dimethylarsinic acid)

Poon et al. [19] reported a study on the use of chitosan and its cross-linked forms with a comparison of activated carbon (AC). Greater crosslinking was found to enhance the sorption capacity toward ROX relative to chitosan, where evidence of competitive adsorption with orthophosphate (10–100 mM) was shown at variable pH values. The role of a different binding mechanism other than the arsenate group of ROX was shown for AC, where minor effects due to competitive binding with phosphate occurred. The role of phenyl interactions with the surface of AC stabilizes the uptake of ROX, as supported by an independent study by Kwon et al. [89]. The favorable role of the graphene adsorption sites of AC with the phenyl group of ROX is in agreement with the limited water solubility of ROX and the role of hydrophobic effects for this system. By contrast, the presence of iron oxide groups play a secondary role for the adsorption of ROX with the surface sites of iron oxide modified AC [18]. In contrast, Kong & Wilson [103] reported on the preparation of iron oxide composites with cellulose, where greater ROX uptake occurs as the iron oxide content increased up to ca. 30 wt.%. The results are in agreement with the hydrophilic character of cellulose versus AC and the role of adsorption via the arsenate group of ROX instead of adsorption via the phenyl moiety. As well, the studies reported in Table 3 provide additional support that arsenate anions form stable complexes with supported iron (hydr)oxide materials for a similar reason. Lunge et al. (see Table 4 in [106]) reported on the uptake of arsenate by a nanomaterial form of Fe_3O_4 (Q_m=154 mg/g) and a biomass supported Fe_3O_4 composite (Q_m=189 mg/g). These results exceed those for iron coated sand (ICS) and chitosan-ICS materials reported by Gupta et al. [113] and further suggest the potential utility of iron oxide composite materials as promising adsorbents.

In contrast to zero-valent iron adsorbents, composite mineral oxides are likely to have greater structural complexity due to the greater number of additive components, heterogeneities, and structural defects. Thus, composite adsorbent materials likely involve a range of tunable adsorptive interactions based on the relative composition and textural properties of the material. Further support for the unique structure (surface area and textural properties) of nanomaterials metal oxides (NMOs) relate to the selection of metal oxides that contain iron, manganese, aluminum, titanium, magnesium, and cerium and their variable uptake capacity with arsenicals [107-113]. The high *surface-to-volume* ratios of NMOs provide abundant surface sites for the uptake of arsenic and other contaminants from aqueous media based on the limited selection of results in Table 3. TiO_2-based materials for arsenic removal may also employ photocatalytic oxidation (PCO) of arsenite and organic arsenic species to less toxic products such as arsenate in conjunction with adsorptive removal [21]. The use of chitosan supports serves a multifunctional role in the stabilization and dispersion of the metal oxide. Composites with supported NMOs have potential use in flow systems to prevent nanoparticle loss to the environment [68]. The variable synthetic preparations and morphology (powders, beads, and granules) of chitosan and its modified forms can be extended to adsorption-based removal of organoarsenicals. The use of combined synthetic methods (surface functionalization, cross-linking, and composite formation) of chitosan can be harnessed to yield tunable physicochemical properties for the controlled uptake of arsenicals.

5.   Conclusion and future outlook

The modification of chitosan via surface functionalization, cross-linking, and composite formation results in adsorbent materials with improved adsorption properties. While cross-linked and functionalized chitosan are well known methods, the formation of composite materials is less well understood. Composite formation can be achieved by facile incorporation of metal oxides, zero-valent iron, metal NPs, and other secondary components that lead to materials with enhanced textural properties and greater sorption capacity toward arsenicals. Further research is required to establish a molecular-level understanding of composites and their unique structure and adsorption property relationships.

Modified chitosan and its composite material forms represent an emerging class of materials in view of their synthetic versatility and structural diversity. Adsorbents with tunable physicochemical properties (textural properties and surface chemistry) that contain chitosan offer wide-ranging opportunities for the development of advanced adsorbent technology for the controlled removal of arsenicals from aquatic environments under different field conditions. Waterborne arsenic contaminants are likely to become an increasingly important issue due to industrial intensification where the demand for *point-of-use* and scalable adsorbent technology is anticipated for the future. Chitosan and its modified forms represent a promising class of adsorbent materials that are positioned to meet this growing demand for the remediation of waterborne arsenical contaminants.

Acknowledgements

LDW acknowledges the support provided by the University of Saskatchewan for this research and the various students and researchers who have contributed to the research cited in this work.

References

[1]   W. M. Alley, R. Alley, *In* High and Dry: Meeting the Challenges of the World's Growing Dependence on Groundwater, NEW HAVEN; Yale University Press, London, UK, 2017, Chapter 13.
https://doi.org/10.12987/yale/9780300220384.001.0001

[2]   E. Merian, M. Anke, M. Ihnat, M. Stoeppler, *In* Elements and Their Compounds in the Environment, Wiley-VCH Verlag GmbH: Weinheim, Germany, 2008, 1321-1364.

[3]   A. Masotti, *In* Arsenic: Sources, Environmental Impact, Toxicity and Human Health – a medical geology perspective, Nova Publishers: New York, USA, 2013.

[4]   S. E. Manahan In *Environmental Chemistry* 10th Edition, CRC Press: Boca Raton, FL., 2017, Chapter 6.

[5]   S. Murcott, Arsenic Contamination in the World, IWA Publishing, London, UK, 2012, Chapter 2.

[6]   C. F. McGuigan, C. L. A. Hamula, S. Huang, S. Gabos, Le Environ. Rev. 18 (2010) 291-307. https://doi.org/10.1139/A10-012

[7]   L. D. Wilson, Chitosan-based Materials and Their Application toward Arsenic Removal from Water, Water Conditioning & Purification International Magazine, 56 (12) (2014) 28-33. www.wcponline.com

[8]   M. Rahim, M. R. H. M. Haris, Application of biopolymer composites in arsenic removal from aqueous medium: A review. J. Radiation Res. Appl. Sci. 8 (2015) 255 -263. https://doi.org/10.1016/j.jrras.2015.03.001

[9]   C. O. Abernathy, R. L. Calderon, W. R. Chappell, *In* Arsenic: Exposure and health effects, Springer Science & Business Media: Hong Kong, Macau, 2012.

[10]   L. Liu, Environment Magazine, 2010. <www.environmentmagazine.org/Archives/Back%20Issues/March-April%202010/made-in-china-full.html>.

[11]   J. Bundschuh, M. I. Litter, F. Parvez, G. Román-Ross, H. B. Nicolli, J-S Jean, C-W Liu, D. López, M. A. Armienta, L. R.G. Guilherme, A. G. Cuevas, L. Cornejo, L. Cumbal, R. Toujaguez, One century of arsenic exposure in Latin America: A review of history and occurrence from 14 countries, Sci. Total Environ. 429 (2012) 2–35. https://doi.org/10.1016/j.scitotenv.2011.06.024

[12]   M. Barlow, *In* Blue Covenant: The global water crisis and the coming battle for the right to water. The New Press. 2010.

[13]   R. Nickson, C. Sengupta, P. Mitra, S. N. Dave, A. K. Banerjee, A. Bhattachary, S. Basu, N. Kakoti, N. S. Moorthy, M. Wasuja, M. Kumar, D. S. Mishra, A. Ghosh, D. P. Vaish, A. K. Srivastava, R. M. Tripathi, S. N. Singh, R. Prasad, S. Bhattachary, P. Deverill, Current knowledge on the distribution of arsenic in groundwater in five states of India, J. Environ. Sci. Health. A. Tox. Hazard. Subst. Environmental Eng. 42 (2007) 1707–1718. https://doi.org/10.1080/10934520701564194

[14]   W. Lepkowski, Arsenic crisis in Bangladesh, Chem. Eng. News, (1998) 27-29. https://doi.org/10.1021/cen-v076n046.p027

[15]   P. Bagla, J. Kaiser, India's spreading health crisis draws global arsenic experts, Science, 274 (1996) 174-175. https://doi.org/10.1126/science.274.5285.174

[16]   Z. Yang, H. Peng, X. Lu, Q. Liu, R. Huang, B. Hu, G. Kachanoski, M. J. Zuidhof, X. C. Le Arsenic Metabolites, Including N-Acetyl-4-hydroxy-m-arsanilic Acid, in Chicken Litter from a Roxarsone-Feeding Study Involving 1600 Chickens. Environ. Sci. Technol., 50 (13) (2016) 6737–6743, and references cited therein. https://doi.org/10.1021/acs.est.5b05619

[17]   G. M. Momplaisir, C.G. Rosal, E.M. Heithmar, Arsenic Speciation Methods for Studying the Environmental Fate of Organoarsenic Animal-Feed Additives, *U.S. EPA, NERL*-Las Vegas, TIM No. 01-11 (2001). www.epa.gov/nerlesd1/chemistry/labmonitor/labresearch.htm

[18]   G. Kyzas, D. Bikiaris, Recent modifications of chitosan for adsorption applications: a critical and systematic review. Mar. Drugs. 13 (2015) 312-337. https://doi.org/10.3390/md13010312

[19]   L. Poon, S. Younus, L. D. Wilson, Adsorption Study of an organo-arsenical with chitosan-based sorbents, J. Colloid Interf. Sci. 420 (2014) 136–144, and references cited therein. https://doi.org/10.1016/j.jcis.2014.01.003

[20]   W. R. Cullen, K. J. Reimer, Arsenic speciation in the environment, Chem. Rev. 89 (1989) 713-764. https://doi.org/10.1021/cr00094a002

[21]   J. Qu, Research progress of novel adsorption processes in water purification: A review, J. Environ. Sci. (China). 20(1) (2008) 1-13. https://doi.org/10.1016/S1001-0742(08)60001-7

[22]   D. Mohan, C. U. Pittman Jr. Arsenic removal from water/wastewater using adsorbents--A critical review, J. Hazard. Mater. 142 (2007) 1–53. https://doi.org/10.1016/j.jhazmat.2007.01.006

[23]   Information on www.rsc.org/education/eic/issues/2007July/HistoricalHighlights-InOrganoarsenicChemistry.asp

[24]   Information on https://pubs.acs.org/cen/government/85/8515gov2.html

[25]   S. Cowen, M. Duggal, T. Hoang, H. A. Al-Abadleh, Vibrational spectroscopic characterization of some environmentally important organoarsenicals - A guide for understanding the nature of their surface complexes. Can. J. Chem., *86* (2008) 942-950. https://doi.org/10.1139/v08-102

[26]   H. R. Guo, H. S. Chang, H. Hu, S. R. Lipsitz, R. R. Monson, Arsenic in drinking water and incidence of urinary cancers, Epidemiol., 8 (1997) 545-550. https://doi.org/10.1097/00001648-199709000-00012

[27]   T. Tsula, A. Babazono, E. Yamamoto, N. Kurumatani, Y. Mino, T. Ogawa, Y. Kishi, H. Ayoama, Ingested arsenic and internal cancer: a historical cohort study followed for 33 years, Amer. J. Epidemiol., 141 (1995) 198-209. https://doi.org/10.1093/oxfordjournals.aje.a117421

[28]   R. A. A. Muzzarelli, *In* Natural Chelating Polymers; Pergamon Press: Oxford, UK, 1973.

[29]   K. M. Varum, O. Smidsrød, O. *In:* S. Dumitriu (Ed.), Polysaccharides – Structure Diversity and Functional Versatility, 2nd ed., Marcel Dekker: New York, 2005, Chapter 26.

[30]   G. Crini, Recent developments in polysaccharide-based materials used as adsorbents in wastewater treatment. Prog. Polym. Sci., 30 (2005) 38–70. https://doi.org/10.1093/oxfordjournals.aje.a117421

[31]   G. Crini, P. Badot, Application of chitosan, a natural aminopolysaccharide, for dye removal from aqueous solutions by adsorption processes using batch studies: A review of recent literature, Prog. Polym. Sci. 33 (2008) 399–447. https://doi.org/10.1016/j.progpolymsci.2007.11.001

[32]   F. Renault, B. Sancey, P.–M. Badot, G. Crini, Chitosan for coagulation/flocculation processes – An ecofriendly approach, Eur. Polym. J. 45 (2009) 1337–1348. https://doi.org/10.1016/j.eurpolymj.2008.12.027

[33]   A. Bhatnagar, M. Sillanpaa, Applications of chitin- and chitosan-derivatives for the detoxification of water and wastewater--a short review, Adv. *Coll. Interface Sci.* 152 (2009) 26–38, and references cited therein. https://doi.org/10.1016/j.cis.2009.09.003

[34]   L. Pontoni M. Fabbricino, Use of chitosan and chitosan-derivatives to remove arsenic from aqueous solutions—a mini review. Carbohydr Res. 356 (2012) 86–92, and references cited therein. https://doi.org/10.1016/j.carres.2012.03.042

[35]   X. Wang, Y. Liu, J. Zheng, Removal of As(III) and As(V) from water by chitosan and chitosan derivatives: a review. Environ Sci Pollut Res. 23 (2016) 13789–13801, and references cited therein. https://doi.org/10.1007/s11356-016-6602-8

[36]   E. B. Denkbas, M. Odabas, Chitosan microspheres and sponges: Preparation and characterization. J. Appl. Polym. Sci., 76 (2000) 1637. https://doi.org/10.1002/(SICI)1097-4628(20000613)76:11<1637::AID-APP4>3.0.CO;2-Q

[37]   K. C.M. Kwok, L. F. Koong, G. Chen, G. McKay, Mechanism of arsenic removal using chitosan and nanochitosan. J. Colloid Interf. Sci. 416 (2014) 1–10. https://doi.org/10.1016/j.jcis.2013.10.031

[38]   H. K. Agbovi, L. D. Wilson, Design of amphoteric chitosan flocculants for phosphate and turbidity removal in wastewater, Carbohydr. Polym. 189 (2018) 360−370, and references cited therein. https://doi.org/10.1016/j.carbpol.2018.02.024

[39]   C. Xue, L. D. Wilson, Design and characterization of chitosan-based composite particles with tunable interfacial properties, Carbohydr. Polym., 132 (2015) 369-77. https://doi.org/10.1016/j.carbpol.2015.06.058

[40]   K.Y. Foo, B.H. Hameed, Insights into the modeling of adsorption isotherm systems. Chem. Eng. J., 156(1) (2010) 2-10. https://doi.org/10.1016/j.cej.2009.09.013

[41]   G. Z. Kyzas, *In* Green Adsorbents, Bentham Science Publishers: 2015, DOI: 10.2174/97816810813661150101. https://doi.org/10.2174/97816810813661150101

[42]   S. N. Kartal, Y. Imamura, Removal of copper, chromium, and arsenic from CCA-treated wood onto chitin and chitosan, Biores. Technol. 96 (2005) 389–392. https://doi.org/10.1016/j.biortech.2004.03.004

[43]   A. J. Verma, S. V. Deshpande, J. F. Kennedy, Metal complexation by chitosan and its derivatives: a review, Carbohydr. Polym., 55 (2004) 77–93. https://doi.org/10.1016/j.carbpol.2003.08.005

[44]   E. Guibal, Interactions of metal ions with chitosan-based sorbents: A review. Sep. Purif. Technol., 38 (2004) 43–74. https://doi.org/10.1016/j.seppur.2003.10.004

[45]   C. Gerente, V. K. C. Lee, P. L. Cloirec, G. McKay, Application of chitosan for the removal of metals from wastewaters by adsorption –mechanisms and models review, Crit. Rev. Environ. Sci. Technol., 37 (2007) 41–127. https://doi.org/10.1080/10643380600729089

[46] W. S. Wan Ngah, L. C. Teong, M. A. K. M. Hanafiah, Adsorption of dyes and heavy metal ions by chitosan composites: a review. Carbohydr. Polym., 83 (2011) 1446–1456. https://doi.org/10.1016/j.carbpol.2010.11.004

[47] D-W. Cho, B-H. Jeon, C-M. Chon, Y. Kim, F. W. Schwartz, E-S. Lee, H. Song. A novel chitosan/clay/magnetite composite for adsorption of Cu(II) and As(V) Chem. Eng. J. 200–202 (2012) 654–662. https://doi.org/10.1016/j.cej.2012.06.126

[48] L. Lalchhingpuii, B.P. Nautiyal, D. Tiwari, S.I. Choi, S-H. Kong; S-M. Lee, Silane grafted chitosan for the efficient remediation of aquatic environment contaminated with arsenic(V). J. Colloid Interf. Sci., 467 (2016) 203–212. https://doi.org/10.1016/j.jcis.2016.01.019

[49] L. Da Sacco, A. Masotti, Chitin and Chitosan as Multipurpose Natural Polymers for Groundwater Arsenic Removal and As_2O_3 Delivery in Tumor Therapy, Mar. Drugs, 8 (2010) 1518-1525. https://doi.org/10.3390/md8051518

[50] K. C. M. Kwok, V. K. C. Lee, C. Gerente, G. McKay, Novel batch reactor design for the adsorption of arsenate on chitosan. *J. Chem. Technol. Biotechnol.* 85 (2010) 1561-68. https://doi.org/10.1002/jctb.2466

[51] W. S. Wan Ngah, S. Fatinathan, Chitosan flakes and chitosan–GLA beads for adsorption of *p*-nitrophenol in aqueous solution. Coll. *Surf. A:* Physicochem. Eng. Aspects, 277 (2006) 214-222. https://doi.org/10.1016/j.colsurfa.2005.11.093

[52] Y.-T Wei, Y.-M. Zheng, J. P. Chen, Enhanced adsorption of arsenate onto a natural polymer-based sorbent by surface atom transfer radical polymerization. J. Colloid Interf. Sci., 356 (2011) 234–239. https://doi.org/10.1016/j.jcis.2010.12.020

[53] R. Brion-Roby, J. Gagnon, J.-S. Deschênes, B. Chabot, Development and treatment procedure of arsenic contaminated water using a new and green chitosan sorbent: kinetic, isotherm, thermodynamic and dynamic studies, Pure Appl. Chem. (2017) http://dx.doi.org/10.1515/pac-2017-0305. https://doi.org/10.1515/pac-2017-0305

[54] R. Brion-Robya, J. Gagnon, J-S Deschênes, B. Chabot Investigation of fixed bed adsorption column operation parameters using a chitosan material for treatment of arsenate contaminated water, J. Environ. Chem. Eng. 6 (2018) 505–511. https://doi.org/10.1016/j.jece.2017.12.032

[55] B. J. Mcafee, W. D. Gould, J. C. Nadeau, A. C. A. Da Costa, Biosorption of metal ions using chitosan, chitin, and biomass of Rhizopus oryzae, Sep. Sci. Technol. 36 (14) (2001) 3207–3222. https://doi.org/10.1081/SS-100107768

[56] Y-T. Wei, Y-M. Zheng, J. P. Chen, Enhanced adsorption of arsenate onto a natural polymer-basedsorbent by surface atom transfer radical polymerization, J. Colloid Interf. Sci. 356 (2011) 234–239. https://doi.org/10.1016/j.jcis.2010.12.020

[57] A. Gupta, V. S. Chauhan, N. Sankararamakrishnan, Preparation and evaluation of iron-chitosan composites for removal of As(III) and As(V) from arsenic contaminated real life groundwater, Water Res. 43 (2009) 3862–3870. https://doi.org/10.1016/j.watres.2009.05.040

[58] S. M. Miller, J. B. Zimmerman, Novel, bio-based, photoactive arsenic sorbent: TiO_2-impregnated chitosan bead, Water Res., 44 (2010) 5722-5729. https://doi.org/10.1016/j.watres.2010.05.045

[59] P. Singh, J. Bajpai, A. K. Bajpai, R.B. Shrivastava, Removal of arsenic ions and bacteriological contamination from aqueous solutions using chitosan nanospheres. Indian J. Chem. Technol., 18 (2011) 403-413.

[60] D. Y. Pratt, L. D. Wilson, J. A. Kozinski, Preparation and sorption studies of glutaraldehyde cross-linked chitosan copolymers, J. Colloid Interf. Sci. 395 (2013) 205-211. https://doi.org/10.1016/j.jcis.2012.12.044

[61] M. V. Dhanapala, K. Subramanian., Modified chitosan for the collection of reactive blue 4, arsenic and mercury from aqueous media, Carbohydr. Polym. 117 (2015) 123–132. https://doi.org/10.1016/j.carbpol.2014.09.027

[62] M. Rahim, M. R. H. M. Haris, Application of biopolymer composites in arsenic removal from aqueous medium: A review. J. Radiation Res. Appl. Sci. 8 (2015) 255 -263. https://doi.org/10.1016/j.jrras.2015.03.001

[63] H. H. dos Santos, C. A. Demarchi, C. A. Rodrigues, J. M. Greneche, N. Nedelko, A. S′lawska-Waniewska, Adsorption of As(III) on chitosan-Fe-crosslinked complex (Ch-Fe). Chemosphere, 82 (2011) 278–283. https://doi.org/10.1016/j.chemosphere.2010.09.033

[64] A. Shekhawat, S. Kahu, D. Saravanan, R. Jugade, Tin(IV) cross-linked chitosan for the removal of As(III), Carbohydr. Polym. 172 (2017) 205–212. https://doi.org/10.1016/j.carbpol.2017.05.038

[65]   I. A. Udoetok, L. D. Wilson, J. V. Headley, Self-Assembled and Cross-Linked
        Animal and Plant-Based Polysaccharides: Chitosan−Cellulose Composites and
        Their Anion Uptake, ACS Appl. Mater. Interfaces, 8 (2016), 33197–33209.
        https://doi.org/10.1021/acsami.6b11504

[66]   M. H. Mahaninia, L. D. Wilson, Phosphate uptake studies of cross-linked chitosan
        bead materials, J. Colloid Interf. Sci., 485 (2017) 201–212.
        https://doi.org/10.1016/j.jcis.2016.09.031

[67]   M. H. Mohamed, L. D. Wilson, Sequestration of Agrochemicals from Aqueous
        Media Using Cross-linked Chitosan-based Sorbents, Adsorption, 22 (2016) 1025–
        1034. https://doi.org/10.1007/s10450-016-9796-7

[68]   J. H. Kwon, L. D. Wilson, Sammynaiken, R. S. Sorptive uptake of selenium with
        magnetite and its supported materials onto activated carbon, J. Colloid Interf. Sci.,
        457 (2015) 388-97. https://doi.org/10.1016/j.jcis.2015.07.013

[69]   L. S. Casey, L. D. Wilson, Investigation of Chitosan-PVA Composite Films and
        their Adsorption Properties, J. Geosci. Environ. Prot., 3 (2015) 78-84.
        https://doi.org/10.4236/gep.2015.32013

[70]   L. Poon, L.D.Wilson, J. V Headley, Chitosan-glutaraldehyde copolymers and their
        sorption properties, Carbohydr. Polym., 109 (2014) 92–101.
        https://doi.org/10.1016/j.carbpol.2014.02.086

[71]   M. H. Mohamed, I. A. Udoetok, R. M. Dimmick, L. D. Wilson, J. V. Headley,
        Fractionation of Carboxylate Anions from Aqueous Solution Using Chitosan
        Cross-Linked Sorbent Materials, RSC Adv., 5 (2015) 82065-82077.
        https://doi.org/10.1039/C5RA13981C

[72]   M. H. Mahaninia, L. D. Wilson, Cross-linked chitosan beads for phosphate
        removal from aqueous solution, J. Applied Polym. Sci., 132 (2015) 42949-42958.

[73]   L. D. Wilson, C. Xue, Studies of Macromolecular Materials Sorbent for Urea
        Capture J. Appl. Polym. Sci. 128 (2013) 667-675.
        https://doi.org/10.1002/app.38247

[74]   M. H. Mahaninia, L. D. Wilson, Modular Cross-Linked Chitosan Beads with
        Calcium Doping for Enhanced Adsorptive Uptake of Organophosphate Anions,
        Ind. Eng. Chem. Res., 55 (45) (2016) 11706–11715.
        https://doi.org/10.1021/acs.iecr.6b02814

[75]   M. H. Mahaninia, L. D. Wilson, A kinetic uptake study of roxarsone using cross-linked chitosan beads, Ind. Eng. Chem. Res. 56 (7) (2017) 1704–1712. https://doi.org/10.1021/acs.iecr.6b04412

[76]   L. D. Wilson, D. Y. Pratt; J. A. Kozinski, Preparation and Sorption Studies of β-Cyclodextrin-Chitosan-Glutaraldehyde Terpolymers, J. Colloid Interf. Sci. 393 (2013) 271-277. https://doi.org/10.1016/j.jcis.2012.10.046

[77]   M. T. Sikder, S. Tanaka, T. Saito, M. Kurasaki. Application of zerovalent iron impregnated chitosan-caboxymethyl-β-cyclodextrin composite beads as arsenic sorbent. J. Environ. Chem. Eng. 2 (2014) 370–376. https://doi.org/10.1016/j.jece.2014.01.009

[78]   Z. K. Elwakeela, E. Guibal. Arsenic(V) sorption using chitosan/Cu(OH)$_2$ and chitosan/CuOcomposite sorbents. Carbohydr. Polym., 134 (2015) 190–204. https://doi.org/10.1016/j.carbpol.2015.07.012

[79]   G. Neeraj, S. K. Raghunandan, P. S. Kumar, H. Cabanac, V. V. Kumar. Adsorptive potential of dispersible chitosan coatediron-oxide nanocomposites toward the elimination of arsenic from aqueous solution. Proc. Saf. Environ. Prot., 104 (2016) 185–195. https://doi.org/10.1016/j.psep.2016.09.006

[80]   D-W. Cho, B-H. Jeon, C-M. Chon, Y. Kim, F. W. Schwartz, E-S. Lee, H. Song, A novel chitosan/clay/magnetite composite for adsorption of Cu(II) and As(V) Chem. Eng. J., 200–202, (2012) 654–662. https://doi.org/10.1016/j.cej.2012.06.126

[81]   F. Su, H. Zhou, Y. Zhang, G. Wang, Three-dimensional honeycomb-like structured zero-valent iron/chitosan composite foams for effective removal of inorganic arsenic in water. J. Colloid Interf. Sci. 478 (2016) 421–429. https://doi.org/10.1016/j.jcis.2016.06.035

[82]   M. (R.) Yazdani, A. Bhatnagar, R. Vahala, Synthesis, characterization and exploitation of nano-TiO2/feldsparembedded chitosan beads towards UV-assisted adsorptive abatement of aqueous arsenic (As), Chem. Eng. J., 316 (2017) 370-382. https://doi.org/10.1016/j.cej.2017.01.121

[83]   A. Anjum, C. K. Seth, M. Datta, Removal of As3O4 Using Chitosan–Montmorillonite Composite: Sorptive Equilibrium and Kinetics Adsorp. Sci. Technol. 31(4) (2013) 303-324. https://doi.org/10.1260/0263-6174.31.4.303

[84] A. Shahzad, W. Miran, K. Rasool, M. Nawaz, J. Jang, S-R. Lim, D. S. Lee, Heavy metals removal by EDTA-functionalized chitosan graphene oxide nanocomposites, RSC Adv., 7 (2017) 9764–9771. https://doi.org/10.1039/C6RA28406J

[85] B. An, H. Kim, C. Park, S-H Lee, J-W. Choi, Preparation and characterization of an organic/inorganic hybrid sorbent (PLE) to enhance selectivity for As(V), J. Hazard. Mater., 289 (2015) 54–62. https://doi.org/10.1016/j.jhazmat.2015.02.029

[86] C. Liu, B. Wang, Y. Deng, B. Cui, J. Wang, W. Chen, S-Y. He. Performance of a New Magnetic Chitosan Nanoparticle to Remove Arsenic and Its Separation from Water. J. Nanomater., Article ID 191829 (2015), http://dx.doi.org/10.1155/2015/191829

[87] L-L. Min, Z-H. Yuan, L-B. Zhong, Q. Liu, R-X. Wu, Y-M. Zheng, Preparation of chitosan based electrospun nanofiber membrane and its adsorptive removal of arsenate from aqueous solution. Chem. Eng. J. 267 (2015) 132–141. https://doi.org/10.1016/j.cej.2014.12.024

[88] M. H. Dehghani, A. Zarei, A. Mesdaghinia, R. Nabizadeh, M. Alimohammadi, M. Afsharnia Response surface modeling, isotherm, thermodynamic and optimization study of arsenic (V) removal from aqueous solutions using modified bentonite-chitosan (MBC), Korean J. Chem. Eng., 34(3) (2017) 757-767. https://doi.org/10.1007/s11814-016-0330-0

[89] J. H. Kwon, L. D. Wilson, R. S. Sammynaiken, Sorptive Uptake Studies of an Arylarsenical with Iron Oxide Composites on an Activated Carbon Support, Materials, 7 (2014) 1880-1898. https://doi.org/10.3390/ma7031880

[90] A. Tawfik A. S. Saleh, M. Tuzen, Chitosan-modified vermiculite for As(III) adsorption from aqueous solution: Equilibrium, thermodynamic and kinetic studies. J. Mol. Liquids 219 (2016) 937–945. https://doi.org/10.1016/j.molliq.2016.03.060

[91] V.M. Boddu, K. Abburib, J.L. Talbottc, E.D. Smitha, R. Haaschd, Removal of arsenic(III) and arsenic (V) from aqueous medium using chitosan-coated biosorbent, Water Res. 42 (2008) 633–642. https://doi.org/10.1016/j.watres.2007.08.014

[92] R. Sharma, N. Singh, A. Gupta, S. Tiwari, S. K. Tiwari, S. R. Dhakate, Electrospun chitosan–polyvinyl alcohol composite nanofibers loaded with cerium for efficient removal of arsenic from contaminated water. J. Mater. Chem. A, 2 (2014) 16669–16677. https://doi.org/10.1039/C4TA02363C

[93] C-Y. Chen, T-H. Chang, J-T. Kuo, Y-F. Chen, Y-C. Chung, Characteristics of molybdate-impregnated chitosan beads (MICB) in terms of arsenic removal from water and the application of a MICB-packed column to remove arsenic from wastewater. Biores. Technol., 219 (2008) 7487–7494. https://doi.org/10.1016/j.biortech.2008.02.015

[94] S. Liu, B. Huang, L. Chai, Y. Liu, G. Zeng, X. Wang, W. Zeng, M. Shang, J. Deng, Z. Zhou, Enhancement of As(V) adsorption from aqueous solution by a magnetic chitosan/biochar composite. RSC Adv., 7 (2017) 10891–10900. https://doi.org/10.1039/C6RA27341F

[95] J. Qi, G. Zhang, H. Li. Efficient removal of arsenic from water using a granular adsorbent: Fe–Mn binary oxide impregnated chitosan bead. Biores. Technol. 193 (2015) 243–249. https://doi.org/10.1016/j.biortech.2015.06.102

[96] M. E. Pena, G. P. Koratis, M. Patel, L. Lippincott, X. Meng, Adsorption of As(V) and As(III) by nanocrystalline titanium dioxide, Water Res., 39 (2005) 2327–2337. https://doi.org/10.1016/j.watres.2005.04.006

[97] E.A. Deliyanni, D.N. Bakoyannnnakis, A.I. Zouboulis, K.A. Matis, Sorption of As(V) ions by akaganeite-type nanocrystals, Chemosphere 50 (2003) 155–163. https://doi.org/10.1016/S0045-6535(02)00351-X

[98] S. R. Kanel, B. Charlet, L. Choi, Removal of As(III) from groundwater by nanoscale zerovalent iron, Environ. Sci. Technol., 39 (2005) 1291–1298. https://doi.org/10.1021/es048991u

[99] B. Dousova, T. Grygar, A. Martaus, L. Fuitova, D. Kolousek, V. Machovi, Sorption of As(V) on aluminosilicates treated with Fe(II) nanoparticles, J. Colloid Interf. Sci., 302 (2006) 424–431. https://doi.org/10.1016/j.jcis.2006.06.054

[100] B. Liu; D. Wang; G. Yu; X. Meng. Adsorption of heavy metal ions, dyes and proteins by chitosan composites and derivatives — A review, J. Ocean Univ. China, 12(3) (2013) 500–508. https://doi.org/10.1007/s11802-013-2113-0

[101] P. Miretzky, A. F. Cirelli, Fluoride removal from water by chitosan derivatives and composites: A review. J. Fluor. Chem., 132 (4) (2011) 231-240. https://doi.org/10.1016/j.jfluchem.2011.02.001

[102] K. J. Reimer, I. Koch, W. R. Cullen, Organoarsenicals. Distribution and transformation in the environment, Metal Ions in Life Sci., 7 (2010) 165-229. https://doi.org/10.1039/9781849730822-00165

[103] D. Kong, L. D. Wilson, Synthesis and characterization of cellulose-goethite composites and their adsorption properties with roxarsone, Carbohyr. Polym., 169 (2017) 282-294. https://doi.org/10.1016/j.carbpol.2017.04.019

[104] T. P. Joshi; G. Zhang; R. Koju, Z. Qi, R. Liu, H. Liu, J. Qu, The removal Efficiency and insight into the mechanism of para-arsanilic acid adsopton on Fe-Mn framework, Sci. Total Environ., 601-602 (2017) 713-722. https://doi.org/10.1016/j.scitotenv.2017.05.219

[105] Y-T. Wei, Y-M. Zheng, J. P. Chen. Uptake of methylated arsenic by a polymeric adsorbent: Process performance and adsorption chemistry, Water Res. 45 (2011) 2290-2296. https://doi.org/10.1016/j.watres.2011.01.002

[106] S. Lunge, S. Singh, A. Sinha, A, Magnetic iron oxide (Fe_3O_4) nanoparticles from tea waste for arsenic removal. J. Magnetism Magnet. Mater., 356 (2014) 21-31. https://doi.org/10.1016/j.jmmm.2013.12.008

[107] Y. C. Sharma, V. Srivastava, V. K. Singh, S. N. Kaul, C. H. Weng, Nano-adsorbents for the removal of metallic pollutants from water and wastewater. Environ. Technol., 30(6) (2009) 583–609. https://doi.org/10.1080/09593330902838080

[108] M. M. Khin, A. S. Nair, V. J. Babu, R. Murugana, S. Ramakrishna, A review on nanomaterials for environmental remediation. Energy Environ. Sci., 5 (2012) 8075-8109. https://doi.org/10.1039/c2ee21818f

[109] L. Cumbal, A.K. Sengupta, Arsenic removal using polymer-supported hydrated iron (III) oxide nanoparticles: role of Donnan membrane effect, Environ. Sci. Technol., 39 (2005) 6508–6515. https://doi.org/10.1021/es050175e

[110] M. Jang, W. F. Chen, F. S. Cannon, Preloading Hydrous Ferric Oxide into Granular Activated Carbon for Arsenic Removal. Environ. Sci. Technol. 42 (2008) 3369–3374. https://doi.org/10.1021/es7025399

[111] B. M. Jovanovi, V. L. Vukašinovi-Peši, Đ. N. Veljović, L. V. Rajaković, Arsenic removal from water using low-cost adsorbents – a comparative study. J. Serb. Chem. Soc., 76 (2011) 1437–1452. https://doi.org/10.2298/JSC101029122J

[112] T. V. Nguyen, S. Vigneswaran, H. H. Ngo, J. Kandasamy, Arsenic removal by iron oxide coated sponge: experimental performance and mathematical models. J. Hazard. Mater. 182 (2010) 723–729. https://doi.org/10.1016/j.jhazmat.2010.06.094

Chitosan-Based Adsorbents for Wastewater Treatment, Ed. Abu Nasar   Materials Research Forum LLC
Materials Research Foundations **34** (2018) 133-160   doi: http://dx.doi.org/10.21741/9781945291753-7

[113] A. Gupta, M. Yunus, N. Sankararamakrishnan, Chitosan-and Iron–Chitosan-
Coated Sand Filters: A Cost-Effective Approach for Enhanced Arsenic Removal,
Ind. Eng. Chem. Res. 52(5) (2013) 2066-2072, and references cited therein.
https://doi.org/10.1021/ie302428z

Chitosan-Based Adsorbents for Wastewater Treatment, Ed. Abu Nasar   Materials Research Forum LLC
Materials Research Foundations 34 (2018) 161-202   doi: http://dx.doi.org/10.21741/9781945291753-8

Chapter 8

Enhancement of Adsorption Capacity of Chitosan by Chemical Modification

Olalekan C. Olatunde[1,2,3], Chidinma G. Ugwuja[1,2,*], Emmanuel I. Unuabonah[1,2,*]

[1]Environmental and Chemical Processes Research Laboratory, Centre for Chemical and Biochemical Research, Redeemer's University, PMB 230, Ede, Osun State, Nigeria

[2]Department of Chemical Sciences, Redeemer's University, PMB 230, Ede, Osun State, Nigeria

[3]Department of Industrial Chemistry, Ekiti State University, Ado-Ekiti, Nigeria

*ugwujac@run.edu.ng, unuabonahe@run.edu.ng

Abstract

Chitosan is a natural biopolymer which has found wide application in fields like medicine, drug delivery and water purification due to its unique chemical and physical properties. Being the second most abundant natural polysaccharide on earth, application of Chitosan for adsorption of pollutants is gaining wide acceptance due to many possible derivatives and properties of Chitosan that can be explored. However, despite its excellent properties, the use of chitosan for adsorption purposes is limited by its poor mechanical property and solubility especially in acidic medium. These limitations could, however, be offset by modifying Chitosan via various chemical routes like the grafting of - sulfur (S), oxygen (O), nitrogen (N) and phosphorus (P) - containing functional moieties and crosslinking reagents. Studies into the mechanism of adsorption of these modified Chitosan materials showed great dependence on pH and on the nature of the incorporated functional group, with chelate formation and electrostatic interaction being the most prominent adsorbent-adsorbate interaction. This chapter seeks to provide an in-depth review on some of these chemical modifications as well as current trends in the engineering of Chitosan for efficient water treatment.

Keywords

Chitosan, Adsorption, Grafting, Crosslinking, Chelate Formation, Electrostatic Interation

Contents

1.   Introduction

There is a huge interest in the development of a cheap adsorbents for the treatment of water, owing to the fact that many of the major problems in the twenty-first century that humanity is faced with are water quality and/ or water quantity issues [1]. Consequently, due to technology advancement and industrial activities, a large and continuously increasing amount of aqueous effluents containing heavy metals and other pollutants are released into water bodies [2]. Over the years, several technologies and processes such as membrane processes, chemical and electrochemical techniques, advanced oxidation processes, coagulation/ flocculation, precipitation and adsorption have been utilized for the purification of water. Nonetheless, each of these processes, with the exception of adsorption, exhibit a number of drawbacks including low efficiency, high capital or operating cost, production of excessive sludge and high cost of maintenance so that some of these methods are unsuitable or even difficult for use by small-scale industries. However, the adsorption process utilizes cheap and readily available adsorbents and it is easy to operate [3] .

Chitosan-Based Adsorbents for Wastewater Treatment, Ed. Abu Nasar   Materials Research Forum LLC
Materials Research Foundations **34** (2018) 161-202   doi: http://dx.doi.org/10.21741/9781945291753-8

The use of biomass as an adsorbent for adsorption processes has attracted a lot of attention. In support of the Sustainable development goals (SDGs) 6, evolution towards a more sustainable society, 'trash to treasure' is a never-ending topic for those in search of profits from recycled wastes. Biomass waste is readily available, biodegradable, low-costs and environmentally friendly, as well as being abundant in nature [4].

For instance, chitosan which is a cationic biopolymer, a derivative of chitin is of great interest as an organic component in composites developed for water treatment. Chitin, a polymer composed of poly $\beta(1-4)$acetyl-D-glucosamine, extracted from shells of crustaceans including shrimp, lobsters, crabs and crayfish is recognized to be the second most abundant biopolymer in nature, next to cellulose [5]. Chitosan and chitin have been employed to remove pollutants via filtration (4%), flocculation (3%), coagulation (4%), precipitation (7%), flotation (1%), membranes filtration (53%), and adsorption (28%) [6-7]. However, chitin is an extremely insoluble material. Therefore, chitosan emerges to be more useful than chitin in several applications, due to the fact that it contains more number of chelating amino groups.

Chitosan or (poly-(1-4)-2-amino-2-deoxy-D-glucose) having a high molecular weight, is an abundantly available and inexpensive biopolymer that possesses several properties as an ideal adsorbent for removing pollutants from wastewater. This biopolymer has unique characteristics which includes hydrophilicity, bifunctionality, non-toxicity, biodegradability, low-cost, biocompatibility, macromolecular structure, abundance in nature, etc and has applications in various fields such as food industry, biotechnology, cosmetics, membrane development, biomedicine and also in adsorption of harmful pollutants [8].

2.   Chemical structure of Chitosan and Chitin

The presence of the amino groups and the hydroxyl groups in chitosan can serve as chelating sites and it is indeed responsible for the uptake of heavy metal cations by a chelating mechanism in near neutral solutions However, in the case of anionic compounds including anionic dyes, metal anions or negatively charged microbial membranes, the adsorption proceeds via electrostatic interaction between the amino groups which can be easily protonated in acidic solutions and the negative compounds [9-10].

Fig. 1 shows the chemical structure of chitosan and chitin. Chitosan is produced in bulk by the alkaline or enzymatic de-acetylation of its parent compound (chitin). The process of de-acetylation involves the elimination of acetyl group from the molecular chain of the parent compound (chitin), leaving behind complete amino groups ($-NH_2$) [11] which gives chitosan a cationic characteristics when dissolved in an acid solution [12]. Degree of de-acetylation is the amount of amino groups formed during the de-acetylation process. In the synthesis process of chitosan from chitin, the acetyl groups of chitin are first hydrolysed and then converted to free amine groups. The degree of de-acetylation or the ratio of de-acetylated to acetylated units is verified by this initial step which can be affected by several factors including concentration of NaOH used in the de-acetylation, temperature and time [11]. In other words, these factors can be used to increase or decrease the degree of de-acetylation. Increasing either the concentration of NaOH, time or temperature can improve the elimination of the acetyl groups from chitin, which results to a variety of chitosan molecules having different properties and applications [13].

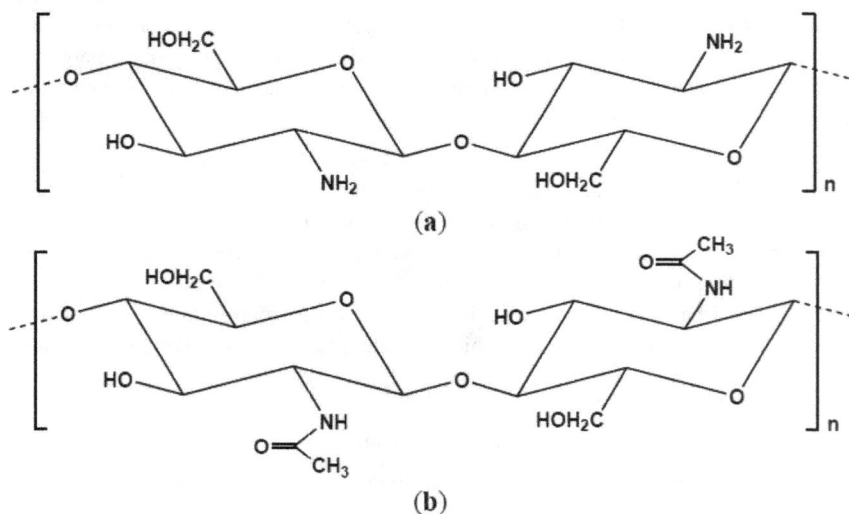

Figure 1 Chemical structure of (a) Chitosan and (b) Chitin (Reproduced from [8]).

Although, various methods have been employed in the determination of the degree of de-acetylation of chitosan, some of these methods are easy to use and inexpensive, while some are more complicated and costly. However, Infra-red spectroscopy and Nuclear Magnetic Resonance (NMR) analysis are the most common ones being used [14-17] In addition, the adsorption capacity of chitosan can be influenced by the degree of deacetylation. High degree of de-acetylation resulting from the presence of large amounts of amino groups can increase adsorption capacity of the chitosan by protonation [18]. For example, chitosan synthesized from locally available shrimps with different degree of de-acetylation was used for the removal of Cd^{2+} and Pb^{2+} [19]. It was observed that the adsorption capacity of the chitosan was affected by its degree of de-acetylation, owing to the fact that, the amount of amino groups available on the material with the high degree of de-acetylation facilitated the adsorption of more metal ions from the aqueous solution [19].

The degree of de-acetylation also has an effect on the chemical, biological and physical properties of the chitosan adsorbent such as its ability to chelate metal ions, self-aggregation, biodegradability, acid-base and electrostatic characteristics [20]. The degree of de-acetylation is usually used to describe chitosan together with other properties such as crystallinity, molecular weight and distribution of amine groups which determine the reaction of chitosan in solution [21]. For instance, low degree of de-acetylation of chitosan refers to the range between 55% and 70% of the N-de-acetylation degree; the 70%-85% of the N-de-acetylation degree is called the medium; the 85%-95% refers to high chitosan while the 95%-100% is called the ultra-high chitosan [22]. Molecular weight less than 50 kDa refers to low molecular weight, the medium molecular weight refers to the molecular weight between 50 kDa and 150 kDa and the molecular weight greater than 150 kDa is called the high molecular weight chitosan [10]. Size exclusion chromatography coupled to an on-line low angle laser light scattering detector and a differential refractive index detector have been utilized for the determination of the molecular weights of chitosan[15, 23] among other methods[14, 24] . In the case of crystallinity, Vakili et al., (2014) stated that molecular weight affects coagulant-flocculant performance of chitosan, tensile strength, crystallinity, solubility and bacteriological properties[25]. However, in the case of solubility, increase in molecular weight reduces solubility[26]. Also, the residual crystallinity of the polymer may have an effect on the accessibility to adsorption sites [27].

3. Forms of Chitosan

3.1 Unmodified Chitosan

The adsorption performances of pristine chitosan in the form of flakes and powder have been studied by many researchers in the past few years.

Chitosan flakes: It is a chitosan derivative obtained from chitin. It is a solid material with high crystallinity that has been used as an adsorbent for dye removal from aqueous solutions by few researchers. Saha et al., [28] was able to use chitosan flakes for the elimination of Reactive Black 5 (RB-5) dye from aqueous solution in a batch system. Its reuse and regeneration potentials by treatment with alkaline after the adsorption process was also highlighted in their study. It was observed that the adsorption capacity of the chitosan flakes increased with decreasing pH. The maximum adsorption capacity (39.5 mg/g) of the material was attained at pH 5.0 while at pH 9.0, the minimum adsorption capacity (12.5 mg/g) was observed. The adsorption of RB-5 dye onto chitosan flakes in a continuous fixed-bed mode column system was also examined by Barron et al., [29]. The results revealed that the process of adsorption was affected by several factors which include bed height, initial dye concentration, particle size and superficial flow velocity. In addition, breakthrough curve studies show that the adsorption of the dye was influenced by mass transfer limitations, possibly because of the intra-particle diffusion [29]. Hadi [30]used chitosan flakes produced from fish shells to remove Methyl Orange from polluted textile wastewater [30]. According to Iqbal et al., [31], the capacity of the chitosan extracted from prawn scales for the adsorption of Acid Yellow 73 dye from water was higher when compared to the chitosan extracted from *Labeorohita*. The result form their analysis also revealed that the removal efficiency of Acid Yellow 73 dye was dependent on pore volume of the adsorbent, pH, surface area and initial dye concentration. Under high initial dye concentration and low pH (3.0-4.0), the high adsorption capacity of the chitosan could be ascribed to the positively charged polymer chain of amino groups which facilitated dye absorption on chitosan [31]. In addition, it was stated by Ignat et al., [32], that the addition of sodium chloride, contact time, pH, temperature and the structure of chitosan influences the adsorption performance of Reactive Red 3 (RR-3) and Direct Brown 95 (DR-95) on chitosan flakes. The highest adsorption values for Reactive Red 3 (RR-3) and Direct Brown 95 (DR-95) was observed to be 151.52 mg/g and 41.84 mg/g at 20 ^{0} C and 50 ^{0} C respectively. It can be elucidated from the result of their analysis that the adsorption capacity decreased at a higher temperature[32].

Chitosan powder: Dotto and Pinto [33], studied the effect of stirring rate during the adsorption process of Food Yellow 3 and Blue 9 dye onto Chitosan powder. According to them, it was stated that the adsorption process occurred through internal and external mass transfer mechanisms and it was chemical in nature. The adsorption capacity of the Chitosan powder for Acid Blue 9 and Food Yellow 3 dyes was increased by 50% and 60% respectively, as the stirring rate increases from 15 rpm to 400 rpm. Hence, stirring rate increased the film diffusivity with subsequent increase in intraparticle diffusivity that ultimately increased adsorption capacity [33]. Zhang and Schiewer [34] showed that As^{5-} could be efficiently removed from aqueous solution using chitosan powder from crab shell.

Although, the various forms of unmodified Chitosan showed good results for the removal of dyes, yet it suffers some drawbacks such as poor stability in water, low mechanical strength, clogging effect, poor solubility in aqueous solution and resistance to intraparticle mass transfer [25-26].

3.2   Modified Chitosan

Modification of chitosan is an effective solution to the several drawbacks earlier mentioned. It generates adsorbents with the desired properties. Modification of chitosan is easier than with other polysaccharides due to the presence of its reactive functional groups including hydroxyl and amino groups and this is achieved through physical and chemical modifications [35].

For instance, in order to decrease crystallinity, increase surface area, improve access to internal sorption sites and increase porosity, which all contributes to the enhancement of the adsorption capacity of chitosan, chitosan flakes are usually subjected to physical modification by converting them into gel beads [25].

3.2.1 Physical modification

Conversion of raw chitosan flakes into beads has been stated to be an important way to enhance its adsorption capability by increasing the surface area and porosity [36-37]. The low-cost biopolymer can be easily conditioned to prepare different physical forms of membranes, films, beads, hydrogels, resins, and fibers [7, 26]. Preparation methods will greatly determine the polymer morphology. The synthesis of different chitosan physical forms is summarized in Fig. 2. Obtaining a desired physical form of the chitosan blends begins with the mixing of the blend components in the liquid form and then adopting the suitable shaping method [38].

Resins (Beads)/ microspheres: Several methods including co-acervation, emulsion and solvent extraction methods have been employed for the synthesis of chitosan beads [38, 40]. The conversion of pristine chitosan flakes into beads increases the adsorption capacity by increasing the porosity, surface area and improves access to internal sorption sites [36]. Comparison between the adsorption capacity of chitosan flakes and beads from 3 different sources, used as adsorbents for removal of Reactive Red 222 (RR 222) dye from wastewater at 30 ^{0}C was carried out by Wu et al [37]. They observed that, because of the higher surface area of the beads, the adsorption rate of chitosan beads was higher than that of chitosan flakes [37]. Similarly, chitosan beads from different sources exhibit higher adsorption capacities by a factor of 2.0-3.8 when compared with chitosan flakes. Thus, chitosan beads are chosen over chitosan flakes as suitable adsorbents for the removal of dyes and heavy metal from aqueous solution [26, 37]. In addition, pristine flakes or powder chitosan are not suitable adsorbents due to their low surface area and non-porosity, which may cause clogging problem in industrial scale column. Developing gel beads from chitosan having large surface area and high porosity, joined with crosslinking to make the beads insoluble in acidic media, could enhance diffusion mechanisms by improving the access to the internal sorption sites, and hydrodynamic performances and improving diffusion mechanisms [41].

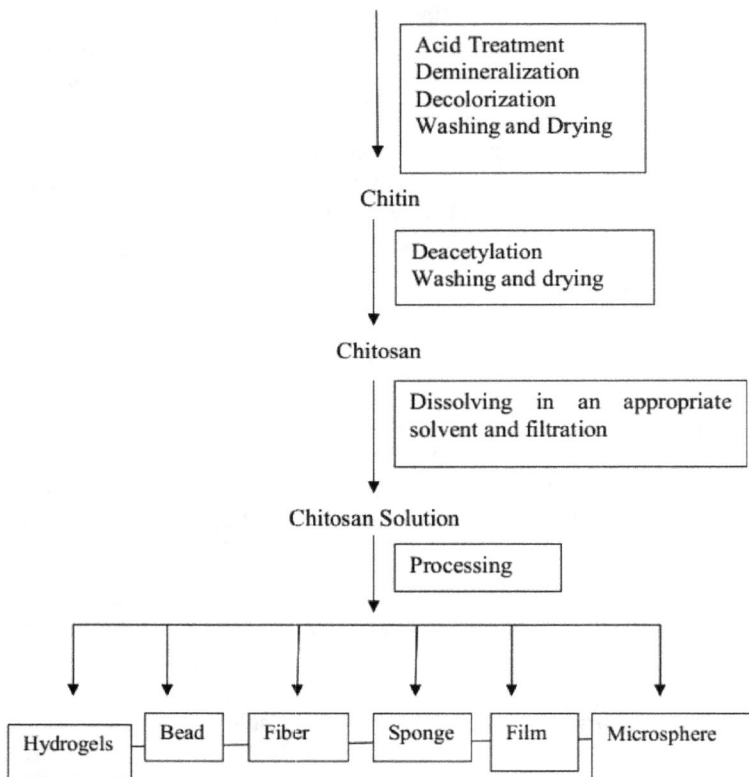

Figure 2 Synthesis of different physical forms of Chitosan obtained from Crustacean shell (Reproduced from [39]).

Chitosan-Based Adsorbents for Wastewater Treatment, Ed. Abu Nasar   Materials Research Forum LLC
Materials Research Foundations **34** (2018) 161-202   doi: http://dx.doi.org/10.21741/9781945291753-8

Hydrogels: Biopolymer hydrogels is an important class of polymeric materials due to their attractive properties for potential application in various fields such as pharmaceutical, environmental and biomedical industries. In literature, several procedures used for preparing hydrogels from chitosan have been proposed [42]. The initial step includes the process of dissolving chitosan in an organic solvent and the formation of emulsions followed by cross-linking of the polymer [43]. Currently, interest in the modification of chitosan and its blends by crosslinking to enhance the hydrogel stability is on the increase. An example is the removal of copper from aqueous solution using chitosan-cellulose hydrogel beads crosslinked with Ethylene glycol diglycidyl ether (EGDE). In the experiment, it was observed that the hydrogel beads material was denser due to the addition of cellulose to chitosan and the crosslinker (EDGE) enhanced the chemical stability of the hydrogel [44]

Films/membranes: Chitosan membranes and films can be prepared by casting technique which involves the casting of chitosan solution mixed with alkali onto a flat surface to evaporate the solvent[38]. Chitosan films are homogenous, flexible and clear with an excellent mechanical and good oxygen barrier properties but low water vapour characteristics. Also, chitosan films are dense and do not have pores. In view of the fact that chitosan degrades before melting, dissolving it in a suitable solvent before casting into films is necessary [38].

Several factors such as source of chitosan, molecular weight, degree of de-acetylation, free amine regenerating mechanism, method of film preparation and most importantly the type of dissolving solvent, affects the morphology of chitosan films, which in turn determines its properties [45]. Acetic acid has been frequently used as a standard solvent for dissolving chitosan for film or membrane production.

Membranes: Chitosan membranes synthesized from various degree of de-acetylation of 75%, 87% and 96% but identical molecular weights were studied by Trung et al., [46]. It was observed that membranes produced from chitosan with higher degree of de-acetylation displayed higher tensile strength and higher elongation at break. Also, membranes from chitosan with 75% degree of de-acetylation exhibited higher water absorption and higher permeability [46]. Similar experiment on chitosan membranes was carried out by Huei and Hwa [47]. In this case chitosan membranes were prepared from different molecular weights. It was observed from the result that chitosan membranes obtained from chitosan with higher molecular weight displayed higher tensile strength compared to those membranes obtained from chitosan with low molecular weights [47].

Fibers: A lot of studies have been carried out on blends of fiber type such as chitosan with polyethyleneoxide [48], collagen [49], polyvinylalcohol [50], alginate [51] and starch [52] for several applications. Fiber type chitosan-based biosorbents are being regarded as excellent adsorbents for the removal or recovery of metal ions, owing to the fact that they can be used in a fixed bed column, allowing easier hydrodynamic transfer and it can bind metal ions quickly [26].

Sponge: Due to chitosan unique characteristics including biocompatibility, non-toxicity, biodegradability and antibacterial activity along with its capability to form film, sponge of chitosan is of high interest in drug carrier system[38]. On the other hand, chitosan solutions do not have the ability to foam alone; therefore it is often mixed with another good foaming polymer such as gelatine, which is also an inexpensive biodegradable polymer with good foaming properties [53]. Gelatin makes good mix with Chitosan and forms polyionic complexes with slower rate of dissolution at the appropriate pH value [54]. Such property could be harnessed for drug release purposes and it can even be applied to a wound while the spongy form absorbs the wound fluid [55].

3.2.2 Chemical modifications of Chitosan

Despite the excellent properties of chitosan as an adsorbent, it suffers some draw backs which are physical and chemical in nature including: (1) its solubility in acidic solutions which limits its use under acidic conditions below pH 5.5 (2) problem of separation when used as powder in fixed bed columns (3) weak mechanical strength [56] (4) low porosity and thermal resistance and (5) low surface area [57-58]. In order to overcome these problems, chemical modification of chitosan provides an easy and convenient way to improve on the adsorptive properties of chitosan. As with physical modification, the choice of chemical modification route, however depends on the final application purpose of the material.

Chemical modification of chitosan can be achieved either via the induction of functional groups into the structure of chitosan [59] or by the modification of the chitosan (CS) structure via the hydroxyl (-OH) and amine ($-NH_2$) groups of the chitosan molecule [60]. Chemical modification can be used to improve the properties of chitosan such as adhesion, bonding characteristics like hydrogen bonding and some physico-chemical properties like permeation and electrostatic charge of its surface [61]. Improved ionic conductivity, mechanical and chemical stability and ion exchange site generation can also be achieved via chemical modification [60]. Though, there are numerous chemical modification techniques for CS, as can be seen in various reviews on this subject [62]. There are also extensive reviews on the adsorption capacity of chitosan and its derivatives for metal ion uptake like with the removal of arsenic [63], mercury [64-66], and even in dye removal [58].

Improving the adsorptive capacity of chitosan via chemical modification usually involves the incorporation of Sulphur (-S), phosphorus (-P), nitrogen (-N) and oxygen (-O) carrying moieties. According to Prashanth and Tharanathan [67], chemical modification of chitosan can be classified into three (3) reaction groups: substitution reactions (e.g acylation, sulfation, metal chelation, cyanoethylation, nitration, phosphorylation and O-/N- carboxyalkylation), chain elongation reactions (e.g crosslinking, graft copolymerization and polymer networks) and depolymerization. Only the first two groups that have so far been explored in enhancing the adsorptive properties of chitosan and they are further considered in this section.

Substitution reaction in chitosan can either take place through the reactions of functional groups containing molecules with one of the amine groups (N-substitution) or through one of the hydroxyl groups (O- substitution) in the chitosan molecule. Substitution reactions usually occur at–OH group located at the C3 or C6 position or the -NH$_2$ on the C2 position on the chitosan skeleton [64]. Protection and de-protection of the 1° amine group is usually required for O-substitution to be effective due to the high reactivity of this group.amine groups, achieved, [68]. On the other hand, graft copolymerization involves the formation of covalent bonds between functional molecules (graft) and chitosan [69]. Grafting enhances numerous properties of chitosan such as chelating and complexation properties, which results in enhanced adsorption properties [70-71]. A review document detailing chemical modifications possible with chitosan are presented by Jayakumar et al [72]

3.2.2.1 S-containing moieties modified chitosan and their adsorption properties

Various methods have been used in the attachment of sulfate groups on the backbone of chitosan. This is achieved either via modification at the amine group (N-sulphonated) or at the hydroxyl group (O-sulphonated) under several reaction conditions including time, reactant concentration and temperature using sulfating agents. The main purpose of incorporating –S containing groups to chitosan is to improve its capacity to hold transition and noble metal ions in aqueous solution via the formation of coordination centres and suppression of the effect of pH that suppresses the influence of competing species on capacity of chitosan. Excellent reviews on Sulphur containing chitosan adsorbents are presented by [73]. Adsorption properties of three marcaptanes modified chitosan-   (6-O-(mercaptoacetat-N-mercaptoacetyl)-chitosan,   N-(2-hyroxy-3-mercaptopropyl) chitosan and (N-(2-hydroxy-3-methyl aminopropyl) chitosan-synthesized via the reaction of epoxy propane were evaluated for Hg^{2+} sorption by Cardenas et al [74]. It was observed that the adsorption of Hg^{2+}on pristine chitosan only occurred between pH 2.5 – 4.5 and the q_{max} (mg/g) varied depending on pH of the solution (357 and 454 mg/g at pH of 2.5 and 4.5 respectively). However, the N-(2-hydroxy-3-mercaptopropyl)-chitosan, showed improved adsorption capacity towards Hg^{2+} that was almost independent of pH (556 and 588 mg/g at pH of 2.5 and 4.5 respectively). The reported adsorption capacity for (6-O-(mercaptoacetat-N-mercaptoacetyl)-chitosan and 1 N-(2-hyroxy-3-mercaptopropyl) at pH 2.5 and 4.5 were 345 and 435 mg/g and 164 and 164 mg/g respectively [74]. Similar trends were observed for the adsorption of Cu(II) on the modified chitosan.

Adsorption of Cu^{2+} and Ni^{2+} ions by Xanthated carboxymethyl chitosan derivative prepared by substitution reactions under basic conditions was reported by Song et al [75]. Though the pristine chitosan showed significant capacity towards Cu^{2+} (108.5 mg/g), when compared with the capacity of the modified chitosan (140.4 mg/g), the capacity of pristine chitosan was significantly low (13.7 mg/g), compared to the capacity of the modified chitosan (110.6 mg/g) for Ni^{2+} adsorption. The improved adsorption capacity of the modified chitosan was due to the presence of metal complexation centers (Sulphur), which enhanced the ability of the adsorbent to form chelates with the metal ions in solution. Study of the effect of xanthation time on the adsorption capacity of the modified chitosan, showed that the higher the xanthation time (corresponding to an increase in the amount of $-S$ atoms incorporated into the chitosan molecule) the higher the adsorption capacity of the adsorbent. This thus confirms the contribution of the $-S$ atoms to the adsorption properties of modified chitosan.

Yong et al [76] reported the synthesis of thiolated chitosan bead via the phase inversion process and it was observed that the amount of S atoms incorporated into the molecule increased with increasing CS_2/chitosan ratio and temperature but decreased with increase in the acidity of the reaction system [76]. Evaluation of the adsorptive properties of the thiolated chitosan on cadmium and copper ions showed that the adsorbed metallic ion species were chemisorbed onto the adsorbent. The q_{max} of the material for Cd^{2+} and Cu^{2+} were 56.2 and 47.2 mg/g respectively. Li et al [77] reported the synthesis of 2.5-Dimercapto-1,3,4-thiodiazole modified chitosan crosslinked with glutaraldehyde and epichlorohydrin [77]. The adsorption capacity of the adsorbent was observed to be significantly affected by the pH of the solution. Evaluation of the adsorption capacity of the adsorbent on Au^{2+}, Pd^{2+} and Pt^{4+} showed a q_{max} of 198.5, 16.2 and 13.8 mg/g respectively at pH of 3.0.

Table 1: List of some modified chitosan, adsorption parameters and their adsorption capacity

Modified chitosan	Adsorbent form	Crosslinking agent	Metal ion	Isotherm Model	Optimum pH	q_{max}	Ref
Thiosemicarbazide chitosan	Framework	None	Cu^{2+}	Langmuir	6	142.8 mg/g	[80]
Carbon disulphide chitosan	Flakes	epichlorohydrin	Cu^{2+}	Langmuir	5	43. 47 mg/g	[79]
Chitosan benzoyl thiourea	Beads	None	^{60}Co $^{153+154}Eu$	Freundlish Langmuir	3.5 8.0	29.47 mg/g 34.54 mg/g	[81]
Thiourea chitosan	Microsphere	Polyethylene glycol	Cu^{2+}	Pseudo second order	5.5	60.6 mg/g	[82]
Thiosemicarbazide chitosan	Hydrogel	None	Pd^{2+} Cd^{2+}	Langmuir	4.26 6.98	325.2mg/g 257.2 mg/g	[83]
Thiosemicarbazide chitosan	Particles	None	Cu^{2+}	Langmuir	3.6	134.0 mg/g	[84]
Thiourea chitosan	Microspheres	Water/oil emulsion	Pt^{4+} Pd^{2+}	Langmuir Langmuir	2.0 2.0	129.9 mg/g 112.4 mg/g	[85]
Ethylenesulfide chitosan	Particles	None	Pb^{2+} Cd^{2+}	Langmuir Langmuir	6.1	1.81 mmol/g 1.94 mmol/g	[86]
Dithiocarbamate chitosan	Particles	None	Pb^{2+} Cu^{2+} Cd^{2+}	Langmuir Langmuir Langmuir	nd	2.24 mmol/g 1.14 mmol/g 0.84 mmol/g	[87]
Thiocarbamoyl chitosan	Particles	glutaraldehyde	Hg^+	Langmuir	2.0	459.5 mg/g	[88]
Thiourea O-carboxymethylated chitosan	Particles	Thiourea/glutaldehyde	Ag^+	Langmuir	4.0	4.0 mg.g	[89]
EDTA-chitosan	Flakes	None	Cu^{2+} Ni^{2+}	Sips Sips	2.1 2.1	63.0 mg/g 71.0 mg/g	[90]
DTPA-chitosan	Flakes	None	Cu^{2+} Ni^{2+}	Sips Sips	2.1 2.1	49.1 mg/g 53.1 mg/g	[90]
Poly(acrylamide)-chitosan	Particles	None	Remacry Red TGL dye	Langmuir	10	309.82	[91]

173

Modified chitosan	Form	Crosslinker	Pollutant	Isotherm Model	Optimum pH	q_{max}	Ref
Diethylenetriamine chitosan	Beads		Acid orange 07	Langmuir-Freundlich		7.29 mmol/g	[92]
			Acid orange 10	Langmuir-Freundlich		4.54 mmol/g	
			Acid red 18	Langmuir-Freundlich		4.94 mmol/g	
			Acid green 25	Langmuir-Freundlich		4.21 mmol/g	
EDTA-chitosan	Particles	None	Nd^{3+}	Langmuir	2.0	74 mg/g	[93]
DTPA-chitosan	Particles	None	Nd^{3+}	Langmuir	2.0	77 mg/g	[93]
2-(diethylamino) salicyaldehyde chitosan	Particles	None	Cu^{2+}	Langmuir	7.0	15.9 mg/g	[94]
Pectin/Carboxymethyl chitosan	Nd	Poly ethylene glycol diglycidyl ether	Pb^{2+}	nd	5	42.0	[95]
Dialdehydes chitosan	Hydrogels	Nd	Hg^{2+}	Nd	5	85%	[96]
			Pb^{2+}			80%	
			Zn^{2+}			35%	
			Co^{2+}			30%	
			Cu^{2+}			20%	
Citric acid chitosan	Flakes	Gltaraldehyde	Cd^{2+}	Freundlich	6	85.4	[97]
O-Carboxymethyl chitosan	Nd	Glutaraldehyde	Crystal violet dye	Langmuir	8.0	239.54	[98]
O-carboxymethyl chitosan	Particles	Glutaraldehyde/thiourea	Hg^{2+}	Langmuir	5.0	6.29 mmol/g	[99]
N, O-carboxymethyl chitosan	Particles	Nd	Congo red dye	Langmuir	7.0	330.62 mg/g	[100]
Carboxymethyl chitosan	Particles	Glutaraldehyde/thiourea	Au^{3+}	Langmuir	4.0	8.32 mmol/g	[101]

			Methylene blue dye	Langmuir	8.0	351 mg/g	[102]
N, O-carboxymethyl chitosan	Particles	-					
N-carboxymethyl chitosan	Particles	-	Pb^{2+}	Langmuir	5.5	421.9 mg/g	[103]
Citric acid chitosan	Flakes	Glutaraldehyde	Pb^{2+}	Langmuir Freundlich	5.0	101.7mg/g	[104]
Carboxylchitosan	Beads	Glutaraldehyde	NO_2^{1-} PO_4^{3-}	Freundlich	7.0 5.0	99.4 mg/g 16.56mg/g	[105]
Poly(acrylic acid) chitosan	particles	None	Remacry Red TGL dye	Langmuir	10	510.74mg/g	[91]
Salicyaldehyde chitosan	Particles	None	Cu^{2+}	Langmuir	7.0	29.2 mg/g	[94]
2, 4-dihydroxybenzaldehyde chitosan	Particles	None	Cu^{2+}	Langmuir	7.0	27.4 mg/g	[94]
Chitosan-Epichlorohydrin-Tripoplyphosphate	Particles	Epichlorohydrin	Cu^{2+} Cd^{2+} Pb^{2+}	Langmuir Langmuir Langmuir	6.0 7.0 5.0	130.72 mg/g 88.75 mg/g 166.94 mg.g	[106]
Chitosan Tripolyphosphate	Beads	Ionotropic	Cu^{2+}	Langmuir	5.0	208.3 mg/g	[107]

Nd- not determined

175

Adsorption properties of thiourea grafted chitosan beads prepared by a preliminary reaction of glutaraldehyde with thiourea was evaluated for Palladium and platinum ion uptake[78]. Epichlorohydrin cross-linked xanthated chitosan was explored for the removal of Cu^{2+} from solution [79]. The adsorption of Cu^{2+} ion was observed to be strongly dependent on pH and temperature, with the maximum adsorption capacity at pH 5.0. The maximum adsorption capacity (q_e) for Cu^{2+} ion uptake was 43.47 mg/g, which was far better than the q_{max} value of 2.0 mg/g reported for the unmodified chitosan. The modified chitosan's ion uptake also showed reverse relationship with respect to pH compared with the pristine chitosan. The adsorption capacity of the pristine chitosan increased with decrease in pH due to exchange of hydrogen ion between the protonated amine groups with metal ions. However, below pH 3.5, there is competition between H^+ ion exchange and metal ion for adsorption sites. Furthermore, with the modification of chitosan with xanthate, the mechanism of adsorption was observed to be by chemisorption of Cu2+ on the xanthated chitosan, associated with complexation.

3.2.2.2   *N*-containing moieties modified Chitosan and their adsorption properties

The main aim of introducing –*N* carrying groups into chitosan structure is to increase the density of electron donating groups that can bind both anions and cations depending on experimental condition. The structure of the modified chitosan has great effect on the adsorption capacity of the adsorbent [108]. An -*N* modified chitosan have been reported by [109]. Ethylenediamine grafted chitosan adsorbent was explored for its adsorption capacity for removal of Eosin Y dye from solution [58]. The modified chitosan showed an improved adsorption capacity with a q_{max} of 294.1 mg/g compared to 70.65 mg/g reported for chitosan. It is has been observed that an increased density of amine group does lead to an increase in the zeta potential of the chitosan adsorbent [110]

Glycidyl trimethyl ammonium chloride chitosan cross-linked with glutaraldehyde was evaluated for the adsorption of reactive Orange 16 dye [111]. The dye uptake was found to be independent of the pH of solution but largely influenced by the concentration of dye. This indicates that the adsorption mechanism is a chemisorption process, with the adsorbed specie attaching itself to the surface of the adsorbent. The q_e of the quaternary adsorbent was 1060 mg/g, corresponding to 75% coverage of the adsorbent's adsorption sites. This modified chitosan showed improved adsorption when compared to most unmodified chitosan reported in the study.

Polyaniline grafted chitosan crosslinked with glutaraldehyde was reported by Karthik & Meenadshi [112]. The modified chitosan was evaluated for the adsorption of Pb^{2+} and Cd^{2+} ions from aqueous solution. The adsorption process was observed be highly dependent on the pH of the solution, with maximum adsorption observed at pH 6.0. At lower pH values reduction in adsorption capacity may be accounted for by the protonation of the amine groups, which limit the availability of the amine groups for attraction of metal ions through the lone pair electrons on the N atoms. The q_{max} for the adsorption of Pb^{2+} and Cd^{2+} was 16.07 and 14.33 mg/g.

Diaminomaleonitrile modified chitosan synthesized via the graft copolymerization process using ceric ammonium nitrate under N_2 atmosphere. The cyano groups of the grafted chitosan was further converted into amidoxime groups by reaction with hydroxylamine hydrochloride[113]. The modified chitosan was explored for the removal of Pb^{2+}, Cd^{2+}, Zn^{2+}, Fe^{2+}, Cu^{2+}, Ni^{2+} and Co^{2+} from aqueous solution. The maximum grafting yield obtained was 84%. The prepared material showed enhanced thermal stability with the main degradation step observed at 388 °C compared to 320 °C for pure chitosan. The reported removal of the metals by the amoxidated chitosan was between 93-98%. The amoxidated chitosan show greater binding ability with the metals [113].

Kondo et al. [114] reported the synthesis of ethylene glycol diglycidyl ether crosslinked chitosan with pyridine groups. The pyridine group was introduced by reacting the crosslinked chitosan with picolinaldehyde which was further reduced by sodium borohydride. The degree of modification obtained was 33.3%. Evaluation of the adsorption capacity of the modified chitosan on Pd^{2+} and Pt^{4+} gave q_e values of 1.55 and 1.35 mmol/g respectively, which were significantly higher than the value obtained for just crosslinked chitosan. The proposed mechanism of adsorption was proposed to be via chelation through the nitrogen atoms on the pyridine and amino groups. The higher adsorption capacity for Pd^{2+} was attributed to the difference in structure between the complex formed between the cations and the adsorbent.

The importance of the introduction $-N$ containing group into chitosan can be confirmed by comparing the work carried out by Azlan et al., [115] and Bhatt et al., [116] who explored the use of crosslinked chitosan for adsorption purposes. Azlan et al., [115] cross linked chitosan using ethylene glycol diglycidyl ether which lead to a reduction in the amount of available amine groups for sorption. This lead to the reduction in the adsorption capacity of the crosslinked chitosan when explored for the removal of acid red 37 and acid blue 25 dyes with q_{max} of 59.52 and 142.86 mg/g respectively compared to 128.21 and 263.15 mg/g of pure chitosan. However in the study by Bhatt et al., [116] ethylenetriaminepentacetic acid was used as a crosslinking agent, leading to an increase in the density of N atoms on the chitosan molecule. Improved adsorption capacity of 192. 3 mg/g was reported compared with the q_{max} of 78 mg/g reported for non-crosslinked chitosan.

3.2.2.3   O-containing moieties modified Chitosan and their adsorptive properties

Introduction of O-containing groups onto chitosan is usually explored in modifying the surface of chitosan for improved adsorption capacity via the formation of highly stable chelate rings with ions of transition metals. It has been suggested that O-containing groups enhances the selectivity of chitosan adsorbents [117]. Carboxyl and hydroxyl containing moieties are major routes to the introduction of $-O$ onto chitosan structure. Chemical modification techniques such as quaternization, grafting, sulphonation, carboxylation, carboxymethylation have been explored in the introduction of $-O$ groups onto chitosan. Numerous reports have been made on the incorporation of O- containing groups [118]

Synthesized glutaradehyde crosslinked chitosan microspheres grafted with pyromellic dianhydride was evaluated for the adsorption of methylene blue and neutral red dyes [119]. The modified chitosan adsorbent showed improve adsorption capacity for the cationic dyes compared to pure chitosan microspheres. The mechanism of adsorption was proposed to be by chemisorption.

In another report by Xing et al. [119], glutaraldehyde crosslinked poly(methacrylic acid)-modified chitosan microspheres was synthesized by graft copolymerization method using potassium persulfate ($K_2S_2O_8$) as the initiator of the reaction[119]. X-ray Photoelectron Spectroscopy (XPS) spectra analysis of crosslinked microsphere and the polymethacrylic acid modified chitosan microsphere showed Carbon:Oxygen ratio of 78:21 and 45:46 respectively, confirming an increase in the number of oxygen atoms due to grafting of polymethacrylic acid to the chitosan skeleton. The adsorption capacity of the modified chitosan was 1001.9 and 473.9 mg/g, for methylene blue and malachite green uptake. This was an increase of ca 10- and 14 fold when compared with 99.1 and 34.2 mg/g reported for crosslinked chitosan. The adsorption capacity of the adsorbent was only affected by pH over a narrow range, with improved sorption capacity observed at high pH.

Konaganti et al., [120] reported the synthesis of chitosan grafted poly(methyl methacrylate) (ChgPMMA), chitosan grafted poly(ethyl methacrylate) (ChgPEMA), chitosan grafted poly(butyl methacrylate) (ChgPBMA) and chitosan grpafted poly(hexyl methacrylate) (ChgPHMA) using potassium persulfate/ascorbic acid as redox initiator. The adsorption capacity of the adsorbents were evaluated for anionic sulfonate dye, Orange-G. The adsorption capacities of the modified chitosan adsorbents were observed to be dependent on the degree of grafting achieved by the synthesis process, with an increase from 72.2 to 91.4 mg/g observed as the percentage grafting increased from 130 to 220 for ChgPEMA adsorbent. A similar trend was also observed for the other poly(alkyl methacrylate) adsorbents. The q_{max} for ChgPMMA, ChgPEMA and ChgPBMA adsorbents and ungrafted chitosan for the removal of Orange-G dye was 94, 91.4, 80.6 and 34 mg/g respectively. The adsorption of anionic dyes on chitosan derivatives was found to greatly influence the surface charge and thus the pH of the environment. The optimum pH range for the adsorption of dye on the poly(alkyl methacrylate) adsorbent was 3.0 - 6.0.

Adsorption of Au^{3+} ions on chitosan and N-carboxymethyl chitosan (NCMC) was reported by Ngah and Liang [121] . The adsorption study showed that the adsorption capacity of the NCMC was not significantly improved by the modification process (q_{max} of ca.31 and 34 mg/g for chitosan and N-carboxylmethyl chitosan, NCMC, respectively). The effect of pH on the sorption capacity of NCMC and Chitosan showed similar pattern, with almost similar adsorption over the studied pH range. This showed that the increase in the positive charge on the NCMC did not contribute significantly to the adsorption of Au^{3+}.

3.2.2.4   *P*-containing moieties modified Chitosan and their adsorptive properties

Incorporation of P-containing groups into chitosan structure results in an increase in adsorption capacity for metal ions and increased resistance towards acid . The increased adsorption capacity is usually due to an increase in the ion-exchange capacity of the adsorbent. Adsorption of metal ions could be through intermolecular or intramolecular with counter ions on the sorbent or through the electrostatic interaction between the positively charged amine groups and the negatively charged tripolyphosphate group [108]. Several phosphorylated chitosan have been reported in literatures [58, 69, 122-123]. Detailed synthesis of phosphorylated chitosan was presented in the review by Jayakumar et al [124]. Even though phosphorylated chitosan exhibits unique adsorption properties, the poor solubility of most of them in water and strong competition by alkaline and alkaline earth metal during adsorption reduces their application, especially when heavy metal recovery is desired[69]. Synthesis of N-methylene phosphonic and N-propyl-N-methylene phosphonate chitosan via reductive N-alkylation reaction using propyl aldehyde was reported by Zuniga et al [125]. P-chitosan was also prepared using phosphorus acid [126-127] and phosphorus oxychloride [128].

Phosphonated chitosan prepared via substitution reaction using a reagent system of orthophosphoric acid-urea in N, N-dimethylformamide was explored for uranium adsorption [129]. Due to the acidic nature of the reaction medium, the amine groups of chitosan are in their ionized form and are capable of forming ionic bonds with free phosphate ions in solution (H_2PO^{4-} and HPO_2^{4-}). Three possible phosphorylation products were proposed viz (1) chitosan phosphate esterified at the hydroxyl group at the C-6 position (2) chitosan pyrophosphates (3) salt linked formed between phosphate groups and amine groups of protonated chitosan. The adsorption of U^{6+} onto P-chitosan in the study was best described by Langmuir model, with q_{max} of 54.95 mg/g.

Phosphorylation of hydrothermally crosslinked chitosan was reported by Dong et al [130]. Sodium tripolyphosphate was used as the P-source, in which crosslinking was facilitated by the adjustment of the pH of the reaction mixture to 5.0. Point of zero charge (pH_{pzc}) of the phosphorylated chitosan was reported to be 6.5, so that at a pH lower than 6.5 the surface of the adsorbent becomes positively charged for optimum anion adsorption and at pH higher than 6.5, the surface becomes negatively charged for optimum cation adsorption. The q_{max} for the adsorption of uranium ion by the phosphorylated chitosan was 409.2 mg/g compared to 200.0 mg/g reported for hydrothermally crosslinked chitosan. The phosphorylated chitosan was observed to also show higher selectivity for the adsorption of U^{6+} in the presence of other competing ions than the crosslinked chitosan.

Phosphorylated chitosan/hydroxyethyl methacrylate interpenetrating polymer network was synthesized via γ-radiation using phosphoric acid [131]. The q_{max} for the adsorption of Cu^{2+} Zn^{2+} and Ca^{2+} ions were 66.3, 57.6 and 48.7 mg/g respectively.

3.2.2.5   Other modified Chitosan

Incorporation of moieties with functional groups other than N, S, O, and P have also been explored in enhancing adsorption capacity of chitosan. Molecularly imprinted chitosan (MIC) have been synthesized to improve the capacity of crosslinked chitosan. MICs are synthesized using the target pollutant as a template, followed by crosslinking and then removal of the template molecule. The template molecules protects the active groups on the chitosan during crosslinking, thus reducing the diminishing effect of crosslinking on the modified chitosan. The selectivity of the MIC is enhanced toward the template molecule after its removal[132]. Synthesis of a polyaminated crosslinked Ni^{2+}-imprinted chitosan beads with ~3 fold sorption capacity for Ni^{2+} was reported by [133]. MICs with enhanced adsorption capacity have also been reported [134-136]

Synthesis of magnetic chitosan is another route to enhancing the adsorptive capacity of chitosan. Comprehensive review on preparation and application of magnetic chitosan was reported by Tong & Chen [56]. Yang et al [137] reported the synthesis of modified chitosan magnetic composite adsorbent with improved adsorption capacity of 758 mg/g for methyl orange dye [137]. Crosslinked magnetic hydroxamated chitosan with 99.2%

removal efficiency for $Cr^{6-}$was reported by Jia et al., [138]. Magnetic chitosan has been used for the adsorption of Zn^{2+} [139], Pb^{2+}, Cu^{2+} [140], Ag^+, Au^{3+} [141]

The use of metal impregnated chitosan for enhanced adsorption of pollutants have also been explored. Tripathi [142] reported the synthesis of molybdenum-impregnated chitosan beads, monocalcium phosphate chitosan beads, with improved adsorption capacity towards Ni^{2+} and Zn^{2+} compare to pristine chitosan [142]. Lanthanum-modified chitosan with 20% impregnation level showed enhanced deflouridation activity with maximum uptake at pH of 6.7 [143]. Mixed rare earths modified chitosan with adsorption capacity of 3.72 mgF^-/g for fluoride uptake was reported by Liang et al., [144]. chitosan crosslinked with Fe^{3+} showed improved adsorption capacity of 131 mg/g for phosphate [145]. Fe-impregnated chitosan[146] is another metal impregnated chitosan explored for fluoride sorption. Only recently, Unuabonah et al., [147] used metal impregnated chitosan composites to disinfect water polluted with *E. coli*, *S. typhi* and *V. Cholerae*.

Surfactant-modified chitosan have also been reported in literatures though with limited research output [148]. The improved adsorption activity of this chitosan is due to the formation of bilayer or multilayers (admicelle) by the surfactant molecule on the chitosan when it is present at a concentration higher than its critical micelle concentration (CMC). This admicelle formed, solubilizes metal ions/cationic dyes by a process called adsolubilization [149-151]. To study the effects of surfactant's hydrophobic tail and ionic head-group on adsorption capacity of surfactant modified chitosan, Lin et al., [152] reported the synthesis of sodium dodecyl sulfate- (SDS), sodium dodecyl sulfonate- (SDOS), dodecyl benzenesulfonic acid sodium salt- (SDBS), dioctyl sulfosuccinate sodium salt- (AOT) and N, N`-ethylene-bis[N(sodium ehtylenesulfonate)-dodecanamide] (DTM-12)-modified chitosan and evaluated their capacity for congo red dye. The q_{max} for the CS/DTM-12, CS/SDOS, CS/SDBS, CS/AOT and CS/SDS were 1732.89, 1539.98, 1637.58, 1766.20 and 1490.65 mg/g respectively.

Crosslinking of chitosan is a method employed for improving the stability of chitosan especially in acidic medium. However, chitosan with improved adsorption capacity can be achieved by using molecules with functional moieties like carboxylate groups and quaternary amines as crosslinks [153]. Chitosan beads crosslinked with epichlorohydrin (EP) and glutaraldehyde (GA) were evaluated for phosphate dianion (HPO_4^{2-}) and *p*-nitrophenolate (PNP) uptake from aqueous solutions. The EP beads showed higher affinity for HPO_4^{2-} than GA beads, while GA showed higher affinity for PNP than EP. The higher adsorption of *p*-nitrophenolate ion on GA was due to the presence of more apolar binding sites, which was favourable for adsorption of the more lipophilic PNP compare to the more hydrophilic HPO_4^{2-} [153]. Application of glutaraldehyde (ALD-CHs) and epichlorohydrin (ECH-CHs) crosslinked chitosan for phosphate removal was reported by Filipkowska et al [154]. ECH-CHs showed higher adsorption capacity (139.4 mg/g) compare to 108.24 mg/g for ALD-CHs and 44.38 mg/g for the pristine chitosan

5. Mechanism of adsorption by modified Chitosan

The mechanism of adsorption of adsorbates by chitosan derivatives have been reported in review papers, experimental and theoretical studies. Despite the large volume of literature on the use of chitosan derivatives as adsorbent, only few studies have taken the pain in elucidating the mechanism involved in the adsorption process. The adsorption of species unto modified chitosan could be via several chemical interactions such as electrostatic attraction, ion-exchange, van der Waal force or chemical bonding which could be either a complex or chelate formation [65]. The adsorption process for a particular adsorbent may not be defined by just one clear cut mechanism, thus several mechanisms may be involved in the uptake of an adsorbate specie by an adsorbent. The functional group incorporated by modification processes most times play a significant role in the mechanism of adsorption. For example, increase in sulphur functionality leads to an increase in charge polarity, leading to increase in attraction toward metal cations due to the partial negative charge on the sulfur [80, 155]. The mechanism of adsorption may also be influenced by factors such as solution composition, speciation of metal ions and pH which affects the polymer protonation and leads to metal cation repulsion.

The formation of chelate by ions with ligands is determined by the availability of its outermost electrons and presence of empty molecular orbitals has described by Person's hard and soft acids and bases theory (HSAB). Two models for metal ion chelation has been proposed: bridge model and pendant model. While the bridge model suggests that metal ions could be bonded through inter- or intra-molecular complexation to amine groups of the same chain or different chain, the pendant model proposes that metal ions are held in pendant mode. The occurrence of any these models in metal ion uptake is determined by nature of metal ion, pH, metal/ligand ratio and presence of additional functional units in the chitosan frame [26, 117]

The study of the adsorption mechanism for the adsorption of Al^{3+} and Pb^{2+} from wastewater by chitosan-tannic acid modified biopolymer was reported by Badawi et al [156]. Adsorption of the metal ions was accounted for by the interaction between polarizable phenolic hydroxyl groups from the tannic acid and the terminal amine on chitosan and the metal ions. The mechanism of adsorption was thus proposed to be by complex formation between the adsorbent and adsorbate as shown in Fig 3. Complex formation was further confirmed by subjecting the Pb^{2+} and Al^{3+} loaded adsorbent to FTIR analysis. The characteristic bands of the adsorbent, which are the stretching vibration due to C-O and C-N observed at 1620 and 1025 cm^{-1} were significantly shifted due to the formation of coordinate bonds between the groups and the metal ions.

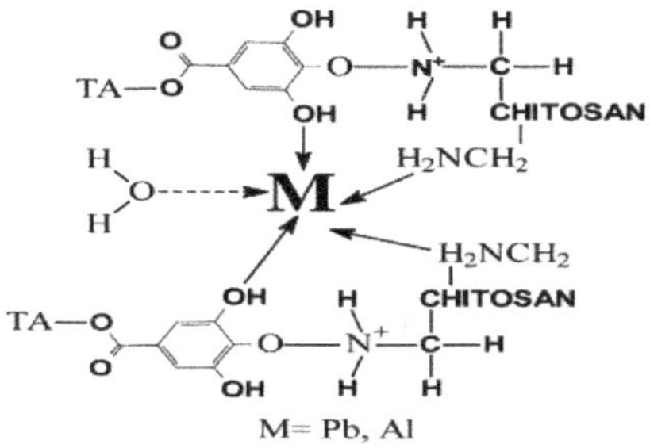

Figure 3: Complex formation by tannic acid modified chitosan and metal ions

Using Fourier Transformed Infrared spectroscopy analysis, Elbarbary & Ghobashy [131] was able to establish the formation of chelates between the adsorbed ions and the hydroxyl, phosphate, amino and carbonyl groups on a phosphorylated chitosan hydroxyethyl methacrylate interpenetrating polymer network adsorbents. Complex formation between adsorbate and adsorbent was also reported by Kannamba & Reddy [79]

In the adsorption of metal anion complexes ion exchange/electrostatic attraction, reduction and metal ion coordination are the mechanism of importance. Interaction between the negatively charged complex and the chitosan is thus controlled by the pH of the solution, since the surface charge/density is determined by the degree of protonation of the groups on the chitosan adsorbent. Though the degree of protonation is determined by the degree of deacetylation, according to Pestov & Bratskaya [117], at pH < 4.5 most studied chitosans are completely protonated. However, below pH of 2, competition between competing anions limits the adsorption capacity. Modification of chitosan for enhanced anion adsorption is thus targeted at reducing its solubility in acidic medium achieved by crosslinking, and enhancing its selectivity and adsorption capacity by introducing functional moieties that will increase the number of electron-donor atoms like nitrogen or moieties with improved affinity [157-159]

Sowmya & Meenakshi [105] and Bhatt et al. [116] studied the mechanism of adsorption of Cr^{6+} using Fourier Transformed Infrared (FTIR), Scanning Electron Microscopy (SEM) and Energy Dispersive X-ray images, X-ray Photoelectron Spectroscopy (XPS) and Electron Spin Resonance (ESR)[74,84]. The adsorption of chromate anion onto the adsorbent was confirmed by FTIR spectra of the loaded adsorbent revealing the presence of the ion as hydrogen chromate and dichromate species. ESR spectra of the loaded adsorbent showed a spectra signal corresponding to exchange coupling between Cr^{3+}-Cr^{3+} ion pairs or Cr^{3+} clusters, showing the reduction of Cr^{6+} to Cr^{3+}. The configuration of the adsorbate-adsorbent interaction was confirmed by the binding energies reported by the XPS analysis. An additional peak at binding energy 288.1 eV resulting from the bond between carbonyl groups and chromium was observed for the loaded adsorbent. The alteration in the N1s spectrum of modified chitosan after loading with chromium suggested that formation of metal-NH_2 complex was not likely the main adsorption route for chromium as only slight changes were observed in this spectrum. The O spectrum of the loaded adsorbent showed two peaks at 531.8 eV due to C=O and 533.1 eV due to $Cr(OH)_3$, instead of the single peak showed by the unmodified and modified chitosan (532.2 eV due to C=O). This was attributed to the splitting of the peak due to C=O group by the Cr-O bond formed. The Cr2p region of the analysis, further proved the reduction of Cr^{6+}. However, the Cr^{3+} was observed to exist in two different species on the surface of the adsorbent with binding energies at 576.9, 579.0, 585.9 and 588.5 eV. The absence of oxygen peaks due to chromates, dichromates expected at ~530 eV confirmed that the chromate ion was adsorbed on the surface in its +3 oxidation state as $Cr(OH)_3$ and Cr^{3+} stabilized by the carboxylate group of the modifier atoms.

Many factors that influence the adsorbate-adsorbent interaction have been reported to also influence to a large extent the adsorption capacity of chitosan derivatives. Factors like solution pH, adsorbate speciation, degree of deacetylation, polymer chain length, metal ion selectivity, conditioning of adsorbent and their physical forms have significant impact on adsorption capacity of chitosan based adsorbents [160]. In the study by Sakkayawong et al, on the sorption of diazo C. I. reactive red dye on chitosan, it was reported that the mechanism of sorption by the sorbent was influenced by the pH of the system [161]. Under acidic conditions, the amine group on chitosan was protonated, while the dye molecules dissociated in solution as shown in the equations 1-3. The adsorption process was thus via electrostatic interaction between the protonated chitosan and the negatively charge dye molecule.

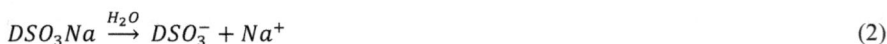

$$R^`- NH_2 + H^+ \rightleftharpoons R - NH_3^+ \tag{1}$$

$$DSO_3Na \xrightarrow{H_2O} DSO_3^- + Na^+ \tag{2}$$

$$R - NH_3^+ + DSO_3^- \rightleftharpoons R - NH_3^{+-}O_3SD \qquad (3)$$

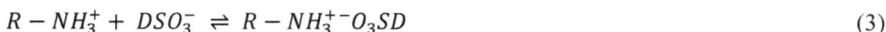

However, under basic condition, the hydroxyl group on chitosan becomes accessible due to its deprotonation as shown in equation 3. Adsorption activity at this point is due to the formation of covalent bond between chitosan and the dye molecule

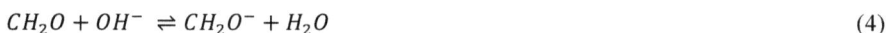

$$CH_2O + OH^- \rightleftharpoons CH_2O^- + H_2O \qquad (4)$$

6.   Recent developments in Chitosan based adsorbent

Since the preliminary work by Muzarrelli in 1969 on the synthesis of chitosan and evaluation of it adsorption capacity for removal of metal ions from solution[162], numerous attempt at improving its capacity and application in adsorption has been the focus in recent researches. Numerous reviews on synthesis of modified chitosan through routes such as crosslinking, grafting and substitution reactions with products exhibiting outstanding adsorption capacities have been presented. However attempts are still ongoing on improving the synthetic routes to modifying chitosan.

Recently, the use of ionic liquids (ILs) in improving the adsorption capacity of chitosan is gaining much focus. ILs are organic salts with organic cations like pyrrolidinium, imdazolium or ammonium derivatives with associated organic anions like CH_3CO^- or inorganic anions like Br^- and BF^{4-} [163]. Advantages of ILs as modifiers for chitosan includes: good thermal stability; formation of bond with other compounds either through, hydrogen bonding, electrostatic attraction, π-π or hydrophobic interactions; ease of tuning their properties through suitable selection of anions and cations and they are green solvents [164-165].

Although, the application of ILs in chitosan modification is relatively new, there is growing research output in this regard. *Sec*-butylammonium acetate and *n*-octylammonium acetate chitosan with improved adsorption capacity for chromium adsorption has been reported [166]. The removal efficiency achieved by the *sec*-butylammonium acetate reached about 99% efficiency. 1-carboxybutyl-3-methylimidazolium chloride chitosan with amphiphilic structure was studied for $Cr_2O_7^{2-}$ and PF_6^- [167]. In order to study the interaction between chitosan and ionic liquid and its effect on Pd^{2+} adsorption, Kumar et al [168] used analytical characterization tools like FT-IR, XRD, SEM and EDAX. It was observed that there was great interaction between Pd^{2+} and Aliquat-336 IL used in modifying the chitosan. Further analysis showed that the Pd^{2+} was adsorbed on the surface of the adsorbent, through electrostatic interaction and ion pair mechanism. The q_{max} achieved by the adsorbent was 187. 61 mg/g, which was more than 20 fold the capacity of unmodified chitosan. ILs modified chitosan have also been reported for adsorption of other adsorbates like: metal ions such as Sr^{2+}, Cs^{2+} [169], and even dyes like malachite green [170]

Despite the advantages of ILs, its relatively high cost makes its application less economical compared to other modifiers. However. processes such as solvent extraction, vacuum treatment and evaporation [171] have been used in ILs recovery.

Chitosan has proven to be material of choice in the removal of metal ions, anions and dye molecules from aqueous solution. The ease of tuning the properties of chitosan also makes it an interesting material for research.. Incorporation of functional moieties not only influence the adsorption capacity of chitosan, but also its mechanism which could be further influenced by the adsorption system parameters like pH, adsorbent form and presence of competing ions. The interest in chemically modified chitosan may not be abating soon with recent focus on the development of ionic liquid modified chitosan with improved adsorption capacity. With this current research direction, the best of chemically modified chitosan adsorbent may be nearer than expected.

References

[1]   R.P. Schwarzenbach, T. Egli, T.B. Hofstetter, U. Von Gunten, B. Wehrli, Global water pollution and human health, Annual Review of Environment and Resources 35 (2010) 109-136. https://doi.org/10.1146/annurev-environ-100809-125342

[2]   M. Mende, D. Schwarz, S. Schwarz, Chitosan–A Natural Adsorbent for Copper Ions, Proceedings of the World Congress on Civil, Structural, and Environmental Engineering (CSEE'16), 2016.

[3]   M. Ilayaraja, S. Sharmilaparveen, R. Sayeekannan, Synthesis and adsorption properties of Chitosan cross linked with phenol–formaldehyde resin for the removal of heavy metals and dyes from water, IOSR-J. Appl. Chem. 7 (2014) 16-26. https://doi.org/10.9790/5736-07211626

[4]   M. Wang, Y. Ma, Y. Sun, S.Y. Hong, S.K. Lee, B. Yoon, L. Chen, L. Ci, J.-D. Nam, X. Chen, Hierarchical Porous Chitosan Sponges as Robust and Recyclable Adsorbents for Anionic Dye Adsorption, Scientific Reports 7(1) (2017) 18054. https://doi.org/10.1038/s41598-017-18302-0

[5]   S. Xie, S. Huang, W. Wei, X. Yang, Y. Liu, X. Lu, Y. Tong, Chitosan Waste-Derived Co and N Co-doped Carbon Electrocatalyst for Efficient Oxygen Reduction Reaction, ChemElectroChem 2(11) (2015) 1806-1812. https://doi.org/10.1002/celc.201500199

[6]   G. Crini, P.-M. Badot, Application of chitosan, a natural aminopolysaccharide, for dye removal from aqueous solutions by adsorption processes using batch studies: A review of recent literature, Progress in Polymer Science 33(4) (2008) 399-447. https://doi.org/10.1016/j.progpolymsci.2007.11.001

[7]   J. Wang, C. Chen, Chitosan-based biosorbents: modification and application for biosorption of heavy metals and radionuclides, Bioresource Technology 160 (2014) 129-141. https://doi.org/10.1016/j.biortech.2013.12.110

[8]   G.Z. Kyzas, D.N. Bikiaris, Recent modifications of chitosan for adsorption applications: a critical and systematic review, Marine Drugs 13(1) (2015) 312-337. https://doi.org/10.3390/md13010312

[9]   E. Guibal, E. Touraud, J. Roussy, Chitosan interactions with metal ions and dyes: dissolved-state vs. solid-state application, World Journal of Microbiology and Biotechnology 21(6-7) (2005) 913-920. https://doi.org/10.1007/s11274-004-6559-5

[10]   R.C. Goy, D.d. Britto, O.B. Assis, A review of the antimicrobial activity of chitosan, Polímeros 19(3) (2009) 241-247. https://doi.org/10.1590/S0104-14282009000300013

[11]   M.R. Hussain, M. Iman, T.K. Maji, Determination of degree of deacetylation of chitosan and their effect on the release behavior of essential oil from chitosan and chitosan-gelatin complex microcapsules, International Journal of Advanced Engineering Applications 6(4) (2013) 4-12.

[12]   M.A. Elgadir, M.S. Uddin, S. Ferdosh, A. Adam, A.J.K. Chowdhury, M.Z.I. Sarker, Impact of chitosan composites and chitosan nanoparticle composites on various drug delivery systems: A review, journal of Food and Drug Analysis 23(4) (2015) 619-629.

[13]   Y. Yuan, B.M. Chesnutt, W.O. Haggard, J.D. Bumgardner, Deacetylation of chitosan: material characterization and in vitro evaluation via albumin adsorption and pre-osteoblastic cell cultures, Materials 4(8) (2011) 1399-1416. https://doi.org/10.3390/ma4081399

[14]   N. Kubota, Y. Eguchi, Facile preparation of water-soluble N-acetylated chitosan and molecular weight dependence of its water-solubility, Polymer Journal 29(2) (1997) 123. https://doi.org/10.1295/polymj.29.123

[15]   M.H. Ottey, K.M. Vårum, O. Smidsrød, Compositional heterogeneity of heterogeneously deacetylated chitosans, Carbohydrate Polymers 29(1) (1996) 17-24. https://doi.org/10.1016/0144-8617(95)00154-9

[16] M. Duarte, M. Ferreira, M. Marvao, J. Rocha, An optimised method to determine the degree of acetylation of chitin and chitosan by FTIR spectroscopy, International Journal of Biological Macromolecules 31(1-3) (2002) 1-8. https://doi.org/10.1016/S0141-8130(02)00039-9

[17] E. El-Nesr, A. Raafat, S.M. Nasef, E. Soliman, E. Hegazy, Chitin and chitosan extracted from irradiated and non-irradiated shrimp wastes (Comparative Analysis Study), Arab Journal of Nuclear Science and Applications 46(1) (2013) 53-66.

[18] J.S. Piccin, M. Vieira, J. Gonçalves, G. Dotto, L.A.d.A. Pinto, Adsorption of FD&C Red No. 40 by chitosan: Isotherms analysis, Journal of Food Engineering 95(1) (2009) 16-20. https://doi.org/10.1016/j.jfoodeng.2009.03.017

[19] J. Unagolla, S. Adikary, Adsorption characteristics of cadmium and lead heavy metals into locally synthesized Chitosan Biopolymer, Tropical Agricultural Research 26(2) (2015).

[20] Q. Li, E. Dunn, E. Grandmaison, M.F. Goosen, Applications and properties of chitosan, Journal of Bioactive and Compatible Polymers 7(4) (1992) 370-397. https://doi.org/10.1177/088391159200700406

[21] S. Jana, A. Saha, A.K. Nayak, K.K. Sen, S.K. Basu, Aceclofenac-loaded chitosan-tamarind seed polysaccharide interpenetrating polymeric network microparticles, Colloids and Surfaces B: Biointerfaces 105 (2013) 303-309. https://doi.org/10.1016/j.colsurfb.2013.01.013

[22] X. He, K. Li, R. Xing, S. Liu, L. Hu, P. Li, The production of fully deacetylated chitosan by compression method, The Egyptian Journal of Aquatic Research 42(1) (2016) 75-81. https://doi.org/10.1016/j.ejar.2015.09.003

[23] A.N. Hernandez-Lauzardo, S. Bautista-Baños, M.G. Velazquez-Del Valle, M. Méndez-Montealvo, M. Sánchez-Rivera, L.A. Bello-Perez, Antifungal effects of chitosan with different molecular weights on in vitro development of Rhizopus stolonifer (Ehrenb.: Fr.) Vuill, Carbohydrate polymers 73(4) (2008) 541-547. https://doi.org/10.1016/j.carbpol.2007.12.020

[24] H. Zhang, S.H. Neau, In vitro degradation of chitosan by a commercial enzyme preparation: effect of molecular weight and degree of deacetylation, Biomaterials 22(12) (2001) 1653-1658. https://doi.org/10.1016/S0142-9612(00)00326-4

[25] M. Vakili, M. Rafatullah, B. Salamatinia, A.Z. Abdullah, M.H. Ibrahim, K.B. Tan, Z. Gholami, P. Amouzgar, Application of chitosan and its derivatives as adsorbents for dye removal from water and wastewater: A review, Carbohydrate Polymers 113 (2014) 115-130. https://doi.org/10.1016/j.carbpol.2014.07.007

[26] E. Guibal, Interactions of metal ions with chitosan-based sorbents: a review, Separation and Purification Technology 38(1) (2004) 43-74. https://doi.org/10.1016/j.seppur.2003.10.004

[27] I. Aranaz, M. Mengíbar, R. Harris, I. Paños, B. Miralles, N. Acosta, G. Galed, Á. Heras, Functional characterization of chitin and chitosan, Current Chemical Biology 3(2) (2009) 203-230.

[28] T.K. Saha, N.C. Bhoumik, S. Karmaker, M.G. Ahmed, H. Ichikawa, Y. Fukumori, Adsorption of methyl orange onto chitosan from aqueous solution, Journal of Water Resource and Protection 2(10) (2010) 898. https://doi.org/10.4236/jwarp.2010.210107

[29] J. Barron-Zambrano, A. Szygula, M. Ruiz, A.M. Sastre, E. Guibal, Biosorption of Reactive Black 5 from aqueous solutions by chitosan: column studies, Journal of Environmental Management 91(12) (2010) 2669-2675. https://doi.org/10.1016/j.jenvman.2010.07.033

[30] A.G. Hadi, Dye removal from colored textile wastewater using synthesized chitosan, Int. J. Sci. Technol. 2(4) (2013) 359-364.

[31] J. Iqbal, F.H. Wattoo, M.H.S. Wattoo, R. Malik, S.A. Tirmizi, M. Imran, A.B. Ghangro, Adsorption of acid yellow dye on flakes of chitosan prepared from fishery wastes, Arabian Journal of Chemistry 4(4) (2011) 389-395. https://doi.org/10.1016/j.arabjc.2010.07.007

[32] M.-E. Ignat, V. Dulman, T. Onofrei, Reactive Red 3 and Direct Brown 95 dyes adsorption onto chitosan, Cellulose Chemistry and Technology 46(5-6) (2012) 357-367.

[33] G. Dotto, L. Pinto, Adsorption of food dyes acid blue 9 and food yellow 3 onto chitosan: Stirring rate effect in kinetics and mechanism, Journal of Hazardous Materials 187(1-3) (2011) 164-170. https://doi.org/10.1016/j.jhazmat.2011.01.016

[34] H. Zhang, S. Schiewer, Arsenic (V) sorption on crab shell based chitosan, Impacts of Global Climate Change2005, pp. 1-7.

[35] G. Ratnamala, K. Brajesh, Biosorption of remazol navy blue dye from an aqueous solution using Pseudomonas putida, International Journal of Science, Environment and Technology 2(1) (2013) 80-89.

[36] P. Miretzky, A.F. Cirelli, Hg (II) removal from water by chitosan and chitosan derivatives: a review, Journal of Hazardous Materials 167(1-3) (2009) 10-23. https://doi.org/10.1016/j.jhazmat.2009.01.060

[37] F.-C. Wu, R.-L. Tseng, R.-S. Juang, Comparative adsorption of metal and dye on flake-and bead-types of chitosans prepared from fishery wastes, Journal of

Hazardous Materials 73(1) (2000) 63-75. https://doi.org/10.1016/S0304-3894(99)00168-5

[38] E.A. El-hefian, M.M. Nasef, A.H. Yahaya, Chitosan physical forms: a short review, Australian Journal of Basic and Applied Sciences 5(5) (2011) 670-677.

[39] N. Shah, R. Mewada, T. Mehta, Chitosan: Development of Nanoparticles, Other Physical Forms and Solubility with Acids, Journal of Nano Research, Trans Tech Publ, 2013, pp. 107-122.

[40] C. Peniche, W. Argüelles-Monal, H. Peniche, N. Acosta, Chitosan: an attractive biocompatible polymer for microencapsulation, Macromolecular Bioscience 3(10) (2003) 511-520. https://doi.org/10.1002/mabi.200300019

[41] E. Guibal, C. Milot, J.M. Tobin, Metal-anion sorption by chitosan beads: equilibrium and kinetic studies, Industrial & Engineering Chemistry Research 37(4) (1998) 1454-1463. https://doi.org/10.1021/ie9703954

[42] J. Berger, M. Reist, J.M. Mayer, O. Felt, R. Gurny, Structure and interactions in chitosan hydrogels formed by complexation or aggregation for biomedical applications, European journal of Pharmaceutics and Biopharmaceutics 57(1) (2004) 35-52. https://doi.org/10.1016/S0939-6411(03)00160-7

[43] S. Kumbar, A. Kulkarni, T. Aminabhavi, Crosslinked chitosan microspheres for encapsulation of diclofenac sodium: effect of crosslinking agent, Journal of Microencapsulation 19(2) (2002) 173-180. https://doi.org/10.1080/02652040110065422

[44] N. Li, R. Bai, Copper adsorption on chitosan–cellulose hydrogel beads: behaviors and mechanisms, Separation and Purification Technology 42(3) (2005) 237-247. https://doi.org/10.1016/j.seppur.2004.08.002

[45] L. Lim, L.S. Wan, The effect of plasticizers on the properties of polyvinyl alcohol films, Drug Development and Industrial Pharmacy 20(6) (1994) 1007-1020. https://doi.org/10.3109/03639049409038347

[46] T.S. Trung, W.W. Thein-Han, N.T. Qui, C.-H. Ng, W.F. Stevens, Functional characteristics of shrimp chitosan and its membranes as affected by the degree of deacetylation, Bioresource technology 97(4) (2006) 659-663. https://doi.org/10.1016/j.biortech.2005.03.023

[47] C.R. Huei, H.-D. Hwa, Effect of molecular weight of chitosan with the same degree of deacetylation on the thermal, mechanical, and permeability properties of the prepared membrane, Carbohydrate polymers 29(4) (1996) 353-358. https://doi.org/10.1016/S0144-8617(96)00007-0

[48] N. Bhattarai, D. Edmondson, O. Veiseh, F.A. Matsen, M. Zhang, Electrospun chitosan-based nanofibers and their cellular compatibility, Biomaterials 26(31) (2005) 6176-6184. https://doi.org/10.1016/j.biomaterials.2005.03.027

[49] Z. Chen, X. Mo, C. He, H. Wang, Intermolecular interactions in electrospun collagen–chitosan complex nanofibers, Carbohydrate Polymers 72(3) (2008) 410-418. https://doi.org/10.1016/j.carbpol.2007.09.018

[50] Y.-T. Jia, J. Gong, X.-H. Gu, H.-Y. Kim, J. Dong, X.-Y. Shen, Fabrication and characterization of poly (vinyl alcohol)/chitosan blend nanofibers produced by electrospinning method, Carbohydrate Polymers 67(3) (2007) 403-409. https://doi.org/10.1016/j.carbpol.2006.06.010

[51] I.-C. Liao, A.C. Wan, E.K. Yim, K.W. Leong, Controlled release from fibers of polyelectrolyte complexes, Journal of Controlled Release 104(2) (2005) 347-358. https://doi.org/10.1016/j.jconrel.2005.02.013

[52] Q. Wang, N. Zhang, X. Hu, J. Yang, Y. Du, Chitosan/starch fibers and their properties for drug controlled release, European Journal of Pharmaceutics and Biopharmaceutics 66(3) (2007) 398-404. https://doi.org/10.1016/j.ejpb.2006.11.011

[53] S. Poole, The foam-enhancing properties of basic biopolymers, International Journal of Food Science & Technology 24(2) (1989) 121-137. https://doi.org/10.1111/j.1365-2621.1989.tb00626.x

[54] D. Thacharodi, K.P. Rao, Collagen-chitosan composite membranes for controlled release of propranolol hydrochloride, International Journal of Pharmaceutics 120(1) (1995) 115-118. https://doi.org/10.1016/0378-5173(94)00423-3

[55] K. Oungbho, B.W. Müller, Chitosan sponges as sustained release drug carriers, International Journal of Pharmaceutics 156(2) (1997) 229-237. https://doi.org/10.1016/S0378-5173(97)00201-9

[56] J. Tong, L. Chen, Review: Preparation and Application of Magnetic Chitosan Derivatives in Separation Processes, Analytical Letters 46(17) (2013) 2635-2656. https://doi.org/10.1080/00032719.2013.807815

[57] A.A. Alhwaige, T. Agag, H. Ishida, S. Qutubuddin, Biobased chitosan hybrid aerogels with superior adsorption: Role of graphene oxide in CO2 capture, RSC Advances 3(36) (2013) 16011-16020. https://doi.org/10.1039/c3ra42022a

[58] M. Vakili, M. Rafatullah, B. Salamatinia, A.Z. Abdullah, M.H. Ibrahim, K.B. Tan, Z. Gholami, P. Amouzgar, Application of chitosan and its derivatives as adsorbents for dye removal from water and wastewater: a review, Carbohydr Polym 113 (2014) 115-30. https://doi.org/10.1016/j.carbpol.2014.07.007

[59] J. Ji, L. Wang, H. Yu, Y. Chen, Y. Zhao, H. Zhang, W.A. Amer, Y. Sun, L. Huang, M. Saleem, Chemical Modifications of Chitosan and Its Applications, Polymer-Plastics Technology and Engineering 53(14) (2014) 1494-1505. https://doi.org/10.1080/03602559.2014.909486

[60] J. Ma, Y. Sahai, Chitosan biopolymer for fuel cell applications, Carbohydrate Polymers 92(2) (2013) 955-975. https://doi.org/10.1016/j.carbpol.2012.10.015

[61] S.K. Shukla, A.K. Mishra, O.A. Arotiba, B.B. Mamba, Chitosan-based nanomaterials: a state-of-the-art review, International Journal of Biological Macromolecules 59 (2013) 46-58. https://doi.org/10.1016/j.ijbiomac.2013.04.043

[62] V.K. Mourya, N.N. Inamdar, Chitosan-modifications and applications: Opportunities galore, Reactive and Functional Polymers 68(6) (2008) 1013-1051. https://doi.org/10.1016/j.reactfunctpolym.2008.03.002

[63] L. Pontoni, M. Fabbricino, Use of chitosan and chitosan-derivatives to remove arsenic from aqueous solutions--a mini review, Carbohydrate Research 356 (2012) 86-92. https://doi.org/10.1016/j.carres.2012.03.042

[64] P. Miretzky, A.F. Cirelli, Hg(II) removal from water by chitosan and chitosan derivatives: a review, Journal of Hazardous Materials 167(1-3) (2009) 10-23. https://doi.org/10.1016/j.jhazmat.2009.01.060

[65] L. Zhang, Y. Zeng, Z. Cheng, Removal of heavy metal ions using chitosan and modified chitosan: A review, Journal of Molecular Liquids 214 (2016) 175-191. https://doi.org/10.1016/j.molliq.2015.12.013

[66] J. Wang, L. Wang, H. Yu, A. Zain Ul, Y. Chen, Q. Chen, W. Zhou, H. Zhang, X. Chen, Recent progress on synthesis, property and application of modified chitosan: An overview, International Journal of Biological Macromolecules 88 (2016) 333-44. https://doi.org/10.1016/j.ijbiomac.2016.04.002

[67] K.H. Prashanth, R. Tharanathan, Chitin/chitosan: modifications and their unlimited application potential—an overview, Trends in Food Science & Technology 18(3) (2007) 117-131. https://doi.org/10.1016/j.tifs.2006.10.022

[68] J. Wang, L. Wang, H. Yu, A. Zain ul, Y. Chen, Q. Chen, W. Zhou, H. Zhang, X. Chen, Recent progress on synthesis, property and application of modified chitosan: An overview, International Journal of Biological Macromolecules 88 (2016) 333-344. https://doi.org/10.1016/j.ijbiomac.2016.04.002

[69] R. Jayakumar, M. Prabaharan, R.L. Reis, J.F. Mano, Graft copolymerized chitosan—present status and applications, Carbohydrate Polymers 62(2) (2005) 142-158. https://doi.org/10.1016/j.carbpol.2005.07.017

[70]   W. Xie, P. Xu, Q. Liu, Antioxidant activity of water-soluble chitosan derivatives, Bioorganic & Medicinal Chemistry Letters 11(13) (2001) 1699-1701. https://doi.org/10.1016/S0960-894X(01)00285-2

[71]   W. Xie, P. Xu, W. Wang, Q. Liu, Preparation and antibacterial activity of a water-soluble chitosan derivative, Carbohydrate Polymers 50(1) (2002) 35-40. https://doi.org/10.1016/S0144-8617(01)00370-8

[72]   R. Jayakumar, M. Prabaharan, R. Reis, J. Mano, Graft copolymerized chitosan—present status and applications, Carbohydrate Polymers 62(2) (2005) 142-158. https://doi.org/10.1016/j.carbpol.2005.07.017

[73]   S.K. Yong, N.S. Bolan, E. Lombi, W. Skinner, E. Guibal, Sulfur-Containing Chitin and Chitosan Derivatives as Trace Metal Adsorbents: A Review, Critical Reviews in Environmental Science and Technology 43(16) (2013) 1741-1794. https://doi.org/10.1080/10643389.2012.671734

[74]   G. Cardenas, P. Orlando, T. Edelio, Synthesis and applications of chitosan mercaptanes as heavy metal retention agent, International Journal of Biological Macromolecules 28(2) (2001) 167-174. https://doi.org/10.1016/S0141-8130(00)00156-2

[75]   Q. Song, C. Wang, Z. Zhang, J. Gao, Adsorption of lead using a novel xanthated carboxymethyl chitosan, Water science and technology : a Journal of the International Association on Water Pollution Research 69(2) (2014) 298-304.

[76]   S.K. Yong, W.M. Skinner, N.S. Bolan, E. Lombi, A. Kunhikrishnan, Y.S. Ok, Sulfur crosslinks from thermal degradation of chitosan dithiocarbamate derivatives and thermodynamic study for sorption of copper and cadmium from aqueous system, Environmental Science and Pollution Research 23(2) (2016) 1050-1059. https://doi.org/10.1007/s11356-015-5654-5

[77]   F. Li, C. Bao, J. Zhang, Q. Sun, W. Kong, X. Han, Y. Wang, Synthesis of chemically modified chitosan with 2, 5-dimercapto-1, 3, 4-thiodiazole and its adsorption abilities for Au (III), Pd (II), and Pt (IV), Journal of Applied Polymer Science 113(3) (2009) 1604-1610. https://doi.org/10.1002/app.30068

[78]   M. Ruiz, A. Sastre, E. Guibal, Pd and Pt recovery using chitosan gel beads. II. Influence of chemical modifications on sorption properties, Separation Science and Technology 37(10) (2002) 2385-2403. https://doi.org/10.1081/SS-120003519

[79]   B. Kannamba, K.L. Reddy, B.V. AppaRao, Removal of Cu(II) from aqueous solutions using chemically modified chitosan, Journal of Hazardous Materials 175(1-3) (2010) 939-48. https://doi.org/10.1016/j.jhazmat.2009.10.098

[80]   M. Ahmad, K. Manzoor, P. Venkatachalam, S. Ikram, Kinetic and thermodynamic evaluation of adsorption of Cu(II) by thiosemicarbazide chitosan, (1879-0003 (Electronic)).

[81]   E. Metwally, S.S. Elkholy, H.A.M. Salem, M.Z. Elsabee, Sorption behavior of 60Co and 152+154Eu radionuclides onto chitosan derivatives, Carbohydrate Polymers 76(4) (2009) 622-631. https://doi.org/10.1016/j.carbpol.2008.11.032

[82]   Y. Zhu, Z.-S. Bai, H.-L. Wang, Microfluidic synthesis of thiourea modified chitosan microsphere of high specific surface area for heavy metal wastewater treatment, Chinese Chemical Letters 28(3) (2017) 633-641. https://doi.org/10.1016/j.cclet.2016.10.031

[83]   M. Li, Z. Zhang, R. Li, J.J. Wang, A. Ali, Removal of Pb(II) and Cd(II) ions from aqueous solution by thiosemicarbazide modified chitosan, International Journal of Biological Macromolecules 86 (2016) 876-84. https://doi.org/10.1016/j.ijbiomac.2016.02.027

[84]   Y.-C. Lin, H.-P. Wang, F. Gohar, M.H. Ullah, X. Zhang, D.-F. Xie, H. Fang, J. Huang, J.-X. Yang, Preparation and copper ions adsorption properties of thiosemicarbazide chitosan from squid pens, International Journal of Biological Macromolecules 95 (2017) 476-483. https://doi.org/10.1016/j.ijbiomac.2016.11.085

[85]   L. Zhou, J. Liu, Z. Liu, Adsorption of platinum(IV) and palladium(II) from aqueous solution by thiourea-modified chitosan microspheres, Journal of Hazardous Materials 172(1) (2009) 439-446. https://doi.org/10.1016/j.jhazmat.2009.07.030

[86]   E.C. da Silva Filho, P.D.R. Monteiro, K.S. Sousa, C. Airoldi, Ethylenesulfide as a useful agent for incorporation on the biopolymer chitosan in a solvent-free reaction for use in lead and cadmium removal, Journal of Thermal Analysis and Calorimetry 106(2) (2011) 369-373. https://doi.org/10.1007/s10973-010-1205-y

[87]   A. Khan, S. Badshah, C. Airoldi, Dithiocarbamated chitosan as a potent biopolymer for toxic cation remediation, Colloids and surfaces. B, Biointerfaces 87(1) (2011) 88-95. https://doi.org/10.1016/j.colsurfb.2011.05.006

[88]   K.C. Gavilan, A.V. Pestov, H.M. Garcia, Y. Yatluk, J. Roussy, E. Guibal, Mercury sorption on a thiocarbamoyl derivative of chitosan, Journal of Hazardous Materials 165(1-3) (2009) 415-26. https://doi.org/10.1016/j.jhazmat.2008.10.005

[89]   L. Wang, R. Xing, S. Liu, H. Yu, Y. Qin, K. Li, J. Feng, R. Li, P. Li, Recovery of silver (I) using a thiourea-modified chitosan resin, Journal of Hazardous Materials 180(1-3) (2010) 577-82. https://doi.org/10.1016/j.jhazmat.2010.04.072

[90] E. Repo, J.K. Warchol, T.A. Kurniawan, M.E.T. Sillanpää, Adsorption of Co(II) and Ni(II) by EDTA- and/or DTPA-modified chitosan: Kinetic and equilibrium modeling, Chemical Engineering Journal 161(1-2) (2010) 73-82. https://doi.org/10.1016/j.cej.2010.04.030

[91] N.K. Lazaridis, G.Z. Kyzas, A.A. vassiliou, D.N. Bikiaris, Chitosan derivatives as biosorbents for basic dyes, Langmuir 23 (2007) 7634-7643. https://doi.org/10.1021/la700423j

[92] Y. Yan, B. Xiang, Y. Li, Q. Jia, Preparation and adsorption properties of diethylenetriamine-modified chitosan beads for acid dyes, Journal of Applied Polymer Science 130(6) (2013) 4090-4098. https://doi.org/10.1002/app.39691

[93] J. Roosen, K. Binnemans, Adsorption and chromatographic separation of rare earths with EDTA- and DTPA-funtionalized chitosan biopolymers, Journal of Materials Chemistry A 2 (2014) 1530-1540. https://doi.org/10.1039/C3TA14622G

[94] H.M. Zalloum, Z. Al-Qodah, M.S. Mubarak, Copper Adsorption on Chitosan-Derived Schiff Bases, Journal of Macromolecular Science, Part A 46(1) (2008) 46-57.

[95] B. Hastuti, D. Siswanta, Mudasir, Triyono, Synthesis and Characterization Pectin-Carboxymethyl Chitosan crosslinked PEGDE as biosorbent of Pb(II) ion, IOP Conference Series: Materials Science and Engineering 299 (2018) 012052. https://doi.org/10.1088/1757-899X/299/1/012052

[96] N.G. Kandile, A.S. Nasr, New hydrogels based on modified chitosan as metal biosorbent agents, International Journal of Biological Macromolecules 64 (2014) 328-33. https://doi.org/10.1016/j.ijbiomac.2013.12.022

[97] H.T.Y. Ly, S. Van Nguyen, in International Conference on Green Technology and Sustainable Development (GTSD). (IEEE, 2016), vol. DOI: 10.1109/GTSD.2016.62, pp. 244-250.

[98] K.S. Manish, Recyclable Crosslinked O-Carboxymethyl Chitosan for Removal of Cationic Dye from Aqueous Solutions, Journal of Waste Water Treatment & Analysis 03(04) (2012). https://doi.org/10.4172/2157-7587.1000138

[99] L. Wang, R. Xing, S. Liu, S. Cai, H. Yu, J. Feng, R. Li, P. Li, Synthesis and evaluation of a thiourea-modified chitosan derivative applied for adsorption of Hg(II) from synthetic wastewater, International Journal of Biological Macromolecules 46(5) (2010) 524-8. https://doi.org/10.1016/j.ijbiomac.2010.03.003

[100] L. Wang, A. Wang, Adsorption properties of congo red from aqueous solution onto N,O-carboxymethyl-chitosan, Bioresource Technology 99(5) (2008) 1403-8. https://doi.org/10.1016/j.biortech.2007.01.063

[101] L. Wang, H. Peng, S. Liu, H. Yu, P. Li, R. Xing, Adsorption properties of gold onto a chitosan derivative, International Journal of Biological Macromolecules 51(5) (2012) 701-4. https://doi.org/10.1016/j.ijbiomac.2012.06.010

[102] L. Wang, Q. Li, A. Wang, Adsorption of cationic dye on N,O-carboxymethyl-chitosan from aqueous solutions: equilibrium, kinetics, and adsorption mechanism, Polymer Bulletin 65(9) (2010) 961-975. https://doi.org/10.1007/s00289-010-0363-1

[103] C. Wang, Q. Song, J. Gao, Investigation of adsorption capacity of N-carboxymethyl chitosan for Pb(II) ions, Water science and technology : a Journal of the International Association on Water Pollution Research 68(8) (2013) 1873-9.

[104] N.V. Suc, H.T.Y. Ly, Lead (II) removal from aqueous solution by chitosan flake modified with citric acid via crosslinking with glutaraldehyde, Journal of Chemical Technology & Biotechnology 88(9) (2013) 1641-1649. https://doi.org/10.1002/jctb.4013

[105] A. Sowmya, S. Meenakshi, Effective removal of nitrate and phosphate anions from aqueous solutions using functionalised chitosan beads, Desalination and Water Treatment 52(13-15) (2013) 2583-2593. https://doi.org/10.1080/19443994.2013.798842

[106] R. Laus, T.G. Costa, B. Szpoganicz, V.T. Favere, Adsorption and desorption of Cu(II), Cd(II) and Pb(II) ions using chitosan crosslinked with epichlorohydrin-triphosphate as the adsorbent, Journal of Hazardous Materials 183(1-3) (2010) 233-41. https://doi.org/10.1016/j.jhazmat.2010.07.016

[107] S.-J. Wu, T.-H. Liou, C.-H. Yeh, F.-L. Mi, T.-K. Lin, Preparation and characterization of porous chitosan–tripolyphosphate beads for copper(II) ion adsorption, Journal of Applied Polymer Science 127(6) (2013) 4573-4580.

[108] M. Ahmad, K. Manzoor, S. Ikram, Versatile nature of hetero-chitosan based derivatives as biodegradable adsorbent for heavy metal ions; a review, International Journal of Biological Macromolecules 105(Pt 1) (2017) 190-203. https://doi.org/10.1016/j.ijbiomac.2017.07.008

[109] S.A. Ali, R.P. Singh, Synthesis and Characterization of a Modified Chitosan, Macromolecular Symposia 277(1) (2009) 1-7. https://doi.org/10.1002/masy.200950301

[110] N. Li, R. Bai, Novel chitosan-cellulose hydrogel adsorbents for lead adsorption, Conference Proceedings, AIChE Annual Meeting, 2004.

[111] S. Rosa, M.C. Laranjeira, H.G. Riela, V.T. Favere, Cross-linked quaternary chitosan as an adsorbent for the removal of the reactive dye from aqueous solutions, Journal of Hazardous Materials 155(1-2) (2008) 253-60. https://doi.org/10.1016/j.jhazmat.2007.11.059

[112] R. Karthik, S. Meenakshi, Removal of Pb(II) and Cd(II) ions from aqueous solution using polyaniline grafted chitosan, Chemical Engineering Journal 263 (2015) 168-177. https://doi.org/10.1016/j.cej.2014.11.015

[113] H. Abdel-Razik, H. Almahy, Recovery of Water from Heavy metals using Chelating Chemcially Modified Chitosan, Int. J. Chem. Sci. 13(4) (2015) 1713-1725.

[114] K. Kondo, R. Eto, M. Matsumoto, Adsorption of Pd and Pt on Chemically Modified Chitosan, Bull Soc Sea Water Sci Jpn 69 (2015) 197-204.

[115] K. Azlan, W.N. Wan Saime, L. Lai Ken, Chitosan and chemically modified chitosan beads for acid dyes sorption, Journal of Environmental Sciences 21(3) (2009) 296-302. https://doi.org/10.1016/S1001-0742(08)62267-6

[116] R. Bhatt, B. Sreedhar, P. Padmaja, Adsorption of chromium from aqueous solutions using crosslinked chitosan-diethylenetriaminepentaacetic acid, Int J Biol Macromol 74 (2015) 458-66. https://doi.org/10.1016/j.ijbiomac.2014.12.041

[117] A. Pestov, S. Bratskaya, Chitosan and Its Derivatives as Highly Efficient Polymer Ligands, Molecules 21(3) (2016) 330. https://doi.org/10.3390/molecules21030330

[118] H. Karaer, İ. Uzun, Adsorption of basic dyestuffs from aqueous solution by modified chitosan, Desalination and Water Treatment 51(10-12) (2013) 2294-2305. https://doi.org/10.1080/19443994.2012.734967

[119] Y. Xing, X. Sun, B. Li, Poly(methacrylic acid)-modified chitosan for enhancement adsorption of water-soluble cationic dyes, Polymer Engineering & Science 49(2) (2009) 272-280. https://doi.org/10.1002/pen.21253

[120] V.K. Konaganti, R. Kota, S. Patil, G. Madras, Adsorption of anionic dyes on chitosan grafted poly(alkyl methacrylate)s, Chemical Engineering Journal 158(3) (2010) 393-401. https://doi.org/10.1016/j.cej.2010.01.003

[121] W.S. Wan Ngah, K.H. Liang, Adsorption of Gold(III) ions intos chitosan and N-carboxymethyl chitosan: Equilibrium studies, Ind. Eng. Chem. Res. 38 (1999) 1411-1414. https://doi.org/10.1021/ie9803164

[122] R. Jayakumar, R.L. Reis, J.F. Mano, Chemisry and Apllications of Phosphorylated Chitin and Chitosan, e-polymers 35 (2006) 1-16.

[123] F. Lebouc, I. Dez, M. Gulea, P.-J. Madec, P.-A. Jaffrès, Synthesis of Phosphorus-Containing Chitosan Derivatives, Phosphorus, Sulfur, and Silicon and the Related Elements 184(4) (2009) 872-889. https://doi.org/10.1080/10426500802715585

[124] R. Jayakumar, N. Selvamurugan, S.V. Nair, S. Tokura, H. Tamura, Preparative methods of phosphorylated chitin and chitosan—An overview, International Journal of Biological Macromolecules 43(3) (2008) 221-225. https://doi.org/10.1016/j.ijbiomac.2008.07.004

[125] A. Zuñiga, A. Debbaudt, L. Albertengo, M.S. Rodríguez, Synthesis and characterization of N-propyl-N-methylene phosphonic chitosan derivative, Carbohydrate Polymers 79(2) (2010) 475-480. https://doi.org/10.1016/j.carbpol.2009.08.011

[126] A. Heras, N.M. Rodríguez, V.M. Ramos, E. Agulló, N-methylene phosphonic chitosan: a novel soluble derivative, Carbohydrate Polymers 44(1) (2001) 1-8. https://doi.org/10.1016/S0144-8617(00)00195-8

[127] V.M. Ramos, Rodrı, x, N.M. guez, Rodrı, x, M.S. guez, A. Heras, E. Agulló, Modified chitosan carrying phosphonic and alkyl groups, Carbohydrate Polymers 51(4) (2003) 425-429. https://doi.org/10.1016/S0144-8617(02)00211-4

[128] D.J. Suchyta, R.J. Soto, M.H. Schoenfisch, Selective monophosphorylation of chitosan via phosphorus oxychloride, Polym Chem 8(16) (2017) 2552-2558. https://doi.org/10.1039/C7PY00123A

[129] A.M.A. Morsy, Adsorptive removal of uranium ions from liquid waste solutions by phosphorylated chitosan, Environmental Technology & Innovation 4 (2015) 299-310. https://doi.org/10.1016/j.eti.2015.10.002

[130] Z.-m. Dong, Y.-f. Qiu, Y. Dai, X.-h. Cao, L. Wang, P.-f. Wang, Z.-j. Lai, W.-l. Zhang, Z.-b. Zhang, Y.-h. Liu, Z.-g. Le, Removal of U(VI) from aqueous media by hydrothermal cross-linking chitosan with phosphate group, Journal of Radioanalytical and Nuclear Chemistry 309(3) (2016) 1217-1226. https://doi.org/10.1007/s10967-016-4722-8

[131] A.M. Elbarbary, M.M. Ghobashy, Phosphorylation of chitosan/HEMA interpenetrating polymer network prepared by gamma-radiation for metal ions removal from aqueous solutions, Carbohydr Polym 162 (2017) 16-27. https://doi.org/10.1016/j.carbpol.2017.01.013

[132] J. Wang, S. Zhuang, Removal of various pollutants from water and wastewater by modified chitosan adsorbents, Critical Reviews in Environmental Science and Technology 47(23) (2017) 2331-2386. https://doi.org/10.1080/10643389.2017.1421845

[133] X. Tang, C. Wang, Adsorption of Ni(II) from Aqueous Solution by Polyaminated Crosslinked Ni(II)-Imprinted Chitosan Derivative Beads, Environmental Engineering Science 30(10) (2013) 646-652. https://doi.org/10.1089/ees.2013.0099

[134] S. Sun, A. Wang, Adsorption properties of carboxymethyl-chitosan and cross-linked carboxymethyl-chitosan resin with Cu(II) as template, Separation and Purification Technology 49(3) (2006) 197-204. https://doi.org/10.1016/j.seppur.2005.09.013

[135] S. Sun, L. Wang, A. Wang, Adsorption properties of crosslinked carboxymethyl-chitosan resin with Pb(II) as template ions, Journal of Hazardous Materials 136(3) (2006) 930-7. https://doi.org/10.1016/j.jhazmat.2006.01.033

[136] Y. Baba, O. kaoru, T. Ohshima, R. Dhakal, Preparation of palladium(II)-imprinted chitosan derivatives and its adsorption properties of precious metals, J. Ion Exchange 18(4) (2007) 226-230. https://doi.org/10.5182/jaie.18.226

[137] D. Yang, L. Qiu, Y. Yang, Efficient Adsorption of Methyl Orange Using a Modified Chitosan Magnetic Composite Adsorbent, Journal of Chemical & Engineering Data 61(11) (2016) 3933-3940. https://doi.org/10.1021/acs.jced.6b00706

[138] L. Jia, J.S. Wang, Q.W. Guo, X.L. Zou, L. Xie, Adsorption of Cr (VI) by Cross-Linked Magnetic Hydroxamated Chitosan, Advanced Materials Research 842 (2013) 175-179. https://doi.org/10.4028/www.scientific.net/AMR.842.175

[139] L. Fan, C. Luo, Z. Lv, F. Lu, H. Qiu, Preparation of magnetic modified chitosan and adsorption of Zn(2)(+) from aqueous solutions, Colloids and surfaces. B, Biointerfaces 88(2) (2011) 574-81. https://doi.org/10.1016/j.colsurfb.2011.07.038

[140] T.V.J. Charpentier, A. Neville, J.L. Lanigan, R. Barker, M.J. Smith, T. Richardson, Preparation of Magnetic Carboxymethylchitosan Nanoparticles for Adsorption of Heavy Metal Ions, ACS Omega 1(1) (2016) 77-83. https://doi.org/10.1021/acsomega.6b00035

[141] A. Donia, A. yousif, A. Atia, M. Elsamalehy, Efficient adsorption of Ag(I) and Au(III) on modified magnetic chitosan with amine functionalities, Desalination and Water Treatment 52 (2014) 2537-2547. https://doi.org/10.1080/19443994.2013.794706

[142] N. Tripathi, G. Choppala, R.S. Singh, P. Srivastava, B. Seshadri, Sorption kinetics of zinc and nickel on modified chitosan, Environ Monit Assess 188(9) (2016) 507. https://doi.org/10.1007/s10661-016-5499-5

[143] S.P. Kamble, S. Jagtap, N.K. Labhsetwar, D. Thakare, S. Godfrey, S. Devotta, S.S. Rayalu, Defluoridation of drinking water using chitin, chitosan and lanthanum-modified chitosan, Chemical Engineering Journal 129(1-3) (2007) 173-180. https://doi.org/10.1016/j.cej.2006.10.032

[144] P. Liang, Y. Zhang, D. Wang, Y. Xu, L. Luo, Preparation of mixed rare earths modified chitosan for fluoride adsorption, Journal of Rare Earths 31(8) (2013) 817-822. https://doi.org/10.1016/S1002-0721(12)60364-0

[145] T. Fagundes, E.L. Bernardi, C.A. Rodrigues, PHOSPHATE ADSORPTION ON CHITOSAN-FeIII-CROSSLINKING: BATCH AND COLUMN STUDIES, Journal of Liquid Chromatography & Related Technologies 24(8) (2001) 1189-1198. https://doi.org/10.1081/JLC-100103441

[146] J. Zhang, N. Chen, T. Zheng, Y. Yu, Q. Hu, C. Feng, A study of the mechanism of fluoride adsorption from aqueous solutions onto Fe-impregnated chitosan, Phys.Chem.Chem.Phys 17 (2015) 12041-12040. https://doi.org/10.1039/C5CP00817D

[147] E.I. Unuabonah, A. Adewuyi, M.O. Kolawole, M.O. Omorogie, O.C. Olatunde, S.O. Fayemi, C. Günter, C.P. Okoli, F.O. Agunbiade, A. Taubert, Disinfection of water with new chitosan-modified hybrid clay composite adsorbent, Heliyon 3(8) (2017) e00379. https://doi.org/10.1016/j.heliyon.2017.e00379

[148] N.A. Negm, H.E. Ali, Modification of heavy metal uptake efficiency by modified chitosan/anionic surfactant systems, Engineering in Life Sciences 10(3) (2010) 218-224. https://doi.org/10.1002/elsc.200900110

[149] M.U. Khobragade, A. Pal, Fixed-bed column study on removal of Mn(II), Ni(II) and Cu(II) from aqueous solution by surfactant bilayer supported alumina, Separation Science and Technology 51(8) (2016) 1287-1298. https://doi.org/10.1080/01496395.2016.1156698

[150] A. Adak, M. Bandyopadhyay, A. Pal, Adsorption of Anionic Surfactant on Alumina and Reuse of the Surfactant-Modified Alumina for the Removal of Crystal Violet from Aquatic Environment, Journal of Environmental Science and Health, Part A 40(1) (2005) 167-182.

[151] D. Das, A. Pal, Adsolubilization phenomenon perceived in chitosan beads leading to a fast and enhanced malachite green removal, Chemical Engineering Journal 290 (2016) 371-380. https://doi.org/10.1016/j.cej.2016.01.062

[152] C. Lin, S. Wang, H. Sun, R. Jiang, Adsorption of anionic dye by anionic surfactant modified chitosan beads: Influence of hydrophobic tail and ionic head-group,

Journal of Dispersion Science and Technology 39(1) (2017) 106-115.
https://doi.org/10.1080/01932691.2017.1298041

[153] M.H. Mahaninia, L.D. Wilson, Cross-linked chitosan beads for phosphate removal from aqueous solution, Journal of Applied Polymer Science (2015) 1-10.

[154] U. Filipkowska, T. Jozwiak, P. Szymczyk, Application of cross-linked chitosan for phosphate removal from aqueous solutions, Progress on Chemistry and Application of Chitin and Its Derivatives XIX (2014) 5-14. https://doi.org/10.15259/PCACD.19.01

[155] M. Ahmad, S. Ahmed, B.L. Swami, S. Ikram, Preparation and characterization of antibacterial thiosemicarbazide chitosan as efficient Cu(II) adsorbent, (1879-1344 (Electronic)).

[156] M.A. Badawi, N.A. Nem, M.T.H. Abou Kana, H.H. Hefni, M.M. Abdel Moneem, Adsorption of aluminum and lead from wastewater by chitosan-tannic acid modified biopolymers Isotherms kinetics thermodynamics and process mechanism, International Journal of Biological Macromolecules 99 (2017) 405-476. https://doi.org/10.1016/j.ijbiomac.2017.03.003

[157] A.V. Pestov, O.V. Koryakova, I.I. Leonidov, Y.G. Yatluk, Gel-synthesis, structure, and properties of sulfur-containing chitosan derivatives, Russian Journal of Applied Chemistry 83(5) (2010) 787-794. https://doi.org/10.1134/S1070427210050058

[158] M. Ruiz, A.M. Sastre, E. Guibal, Palladium sorption on glutaraldehyde-crosslinked chitosan, Reactive and Functional Polymers 45(3) (2000) 155-173. https://doi.org/10.1016/S1381-5148(00)00019-5

[159] P. Chassary, T. Vincent, J. Sanchez Marcano, L.E. Macaskie, E. Guibal, Palladium and platinum recovery from bicomponent mixtures using chitosan derivatives, Hydrometallurgy 76(1) (2005) 131-147. https://doi.org/10.1016/j.hydromet.2004.10.004

[160] C. Gerente, V.K.C. Lee, P.L. Cloirec, G. McKay, Application of Chitosan for the Removal of Metals From Wastewaters by Adsorption—Mechanisms and Models Review, Critical Reviews in Environmental Science and Technology 37(1) (2007) 41-127. https://doi.org/10.1080/10643380600729089

[161] N. Sakkayawong, P. Thiravetyan, W. Nakbanpote, Adsorption mechanism of synthetic reactive dye wastewater by chitosan, Journal of colloid and interface science 286(1) (2005) 36-42. https://doi.org/10.1016/j.jcis.2005.01.020

[162] G.Z. Kyzas, D.N. Bikiaris, Recent modifications of chitosan for adsorption applications: a critical and systematic review, Marine drugs 13(1) (2015) 312-37. https://doi.org/10.3390/md13010312

[163] S.S. Silva, J.F. Mano, R.L. Reis, Ionic liquids in the processing and chemical modification of chitin and chitosan for biomedical applications, Green Chemistry 19(5) (2017) 1208-1220. https://doi.org/10.1039/C6GC02827F

[164] K. Ghandi, A Review of Ionic Liquids, Their Limits and Applications, Green and Sustainable Chemistry 04(01) (2014) 44-53. https://doi.org/10.4236/gsc.2014.41008

[165] Y. Liu, Y. Liu, T. Huo, X. Wu, J. Wei, P. Dong, D. Di, J. Wang, Y. Sun, Effect of the ionic liquid group in novel interpenetrating polymer networks on the adsorption properties for oleuropein from aqueous solutions, New Journal of Chemistry 39(12) (2015) 9181-9190. https://doi.org/10.1039/C5NJ01475A

[166] K.P. Eliodorio, V.S. Andolfatto, M.R.G. Martins, B.P. de Sá, E.R. Umeki, A. de Araújo Morandim-Giannetti, Treatment of chromium effluent by adsorption on chitosan activated with ionic liquids, Cellulose 24(6) (2017) 2559-2570. https://doi.org/10.1007/s10570-017-1264-3

[167] Y. Wei, W. Huang, Y. Zhou, S. Zhang, D. Hua, X. Zhu, Modification of chitosan with carboxyl-functionalized ionic liquid for anion adsorption, International Journal of Biological Macromolecules 62 (2013) 365-9. https://doi.org/10.1016/j.ijbiomac.2013.09.020

[168] A.S. Kumar, S. Sharma, R.S. Reddy, M. Barathi, N. Rajesh, Comprehending the interaction between chitosan and ionic liquid for the adsorption of Palladium, International Journal of Biological Macromolecules 72 (2015) 633-9. https://doi.org/10.1016/j.ijbiomac.2014.09.002

[169] L. Lupa, R. Voda, A. Popa, Adsorption behavior of cesium and strontium onto chitosan impregnated with ionic liquid, Separation Science and Technology 53(7) (2017) 1107-1115. https://doi.org/10.1080/01496395.2017.1313274

[170] F. Naseeruteen, N.S.A. Hamid, F.B.M. Suah, W.S.W. Ngah, F.S. Mehamod, Adsorption of malachite green from aqueous solution by using novel chitosan ionic liquid beads, International Journal of Biological Macromolecules 107(Pt A) (2018) 1270-1277.

[171] D. Ishii, C. Ohashi, H. Hayashi, Facile enhancement of the deacetylation degree of chitosan by hydrothermal treatment in an imidazolium-based ionic liquid, Green Chemistry 16(4) (2014) 1764-1767. https://doi.org/10.1039/c3gc41852a

Chitosan-Based Adsorbents for Wastewater Treatment, Ed. Abu Nasar   Materials Research Forum LLC
Materials Research Foundations **34** (2018) 203-229   doi: http://dx.doi.org/10.21741/9781945291753-9

Chapter 9

Role of Surfactants in Enhancing the Biosorption Capacity of Chitosan

Preeti Pal[1], Anjali Pal[2]*

[1]School of Environmental Science and Engineering, Indian Institute of Technology, Kharagpur, West Bengal - 721302, India

[2]Civil Engineering Department, Indian Institute of Technology, Kharagpur, West Bengal - 721302, India

*anjalipal@civil.iitkgp.ac.in

Abstract

This chapter highlights the properties and applications of chitosan and its derivatives for removal of heavy metals and dyes from aqueous solutions. It also explains the role of anionic surfactant (AS), sodium dodecyl sulfate (SDS) for enhancing the removal capacity of the chitosan. Different types of modifications can be done over the chitosan beads among which SDS modification is one of the best ways to modify the chitosan beads surface to obtain the maximum removal of heavy metals and dyes from the aqueous solutions. While SDS above its CMC is applied to the chitosan beads for the surface modification it is expected to form a bilayer over the surface and make it feasible to capture the contaminants preferably cations and cationic dyes, as the surface is negatively charged. This chapter compiles some of the recently published studies which show the application of SDS for making the aqueous solution free of contaminants.

Keywords

Marine Waste, Chitin and Chitosan, Anionic Surfactant, Sodium Dodecyl Sulfate, Heavy Metal, Dye, Adsorption

Contents

1.   Introduction of Chitin and Chitosan

Chitosan belongs to the family of polysaccharides which is obtained by partial deacetylation of chitin. Chitin is one of the most abundant renewable resources on Earth after cellulose. According to the report of The Hindu [1], 50 - 70% of raw material generated from sea food goes as waste. In India, along with Andhra Pradesh (AP) coast, approximately 2.5×10^5 tons marine food, including 3.0×10^4 tons of shrimp and 7.0×10^3 tons of crabs is produced annually. The report also gives the clue that nearly 8.5×10^6 tons of waste/year is generated by the shellfish processing industries and more than 1×10^5 tons of industrial waste is generated from shrimp processing activities. Munoz et al. [2] reported recently that general-purpose chitosan produced from Mahtani Chitosan Private Limited (on the coast of Gujarat) in India is 50 tons annually. Chitosan produced in India is used mainly for agricultural purpose, whereas, European producers entirely focuses on the good quality chitosan which can be used in the medical sector [2]. Chitosan contributes as a part of our food supply. To apply chitosan and its derivative in biomedical and pharmaceutical purpose, the purity of chitosan is of utmost importance. However, presently available usage is still less as compared to its potential, as chitosan has a wide range of application fields [3]. Chitosan is used in environmental remediation as it has the ability to remove pollutants from soil and water. It is used in sorption of inorganic pollutants such as heavy metals [4–11], organic pollutants like dyes [6,12–16], residue oil [17–19], pesticides and herbicides from agriculture-based wastewater [20–23] etc.

2.   Chitosan: waste to wealth

Yan and Chen [3] recently published a research paper where the researchers described that seafood waste is potentially very rich in protein content (20-40 %), calcium carbonate (20-50 %) and chitin (15-40 %). Seafood waste can be converted into a protein-rich feed for animals and can replace soya bean feed, which is less in protein content. Another very important commercial product that can be extracted from the crustacean shells is calcium carbonate. It is also reported by Munoz et al. [2] that, the production of 1 kg chitin requires 33 kg (wet weight) shrimp shells which resulted into the production of 1.5 kg calcium salts/kg chitin. Calcium carbonate is used in the pharmaceutical industries [3]. Authors also described some of the very important aspects which should be taken care of to improve the values of the crustacean waste. The production of chitin and chitosan from the waste materials uses corrosive and hazardous chemicals which cause the degradation of most of the valuable contents of the crustacean shells. Moreover, for the production of 1 kg of chitosan from shrimp shells, more than 1 ton of water is required which increase the cost of chitin up to \$200 per kg. Despite using the cheap starting material the cost of the final product is very high and that too with the use of non-environment friendly chemicals [3]. For the application of chitosan and its derivatives in biomedical and pharmaceuticals, the purity of chitosan matters which again increases the cost of the high-quality chitin and chitosan. Extremely pure chitosan is required for biomedical applications like tissue engineering and drug delivery. Many grades of chitosan are available based on its application. Wastewater treatment is the area of environmental remediation where chitosan is widely used nowadays [10,24].

2.1   Chitosan: synthesis and properties

Chitin can be used as a starting material to produce chitosan with a different degree of deacetylation. The production of chitosan from crustacean shells involves the basic steps: (i) pretreatment i.e., decalcification, (ii) deproteinization, (iii) demineralization, (iv) decoloration, and (v) deacetylation [25–27]. Each step was followed by thorough washing with water. A patent by Bristow [26] provided a detailed process for manufacturing medical grade chitosan without the need of decoloration step which is one of the most important steps in chitosan manufacturing processes. Exclusion of the decoloration step exempts the use of $KMnO_4$ and oxalic acid, which are the essential reagents used in this step [26]. A simplified flowchart of the general processes followed in the production of chitosan is shown in Fig. 1. The experimental conditions are also mentioned. Chitosan manufacturing process uses harsh chemicals like HCl and NaOH in highly concentrated form. The washing, which is mandatory after every use of acid or alkali, requires a huge quantity of water. Consumption of water is so large in the

Chitosan-Based Adsorbents for Wastewater Treatment, Ed. Abu Nasar   Materials Research Forum LLC
Materials Research Foundations **34** (2018) 203-229   doi: http://dx.doi.org/10.21741/9781945291753-9

manufacturing process of chitosan that, ultimately the cost of the end product is a lot more than the raw material [3]. For instance, Munoz et al. [2] reported that Mahtani Chitosan Pvt. Ltd. (Gujarat) uses 250 L of water along with 1.4 kg chitin, 5.18 kg NaOH, 1.06 kWh energy, 31 MJ wood fuel to produce 1 kg of chitosan. The used chemicals are not only detrimental to the environment, but they destroy the other components which could be extracted from the crustacean shells. A recent report of Yan and Chen [3] emphasized that, the proteins, calcium carbonate, which are also extractable from the crustacean shells, are lost due to the use of those chemicals. Chitosan can also be extracted enzymatically as shown in Fig. 1. This method is environment friendly. This also reduces the use of chemicals, and at the same time it involves less cost in the process.

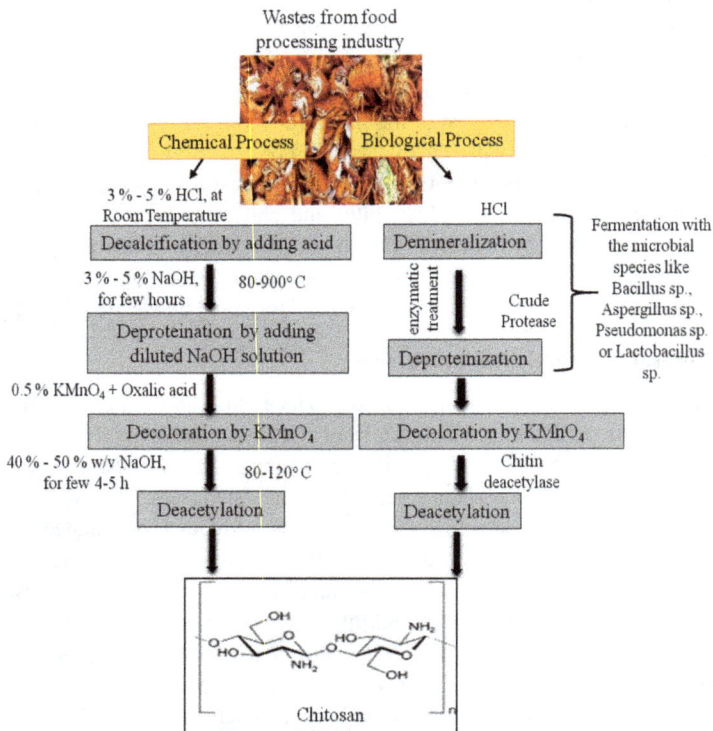

Figure 1. Flowchart of the general processes followed for the production of chitosan.

Chitin and chitosan both are attractive polymers because of their multidimensional properties [28]. Chitin is a white, hard, inelastic, nitrogenous, polysaccharide, which is made up of randomly distributed β-(1-4)-linked D-glucosamine and N-acetyl-D-glucosamine. Chitosan is naturally abundant and renewable polymer having properties such as biodegradability, biocompatibility, and non-toxicity. It can be used as a potential adsorbent. Chitosan and cellulose have similarity in the structure except that, chitosan has –NH₂ groups at C-2. Due to this, the reaction of chitosan is considerably more versatile than cellulose. Chitosan has unique properties such as solubility in various media, polyelectrolytic behavior, polyoxy salt formation, ability to form films, metal chelation, optical and structural characteristics etc. Chemical properties of chitosan depict that, it is a linear polyamine having active –NH₂ and -OH groups which can chelate many metal ions.

2.2   Chitosan for heavy metal removal

With the advancement of lifestyle, more pollutants are emanated in the environment. In the present era of technology, the industries which involve in the production of our daily stuff deliberately or inadvertently emanate the life-threatening contaminants in water, soil, and air. Wastewater coming out from different industries like zinc mining, smelting, paint manufacturing, cosmetics, battery manufacturing, alloys manufacturing, pesticides and other chemical industries contains heavy metals such as Cd, Hg, Zn, As, Fe, Mn, Mg, Co, Cr, Cu, and many other contaminants [29,30]. The contaminants can pose risks to human beings who work regularly on the contaminated site [6,31]. Some of the above metals/metalloids (essential elements) are required in trace amounts in the body for specific functions. But most of the elements mentioned are toxic to human [31] even at very low concentrations [6]. The contaminants are transferred from one place to another with rain water and ultimately they reach the water bodies. This cycle continues in an endless process. It is the prime necessity to make the discharging water free of all contaminants so that it will not pollute the rivers, ponds or other water bodies. Biosorption using chitosan is also independent of the physical form of the adsorbent. According to the applications, the physical form of the chitosan may be selected. Figure 2 shows some of the far and wide used forms of chitosan. For example, the recent research of Abdelaal et al. [32], described the successful preparation of the different derivatives of chitosan by treating it with ethyl cellulose, cellulose triacetate. The authors also modified the chitosan with glycidyltrimethyl ammonium chloride, phthalic anhydride, and succinic acid derivatives. All the obtained derivatives were then crosslinked with glutaraldehyde (GLA) and analyzed for the removal of transition metals from aqueous solutions [32]. Low molecular weight chitosan derivative of pyruvic acid was prepared by Boamah et al. [33] and thiocarbamoyl chitosan (TC-Chitosan) was prepared by Chauhan et al. [7]. Pyruvic acid derivative prepared by reaction of chitosan with pyruvic acid and then this derivative was crosslinked with GLA in order to increase the stability of the chitosan against acid or alkali or other chemicals [33]. Cd and Cr removal from wastewater emanating from electroplating industries were performed by the prepared adsorbent thiocarbamoyl chitosan (TC-Chitosan) [7]. Chen et al. [34,35] presented in their studies the preparation of novel type of composites of thiourea-modified magnetic ion-imprinted chitosan-TiO_2 and carbon disulfide-modified magnetic ion-imprinted chitosan-Fe(III), respectively. The former was named as MICT and later named as MMIC-Fe(III) composites. The prepared composites were used for the removal of Cd [34,35], tectracycline [35] and 2,4-dichlorophenol (2,4-DCP) [34]. Various other chitosan derivatives which can be used for the effective removal of contaminants from aqueous solutions are documented. These are poly (itaconic acid) grafted chitosan adsorbents [36], thiosemicarbazide-chitosan powder [5], surfactant-modified chitosan beads [8,15,37,38], eskom fly ash-chitosan composite [39] etc. Li et al. [4] prepared chitosan/sulfydryl-functionalized graphene oxide composites by the technique of covalent modification and electrostatic self-assembly. Surface imprinting technique was used by Lu et al. [40] for the preparation of ion-imprinted carboxymethyl-chitosan functionalized silica gel sorbent for removal of Cd. Biocomposite preparation by intercalation between chitosan and vermiculite using ultrasound irradiation was reported by Padilla-Ortega et al. [41]. Heidari et al. [42] reported the adsorptive removal of Cd, Pb and Ni metal ions from wastewater using chitosan-methacrylic acid (chitosan-MAA) nanoparticles.

3.   Structure and properties of surfactants

The surfactants are surface active agents which reduce the surface tension between two liquids or between a liquid and a solid. Surfactants have many fold applications because of their cleaning abilities, froth forming properties and micellization above critical micelle concentration (CMC). Surfactants consist of two parts: one is the hydrophobic tail and the other is the hydrophilic head. The tail part of most surfactants is similar, and it consists of a hydrocarbon chain. The tail part can be branched, linear, or aromatic. Surfactants which have two tails are called double-chained. On the basis of their polar head groups, surfactants are classified majorly as, non-ionic, cationic, anionic, and amphoteric or zwitterionic (as shown in Fig. 3). General structures those are formed by the surfactants when they come in contact with water are presented in Fig. 4. When surfactants are present at a lower concentration in aqueous solution they are present in the monomeric form. However, with an increase in concentration, they arrange themselves in the form of bilayer or micelle. Micelle formation in the solution reduces the system's free energy and the extreme point at which this condition exists is called as CMC. This

Chitosan-Based Adsorbents for Wastewater Treatment, Ed. Abu Nasar   Materials Research Forum LLC
Materials Research Foundations **34** (2018) 203-229   doi: http://dx.doi.org/10.21741/9781945291753-9

character of the surfactants gives them the antibacterial property, detergency and solubilization property [43]. Surfactants have taken their place in many of the industries including pharmaceutical, cosmetics, sanitization, textile, mining, leather industry etc. In pharmaceutical industry surfactants increase the efficiency of the drug/ingredient by either directly binding to the drug or it can increase the adsorption/absorption and partition of the drugs between the hydrophobic and hydrophilic components of the organisms [44].

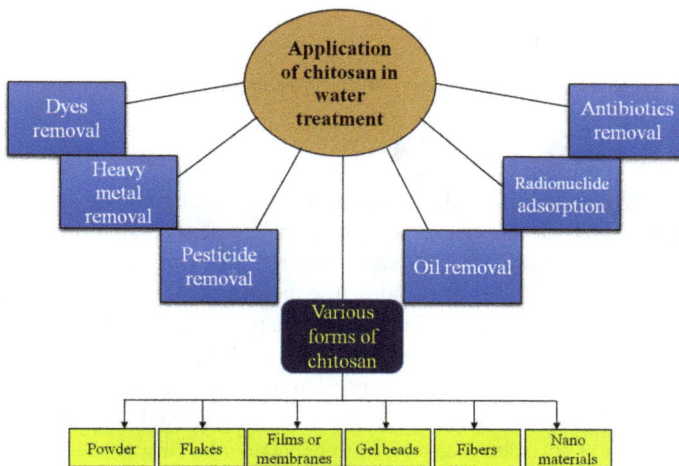

Figure 2. Applications of chitosan in water treatment, and its different forms.

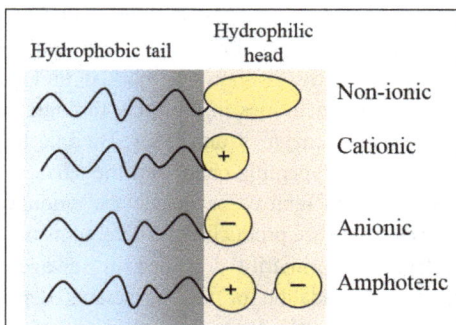

Figure 3. Classification of surfactants on the basis of their head group.

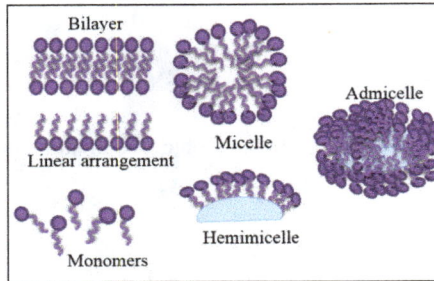

Figure 4. General structures formed by the surfactants when they come in contact with water.

Among all the detergents anionic surfactants (AS) are the oldest and most common types of detergent which are the major component in the detergent formulation. The global production of surfactants in 2006 was 12.5×10^6 tons, and the use of surfactants in Western Europe in 2007 was 3×10^6 tons. The production of surfactants is increasing day by day [43]. It is reported that 25% of total consumption of surfactants are used in Europe only, followed by North America (United States and Canada (22%)), and China (18%). The highest growth rate in the consumption is expected in China and other Asian countries [45]. The major use of surfactants is in the household activities and cosmetics. The application of surfactants also leads to generate a lot of waste surfactants which are being discharged into the water bodies. As a result, surfactants are going to the water, air, and soil which cause the unnecessary involvement of the surfactant to the ecosystem. Surfactants have the property to denature the proteins and DNA which makes it an important material to be used in biotechnology applications and research fields [44,46]. Once surfactant goes into the body, it shows a marked biological activity. Anionic surfactants can hinder the normal activities of the body through its binding to bioactive macromolecules such as peptides, enzymes, and DNA. Binding to proteins and peptides may change the folding of the polypeptide chain and the surface charge of a molecule [44]. Therefore, it is necessary to remove or reduce the amount of surfactants in the wastewater of the industries which are producing the detergents or using the surfactant as a raw material. Surfactants are primarily degraded in sewage treatment plants and through microbial activity. Degradation of the surfactants depends on the chemical properties of the surfactants as well as on the physicochemical conditions of the environment. Recent research of Pal et al. [16] has shown the utilization of chitosan beads for the removal of surfactants from the water. Some of the reports also suggested the productive use of surfactants in the removal of heavy metals by using the technique called micellar enhanced ultrafiltration [47–50]. The present article gives the compilation of recent researches which came up with the utilization of surfactants for the modification of the adsorbents (especially chitosan) which are used for the removal of contaminants from water and wastewater.

4. Interaction of surfactants with Chitosan

The modification or the interaction of the surfactants to any surface depends on the charge of the material. If the adsorbent surface is positively charged, then it can be linked with the anionic surfactants and if the surface is negatively charged then it can be modified with the cationic surfactants. Chitosan is a cationic polymer which can be easily modified with sodium dodecyl sulfate (SDS). However, to improve the adsorption capacity of the chitosan it is important to know the interaction of chitosan with the surfactants. Chitosan is biopolycation in acidic conditions (pK_a=6.5) and the fraction of N-acetyl glucosamine in chitosan determines its degree of deacetylation (DA) [51]. The complex formation in the system of oppositely charged polyelectrolytes and surfactants increases the entropy due to the release of small counter ions and water molecules. The surfactant binding to the polymer chain can be cooperative, non-cooperative or anti-cooperative in nature. The cooperative binding takes place if the completion of the binding process occurs in a narrow concentration range (above CMC). Anti-coopertaive binding occurs between the critical aggregation concentration (CAC) and CMC, and non-cooperative binding is observed when the concentration of surfactant is below CAC [46,52,53]. The complexes are useful in many applications due to the modification in their rheological properties and incorporation of new active agents. Applications such as in the field of cosmetics and pharmaceutical, drug delivery have the use of SDS/chitosan complexes because of their nontoxicity and degradable nature [51,53–55]. When surfactants are present in the form of aggregates an increase in the solubilization capacity (due to the presence of hydrophobic domains) as well as in the turbidity occurs.

The interactions between surfactant and polymer can be divided into "vertical" and "horizontal" forces. The vertical force between the surfactant and polymer is mostly electrostatic force but hydrophobic forces and hydrogen bonding also play a role. The horizontal interaction is among the surfactant tails which is mainly due to the hydrophobic and dispersion forces [51,56]. Wei and Hudson [56] reported that the degree of deacetylation (DA) and ionic strength also affects the interaction with the SDS. The authors concluded that an increase in DA decreases the co-operativity of the binding process which increases the spacing among the surfactant tails but it does not affect the vertical surfactant/glucosammonium interaction. Increase in ionic strength does not affect the co-operativity of the process. This indicates the electrostatic nature of the vertical interaction and hydrophobic origin of the co-operativity [56]. The complex mixture of surfactants and polyelectrolyte limits their use because of less solubility. Hence, efforts have been made to exploit their properties in various forms. When chitosan is added to the surfactant solution, or vice-versa (to make capsules), ionic cross-linking of the polymer chains by the surfactant micelles occurs. Chitosan–SDS emulsion capsules are not stable towards environmental stresses (temperature, pH or salt). The stability of these complexes can be largely enhanced by covering them with the layer of negatively charged biopolymers such as pectin. The structure of such complexes can be modified easily especially by varying the natural conditions of the materials (like pH, temperature). Modified forms can be exploited for the purpose of delivery, sequestration, water-based applications, catalyst preparation, etc. due to their biocompatibility [51]. It is evidenced that apart from chitosan, other materials like tea waste [57], iron oxide [58], iron humate [59], chitosan beads [8,15,37,38], alumina [60–64], wheat straw [65], laterite soil [66], etc. can also be modified using surfactants to improve the adsorption capacity of the material. Some of the recent researches by Pal and Pal [8,38], Das and Pal [15], Pal et al. [16], Kongarapu et al. [37] have suggested that the associative interactions between oppositely charged surfactants (such as AS) and polyelectrolyte chitosan gel beads lead to a modified scaffold, which is useful in environmental applications. Modification method is described in the next section.

5. Modification of Chitosan beads using surfactants

Chitosan is a widely used adsorbent for the removal of heavy metals. Among various forms of the chitosan, beads are now coming up with their advantage in diverse fields. A convenient method for the preparation of chitosan beads was reported by Mitani et al. [67]. The method is very popular because of its ease and simplicity. The process involves the mixing of 3 g (w/v) chitosan powder per 250 mL solution containing the 7 % glacial acetic acid. The prepared chitosan solution is then stirred for 4-5 h to make the solution bubble-free. The same solution was then added drop-wise to the coagulating mixture containing MeOH:NaOH:H$_2$O (5:1:4) (wt/wt). This leads to bead formation. Prepared beads were washed thoroughly to bring the pH to neutral. This was followed by the modification of the chitosan (CS) beads by an AS, SDS, in a simple way [8,15,37,38]. In this method, the chitosan beads were dipped in the SDS solution at a particular concentration and left undisturbed for 3-5 days to adsorb maximum SDS. The SDS solution was prepared separately with different concentrations as required. After equilibrium (3-4 days) the SDS remaining in the solution was measured by

Chitosan-Based Adsorbents for Wastewater Treatment, Ed. Abu Nasar   Materials Research Forum LLC
Materials Research Foundations 34 (2018) 203-229   doi: http://dx.doi.org/10.21741/9781945291753-9

spectrophotometric method [68]. The modified chitosan beads were then utilized for the removal of heavy metals or dyes from aqueous solutions and for other applications. For instance, Kongarapu et al. [37] used modified chitosan beads for removal of Ni^{2+} followed by the formation Ni@NiO core-shell nanoparticles for 4-nitrophenol reduction. Das and Pal [15] also utilized SDS for modification of chitosan beads with the concentrations of 600 and 3000 mg/L for the removal of malachite green (MG). Pal and Pal [8,38] exploited the SDS (concentration = 3000 mg/L) modified chitosan (SMCS) beads for the removal of Cd^{2+} and Pb^{2+}. Scanning electron microscopic (SEM) images of CS and SMCS beads are shown in Fig. 5 (a) and (b), respectively. SDS flakes can be seen clearly over the SMCS beads.

Synergistic effects on the adsorption of SDS and crystal violet (CV) was also reported by Pal et al. [16]. Lin et al. [69] reported recently the chitosan hydrogel beads modification by anionic surfactants (SDS, sodium dodecyl sulfonate (SDOS), sodium dodecylbenzenesulfonate (SDBS), dioctyl sulfosuccinate sodium salt (AOT), and N,N'-ethylene-bis[N (sodium ethylenesulfonate)-dodecanamide] (DTM-12)) and their use for adsorptive removal of congo red (CR) from aqueous solutions. It is one of the emerging techniques to modify the adsorbent material with the surfactant to improve the adsorption properties of the material. For example, attapulgite which is a crystalline hydrated magnesium aluminum silicate has permanent negative charges on its surface and has a fibrous morphology, which enables it to be modified by cationic surfactants to enhance contaminant retention. Asgari et al. [70] used a cationic surfactant, hexadecyltrimethyl ammonium bromide (HDTMA) to functionalize pumice for removal of fluoride from drinking water [70]. The surfactant-modified pumice resulted in the removal of over 96% of fluoride from a solution having concentration 10 mg/L. The adsorbent dose applied was 0.5 g/L, medium pH was 6.0 and the contact time was 30 min. The maximum adsorption capacity obtained was 41.0 mg/g [70].

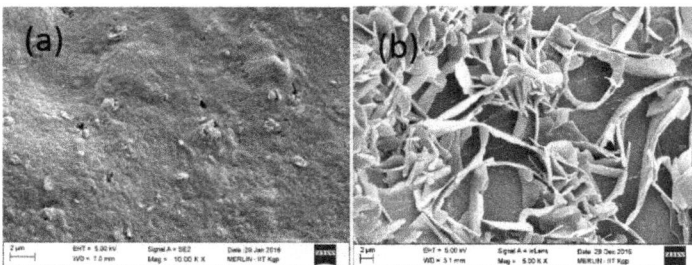

Figure 5. SEM images of CS and SMCS beads.

6.   Removal of heavy metals and dyes using surfactant-modified Chitosan beads

The complex formation between the surfactant and polyelectrolyte occurs when they are oppositely charged. Chitosan in complexing with the surfactants shows the synergistic effect which may be helpful in the food industry, drug delivery, textile industry [71] and for the removal of other contaminants like dyes [16]. The first report of the synergistic effect of the SDS on crystal violet (CV) removal from water was published by Pal et al. [16]. The experiments were conducted to remove the SDS from aqueous solution using the CS beads and it was found that with the dose of 0.5 g/L, 95-99% of SDS was removed from the solution having a concentration in the range of 10-20 mg/L. Maximum adsorption capacity obtained was 76.9 mg/g. Moreover, crystal violet removal was also analyzed when both are present together. Interesting results were obtained revealing that, both the adsorbates have a synergistic effect on the removal of each other. For instance, at 10 mg/L initial concentration of CV and SDS, while present together, the percent removal of CV was 100% with the contact time of 24 h while it was 60% in absence of SDS. Similar results were obtained in case of SDS removal in presence of CV. The SDS removal (C_0=10 mg/L) in presence of CV was 100%, however, it was only 70% in absence of CV. This type of synergism is very beneficial for the treatment of industrial wastewater [16].

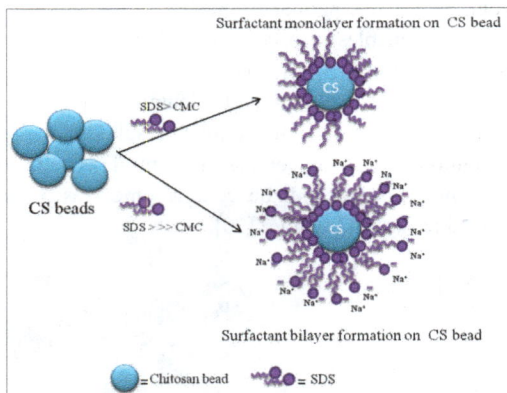

Figure 6. Schematic representation of monolayer and bilayer formation on CS beads.

Kongarapu et al. [37] have published a recent research where SDS was used for the surface modification of CS beads. The surfactant-modified chitosan beads were used for the successful removal of Ni^{2+} and later nickel loaded beads were used for the synthesis of Ni@NiO catalyst to be used for the 4-nitrophenol (4-NP) reduction. The concentration of SDS used for the modification of CS beads was 1500 (pre-micellar) and 6000 (post-micellar) mg/L and the surfactant-modified chitosan beads were designated as PRECS and POSTCS beads, respectively. Adsorption of SDS occurred on the CS beads to create SDS bilayer (admicelle). These beads were called as the POSTCS beads. Whereas, when the SDS monolayer are formed on the CS beads (hemi micelle), those beads were named as PRECS beads. A schematic of the monolayer and bilayer formation on CS beads is presented in Fig. 6.

CS, PRECS and POSTCS beads were evaluated for their adsorption capacity for Ni^{2+} removal. It was found that the adsorption capacity of POSTCS beads for Ni^{2+} removal was 37.82 mg/g, for PRECS beads it was 18.56 mg/g and for normal CS beads, it was 10.0 mg/g. The enhanced removal of Ni^{2+} by POSTCS beads was due to the presence of surfactant bilayer which is believed to provide the adsolubilization of Ni^{2+} due to the electrostatic interaction between the anionic head group of SDS and the positively charged nickel ion. The POSTCS beads which were highly loaded with Ni^{2+} were used for the Ni@NiO catalyst preparation in presence of sodium borohydride ($NaBH_4$) which reduced the Ni^{2+} to form Ni(0). The average core size of Ni(0) was found to be ~100 nm and the shell thickness was 25 nm as predicted by the TEM analysis. Catalytic reduction of 4-NP to 4-aminophenol (4-AP) followed four steps which are adsorption of reactant molecules (i.e. 4-NP and $NaBH_4$) to the catalyst surface, surface complexation at the active site of the catalyst, reduction of 4-NP to 4-AP in presence of $NaBH_4$, and desorption of 4-AP from the catalyst surface. Effect of dose, contact time, initial concentration of the 4-NP, temperature was evaluated and it was found that at lower concentration the reaction followed zero order kinetics and at higher concentration (1.0×10^{-4} M) the reaction followed first order kinetics. The study on the effect of temperature (293, 303 and 313 K) was done at 4-NP concentration of 1×10^{-4} M with the catalyst dose of 2.0 g/L and $NaBH_4$ concentration 0.1 M. It was observed that the first order rate constant decreased with increase in temperature. The activation energy of the catalytic reaction was found to be negative in the present study. The turn over number (TON) and turn over frequency (TOF) of the catalyst at a dose of 2.0 g/L with the 4-NP concentration in presence of 0.1 M $NaBH_4$ was found to be 3.0115×10^{19} molecules/g and 5.019×10^{16} molecules/g/s, respectively [37].

Chitosan–surfactant–core–shell (CSCS), surfactant–polyelectrolyte-complex (SPEC) and CS–SDS composite material (CSC) were prepared by Das and Pal [15] for the adsorptive removal of MG. The complex forming ability of surfactant with counter ion polyelectrolyte (i.e., chitosan) is utilized to fabricate the above said adsorbents. The CSCS beads which were modified with the 3000 mg/L SDS concentration were found to give the best adsorption capacity for MG which is 360 mg/g. The SDS concentration used for the preparation of adsorbent was >>CMC which is believed to form bilayer over the surface of chitosan beads. With 200-350 mg/L initial concentration of MG, the removal was 50–60% only due to the limited number of active sites on the adsorbent, whereas with the initial concentration of 10 and 100 mg/L 80–90 % removal was achieved. This idea of a modification of CS beads with the SDS concentration>>CMC was also followed by Pal and Pal [8] in their recent research where authors have reported the visual change (Fig. 7 (a, b)) in the physical appearance of SMCS beads after Cd^{2+} removal. SMCS beads after Cd^{2+} removal becomes white and opaque, while the CS beads are more transparent.

Figure 7. Physical appearance of SMCS beads before (a) and after (b) Cd^{2+} removal.

The comparative study of CS and SMCS beads for Cd^{2+} removal from synthetic wastewater showed that SMCS beads showed 3-fold higher adsorption capacity than that of CS beads. For instance, the adsorption capacity of the SMCS beads was 108.72 mg/g for 100 mg/L of Cd^{2+}, while it was 37.57 mg/g in case of CS beads with the dose of 0.45 g/L at a contact time of 10 h. Fourier transformed infrared (FTIR) and SEM analysis showed a successful modification of CS beads. FTIR spectra showed the emergence of some new peaks as well as shifting of some peaks due to the ionic interaction of the SDS bilayer and Cd^{2+}. Isotherm data followed Langmuir adsorption model which gives the clue that the adsorption is monolayer which occurred due to the ionic interaction or chemisorption. Kinetic data complied with pseudo-second-order model. The maximum adsorption capacity or Langmuir adsorption capacity (Q_{max}) of SMCS beads obtained for Cd^{2+} removal was 125 mg/g. The above prepared SMCS beads could also be used for the removal of Pb^{2+} in the range of 10-100 mg/L as reported by Pal and Pal [38]. Apart from the Pb^{2+} analysis by AAS, the confirmation was also done by SEM and FTIR spectrum analysis which gave the assurance of Pb^{2+} adsorption onto SMCS beads. Authors explained that the removal efficiency using SMCS beads was similar to CS beads at lower concentrations (10-30 mg/L) of lead. However, at higher concentration of Pb^{2+} (30-100 mg/L), the removal percentage by SMCS beads was 3 times higher than that of CS beads. Adsorption of Pb^{2+} on to SMCS beads showed to follow pseudo-second-order kinetics and equilibrium data complied with the Langmuir isotherm model. Isotherm study here was done by varying the dose of adsorbent from 0.09 to 1.13 g/L, and the maximum adsorption capacity obtained was 100.0 mg/g.

Some of the studies are also reported which showed the complexing of chitosan with the surfactants by mixing it in the solution phase and then reformed the hydrogel beads. This also resulted in the increment of the adsorption capacity of the adsorbent due to the addition of surfactants. Chatterjee et al. [72] reported the enhanced adsorption capacity of chitosan beads impregnated by a cationic surfactant (CTAB; cetyl trimethyl ammonium bromide) for CR removal. The ratio of CTAB:CS was optimized as 0.05:1 (wt %) which increased the adsorption capacity of CS/CTAB beads 2.2 times to that of CS beads. For instance, the adsorption capacity for CS beads was 162.3 mg/g (without CTAB impregnation) which increased to 352.5 mg/g when it was mixed with 0.05% CTAB when the initial concentration of CR was taken as 500 mg/L. Isotherm data showed a better fit to the Sips isotherm model and kinetic data followed pseudo-second-order rate model. Langmuir adsorption capacity obtained for CR removal by CS/CTAB was 385.9 mg/g whereas it was 182.7 mg/g for CS beads. Authors explained that this enhancement of adsorption capacity may be partly due to hydrophobic interactions between the hydrophobic tail of CTAB and hydrophobic moieties of CR. Also, the increase in porosity contributes to the increase in removal of CR [72]. This is also supported by the study of Seredych et al. [73] which reports that the amount of adsorbate adsorbed on to the adsorbent depends on its porosity [72,73].

Chatterjee et al. [74] prepared the chitosan hydrogel beads by gelation with the SDS (designated as CSB), which were used for the removal of methylene blue (MB) from aqueous solutions with the initial concentration of 100 mg/L. The concentration of SDS was varied to get the maximum adsorption capacity with the CSB and it was observed that the optimum equilibrium adsorption capacity was maximum while 4 g/L of SDS was used for ionic gelation with chitosan beads (CB). The adsorption isotherm data best fitted with the Sips isotherm model, which resulted in the maximum adsorption capacity of 226.24 mg/g; which was higher than that of CB (Q_{max}=99.01 mg/g). All the experiments were performed with the adsorbent dose of 20.0 g/L (wet weight of the beads) at 30°C keeping pH at 6.0 [74]. Recently, Lin et al. [69] reported the adsorptive removal of CR using chitosan beads modified with different anionic surfactants viz., SDS, SDOS, SDBS, AOT, and DTM-12. The comparative results for CS and all surfactant-modified beads revealed that CS beads had much less adsorption efficiency compared to that of modified beads. High adsorption capacity was observed for CS/AOT beads and CS/DTM-12 beads which are having two hydrophobic tails. Adsorption isotherm studies showed that Sips isotherm model was better followed and the adsorption process was heterogeneous. Authors suggested that the surfactant concentration, surfactant ionic head-group, and surfactant hydrophobic tail were crucial for the effective adsorption of CR from aqueous solution. Effect of the ionic head group was evaluated by modifying the CS beads with SDBS, SDOS and SDS, and it was found that the adsorption capacities for CR removal were in the order of SDBS/CS beads >SDOS/CS beads >SDS/CS beads. Despite the same tail length, the adsorption capacity differs probably due to the miceller solubilization. Effect of the hydrophobic tail was examined using AOT (with single head-group and double tails per molecule) and DTM-12 (with two head-groups and two tails per molecule). The AOT/CS and DTM-12/CS beads both showed higher adsorption capacities which were 1783.44 and 1732.89 mg/g, respectively. This increment in the adsorption capacity than that of SDOS/CS, SDS/CS beads was due to the presence of more than one hydrophobic tails [69]. Efforts have been made to compile up the studies which came up recently to improve the chitosan adsorption capacity by surfactant modification with it. Table 1 shows the comparison in adsorption capacity with the raw material and after modification with a surfactant.

Table 1.   Adsorption capacities of some of the raw materials and their surfactant-modified forms

Raw Material	Designated names of the adsorbents prepared by surfactant modification	Contaminant removed	Adsorption capacity (mg/g)	Ref.
Chitosan beads	CS beads	Ni^{2+}	10.0	[37]
Surfactant-modified chitosan beads	PRECS beads		18.56	
	POSTCS beads		37.82	
Chitosan beads	CS beads	MG	171.35	[15]
Surfactant-modified chitosan beads	CSC		239.47	
	SPEC		352.52	
	CSCS		359.42	
	SMCS beads	Cd^{2+}	125.0	[8]
	SMCS beads	Pb^{2+}	100.0	[38]
Chitosan beads	CS beads	CR	162.32	[72]

Surfactant-modified chitosan beads	CS/CTAB beads	CR	352.5	
Chitosan beads	CB	MB	99.01	[74]
Surfactant-modified chitosan beads	CSB		226.24	
Surfactant-modified chitosan beads	CS/SDS beads	CR	1490.65	[69]
	CS/SDOS beads		1539.98	
	CS/SDBS beads		1637.58	
	CS/AOT beads		1766.20	
	CS/DTM-12 beads		1732.89	
Alumina	Surfactant-modified alumina (SMA)	CV	111.6	[60]
		Phenol	6.64	[61]
		Mn^{2+}	2.04	[63]
		Cu^{2+}	9.84	[64]
		Ni^{2+}	6.87	
Raw wheat straw	Wheat straw	MB	55.0	[65]
Surfactant-modified wheat straw	SDS modified wheat straw		126.6	
Tea waste	CTAB-TW	CR	106.4	[57]
Laterite soil	SDS-modified laterite soil	Cu^{2+}	185.0	[66]
Hydrotalcite-iron oxide magnetic organocomposite	HT-DS/Fe	MB	110.05	[58]
	HT-DSB/Fe		94.69	

7. Conclusions

Surfactants have a very important role in our day to day life. Because of their immense use, they are becoming part of wastewater inadvertently along with the heavy metals and

Chitosan-Based Adsorbents for Wastewater Treatment, Ed. Abu Nasar   Materials Research Forum LLC
Materials Research Foundations **34** (2018) 203-229   doi: http://dx.doi.org/10.21741/9781945291753-9

dyes. Here, we have summarized some of the recent researches which focused on the smart use of surfactant for the modification of chitosan and some other biomaterials and their subsequent use for contaminant removal from aqueous solutions. It has been found that the micelle-forming capability of the surfactants shows a synergistic effect for the removal of contaminants. Also, above the CMC, surfactant form bilayer/admicelles over the solid surfaces which are able to solubilize the contaminants from the aqueous solutions.

Acknowledgements

The authors are thankful to IIT Kharagpur for providing facilities to carry out this research.

References

[1]   Information on http://www.thehindu.com/todays-paper/tp-national/tp-andhrapradesh/fishery-waste-offers-huge-business-potential/article2561395.ece

[2]   I. Munoz, C. Rodriguez, D. Gillet, B. M. Moerschbacher, Life cycle assessment of chitosan production in India and Europe, Int. J. Life Cycle Assess. (2017) 1–10. https://doi.org/10.1007/s11367-017-1290-2

[3]   N. Yan, X. Chen, Don't waste seafood waste, Nature. 524 (2015) 155–157. https://doi.org/10.1038/524155a

[4]   X. Li, H. Zhou, W. Wu, S. Wei, Y. Xu, Y. Kuang, Studies of heavy metal ion adsorption on chitosan/sulfydryl-functionalized graphene oxide composites, J. Colloid Interface Sci. 448 (2015) 389–397. https://doi.org/10.1016/j.jcis.2015.02.039

[5]   M. Li, Z. Zhang, R. Li, J.J. Wang, A. Ali, Removal of Pb(II) and Cd(II) ions from aqueous solution by thiosemicarbazide modified chitosan, Int. J. Biol. Macromol. 86 (2016) 876–884. https://doi.org/10.1016/j.ijbiomac.2016.02.027

[6]   H.K. No, S.P. Meyers, Application of chitosan for treatment of wastewaters, Rev. Environ. Contam. Toxicol. 163 (2000) 1–28. https://doi.org/10.1007/978-1-4757-6429-1_1

[7]   D. Chauhan, M. Jaiswal, N. Sankararamakrishnan, Removal of cadmium and hexavalent chromium from electroplating waste water using thiocarbamoyl chitosan, Carbohydr. Polym. 88 (2012) 670–675. https://doi.org/10.1016/j.carbpol.2012.01.014

[8]   P. Pal, A. Pal, Surfactant-modified chitosan beads for cadmium ion adsorption, Int. J. Biol. Macromol. 104 (2017) 1548–1555. https://doi.org/10.1016/j.ijbiomac.2017.02.042

[9]   E. Igberase, P. Osifo, Equilibrium, kinetic, thermodynamic and desorption studies of cadmium and lead by polyaniline grafted cross-linked chitosan beads from aqueous solution, J. Ind. Eng. Chem. 26 (2015) 340–347. https://doi.org/10.1016/j.jiec.2014.12.007

[10] S.M. Nomanbhay, K. Palanisamy, Removal of heavy metal from industrial wastewater using chitosan coated oil palm shell charcoal, Electron. J. Biotechnol. 8 (2005) 43–53. https://doi.org/10.2225/vol8-issue1-fulltext-7

[11] P.O. Osifo, The use of chitosan beads for the adsorption and regeneration of heavy metals, North-West University, 2007.

[12] W.S. Wan Ngah, L.C. Teong, M.A.K.M. Hanafiah, Adsorption of dyes and heavy metal ions by chitosan composites: A review, Carbohydr. Polym. 83 (2011) 1446–1456. https://doi.org/10.1016/j.carbpol.2010.11.004

[13] G. Crini, P.M. Badot, Application of chitosan, a natural aminopolysaccharide, for dye removal from aqueous solutions by adsorption processes using batch studies: A review of recent literature, Prog. Polym. Sci. 33 (2008) 399–447. https://doi.org/10.1016/j.progpolymsci.2007.11.001

[14] K. Azlan, W.N. Wan Saime, L. Lai Ken, Chitosan and chemically modified chitosan beads for acid dyes sorption, J. Environ. Sci. 21 (2009) 296–302. https://doi.org/10.1016/S1001-0742(08)62267-6

[15] D. Das, A. Pal, Adsolubilization phenomenon perceived in chitosan beads leading to a fast and enhanced malachite green removal, Chem. Eng. J. 290 (2016) 371–380. https://doi.org/10.1016/j.cej.2016.01.062

[16] A. Pal, S. Pan, S. Saha, Synergistically improved adsorption of anionic surfactant and crystal violet on chitosan hydrogel beads, Chem. Eng. J. 217 (2013) 426–434. https://doi.org/10.1016/j.cej.2012.11.120

[17] A.L.Ã. Ahmad, S. Sumathi, B.H. Hameed, Adsorption of residue oil from palm oil mill effluent using powder and flake chitosan: Equilibrium and kinetic studies, Water Res. 39 (2005) 2483–2494. https://doi.org/10.1016/j.watres.2005.03.035

[18] P.F. Rupani, P.S. Rajeev, M.H. Irahim, N. Esa, Review of current palm oil mill effluent (POME) treatment methods: Vermicomposting as a sustainable practice, World Appl. Sci. J. 11 (2010) 70–81.

[19] A.R. Gentili, M.A. Cubitto, M. Ferrero, M.S. Rodriguéz, Bioremediation of crude oil polluted seawater by a hydrocarbon-degrading bacterial strain immobilized on chitin and chitosan flakes, Int. Biodeterior. Biodegradation. 57 (2006) 222–228. https://doi.org/10.1016/j.ibiod.2006.02.009

[20] S. Moradi Dehaghi, B. Rahmanifar, A.M. Moradi, P.A. Azar, Removal of permethrin pesticide from water by chitosan–zinc oxide nanoparticles composite

as an adsorbent, J. Saudi Chem. Soc. 18 (2014) 348–355.
https://doi.org/10.1016/j.jscs.2014.01.004

[21] B. Rahmanifar, S. Moradi Dehaghi, Removal of organochlorine pesticides by chitosan loaded with silver oxide nanoparticles from water, Clean Technol. Environ. Policy. 16 (2013) 1781–1786. https://doi.org/10.1007/s10098-013-0692-5

[22] M. Agostini de Moraes, D.S. Cocenza, F. da Cruz Vasconcellos, L.F. Fraceto, M.M. Beppu, Chitosan and alginate biopolymer membranes for remediation of contaminated water with herbicides, J. Environ. Manage. 131 (2013) 222–227. https://doi.org/10.1016/j.jenvman.2013.09.028

[23] R.T.A. Carneiro, T.B. Taketa, R.J. Gomes Neto, J.L. Oliveira, E.V.R. Campos, M.A. de Moraes, C.M.G. da Silva, M.M. Beppu, L.F. Fraceto, Removal of glyphosate herbicide from water using biopolymer membranes, J. Environ. Manage. 151 (2015) 353–360. https://doi.org/10.1016/j.jenvman.2015.01.005

[24] A. Rathinam, B. Maharshi, S.K. Janardhanan, R.R. Jonnalagadda, B.U. Nair, Biosorption of cadmium metal ion from simulated wastewaters using Hypnea valentiae biomass: A kinetic and thermodynamic study, Bioresour. Technol. 101 (2010) 1466–1470. https://doi.org/10.1016/j.biortech.2009.08.008

[25] A.G. Hadi, Synthesis of chitosan and its use in metal removal, Chem. Mater. Res. 3 (2013) 22–26.

[26] J. Bristow, U.S. Patent 8,318,913 (2012).

[27] P.K. Dutta, M.N. V. Ravikumar, J. Dutta, Chitin and chitosan for versatile applications, J. Macromol. Sci. Part C Polym. Rev. 42 (2002) 307–354. https://doi.org/10.1081/MC-120006451

[28] C.K.S. Pillai, W. Paul, C.P. Sharma, Chitin and chitosan polymers: Chemistry, solubility and fiber formation, Prog. Polym. Sci. 34 (2009) 641–678. https://doi.org/10.1016/j.progpolymsci.2009.04.001

[29] S.E.A. Sharaf El-Deen, F.S. Zhang, Immobilisation of TiO_2-nanoparticles on sewage sludge and their adsorption for cadmium removal from aqueous solutions, J. Exp. Nanosci. 11 (2016) 239–258. https://doi.org/10.1080/17458080.2015.1047419

[30] A.K.A. Rathi, S.A. Puranik, Chemical industry wastewater treatment using adsorption, Ind. Res. 61 (2002) 53–60.

[31] N. Sankararamakrishnan, A.K. Sharma, R. Sanghi, Novel chitosan derivative for the removal of cadmium in the presence of cyanide from electroplating wastewater, J. Hazard. Mater. 148 (2007) 353–359. https://doi.org/10.1016/j.jhazmat.2007.02.043

[32] M.Y. Abdelaal, T.R. Sobahi, H.F. Al-Shareef, Modification of chitosan derivatives of environmental and biological interest: A green chemistry approach, Int. J. Biol. Macromol. 55 (2013) 231–239. https://doi.org/10.1016/j.ijbiomac.2013.01.013

[33] P.O. Boamah, Y. Huang, M. Hua, Q. Zhang, Y. Liu, J. Onumah, W. Wang, Y. Song, Removal of cadmium from aqueous solution using low molecular weight chitosan derivative., Carbohydr. Polym. 122 (2015) 255–264. https://doi.org/10.1016/j.carbpol.2015.01.004

[34] A. Chen, G. Zeng, G. Chen, X. Hu, M. Yan, S. Guan, C. Shang, L. Lu, Z. Zou, G. Xie, Novel thiourea-modified magnetic ion-imprinted chitosan/TiO$_2$ composite for simultaneous removal of cadmium and 2,4-dichlorophenol, Chem. Eng. J. 191 (2012) 85–94. https://doi.org/10.1016/j.cej.2012.02.071

[35] A. Chen, C. Shang, J. Shao, Y. Lin, S. Luo, J. Zhang, H. Huang, M. Lei, Q. Zeng, Carbon disulfide-modified magnetic ion-imprinted chitosan-Fe(III): A novel adsorbent for simultaneous removal of tetracycline and cadmium, Carbohydr. Polym. 155 (2017) 19–27. https://doi.org/10.1016/j.carbpol.2016.08.038

[36] G.Z. Kyzas, P.I. Siafaka, D.A. Lambropoulou, N.K. Lazaridis, D.N. Bikiaris, Poly (itaconic acid)-grafted chitosan adsorbents with different cross-linking for Pb(II) and Cd(II) uptake, Langmuir. 30 (2014) 120–131. https://doi.org/10.1021/la402778x

[37] R.J. Kongarapu, P. Mahamallik, A. Pal, Surfactant modification of chitosan hydrogel beads for Ni@NiO core-shell nanoparticles formation and its catalysis to 4-nitrophenol reduction, J. Environ. Chem. Eng. 5 (2017) 1321–1329. https://doi.org/10.1016/j.jece.2017.02.017

[38] P. Pal, A. Pal, Enhanced Pb^{2+} removal by anionic surfactant bilayer anchored on chitosan bead surface, J. Mol. Liq. 248 (2017) 713–724. https://doi.org/10.1016/j.molliq.2017.10.103

[39] S. Pandey, S. Tiwari, Facile approach to synthesize chitosan based composite-characterization and cadmium(II) ion adsorption studies, Carbohydr. Polym. 134 (2015) 646–656. https://doi.org/10.1016/j.carbpol.2015.08.027

[40] H. Lü, H. An, Z. Xie, Ion-imprinted carboxymethyl chitosan-silica hybrid sorbent for extraction of cadmium from water samples, Int. J. Biol. Macromol. 56 (2013) 89–93. https://doi.org/10.1016/j.ijbiomac.2013.02.003

[41] E. Padilla-Ortega, M. Darder, P. Aranda, R. Figueredo Gouveia, R. Leyva-Ramos, E. Ruiz-Hitzky, Ultrasound assisted preparation of chitosan-vermiculite bionanocomposite foams for cadmium uptake, Appl. Clay Sci. 130 (2016) 40–49. https://doi.org/10.1016/j.clay.2015.11.024

[42] A. Heidari, H. Younesi, Z. Mehraban, H. Heikkinen, Selective adsorption of Pb(II), Cd(II), and Ni(II) ions from aqueous solution using chitosan-MAA

nanoparticles, Int. J. Bol. Macromol. 61 (2013) 251–263.
https://doi.org/10.1016/j.ijbiomac.2013.06.032

[43] G.G. Ying, Fate, behavior and effects of surfactants and their degradation products in the environment, Environ. Int. 32 (2006) 417–431.
https://doi.org/10.1016/j.envint.2005.07.004

[44] T. Ivanković, J. Hrenović, Surfactants in the environment, Arh. Hig. Rada Toksikol. 61 (2010) 95–110. https://doi.org/10.2478/10004-1254-61-2010-1943

[45] Information on https://www.ihs.com/products/chemical-surfactants-scup.html.

[46] E. Guzmán, S. Llamas, A. Maestro, L. Fernández-Peña, A. Akanno, R. Miller, F. Ortega, R.G. Rubio, Polymer-surfactant systems in bulk and at fluid interfaces, Adv. Colloid Interface Sci. 233 (2016) 38–64.
https://doi.org/10.1016/j.cis.2015.11.001

[47] K. Xu, G. Zeng, J. Huang, J. Wu, Y. Fang, G. Huang, J. Li, B. Xi, H. Liu, Removal of Cd^{2+} from synthetic wastewater using micellar-enhanced ultrafiltration with hollow fiber membrane, Colloids Surfaces A Physicochem. Eng. Asp. 294 (2007) 140–146. https://doi.org/10.1016/j.colsurfa.2006.08.017

[48] R.S. Juang, Y.Y. Xu, C.L. Chen, Separation and removal of metal ions from dilute solutions using micellar-enhanced ultrafiltration, J. Memb. Sci. 218 (2003) 257–267. https://doi.org/10.1016/S0376-7388(03)00183-2

[49] V.D. Karate, K.V. Marathe, Simultaneous removal of nickel and cobalt from aqueous stream by cross flow micellar enhanced ultrafiltration, J. Hazard. Mater. 157 (2008) 464–471. https://doi.org/10.1016/j.jhazmat.2008.01.013

[50] J.H. Huang, G.M. Zeng, C.F. Zhou, X. Li, L.J. Shi, S.B. He, Adsorption of surfactant micelles and Cd^{2+}/Zn^{2+} in micellar-enhanced ultrafiltration, J. Hazard. Mater. 183 (2010) 287–293. https://doi.org/10.1016/j.jhazmat.2010.07.022

[51] L. Chiappisi, M. Gradzielski, Co-assembly in chitosan-surfactant mixtures: Thermodynamics, structures, interfacial properties and applications, Adv. Colloid Interface Sci. 220 (2015) 92–107. https://doi.org/10.1016/j.cis.2015.03.003

[52] T. Nylander, Y. Samoshina, B. Lindman, Formation of polyelectrolyte-surfactant complexes on surfaces, Adv. Colloid Interface Sci. 123–126 (2006) 105–123.
https://doi.org/10.1016/j.cis.2006.07.005

[53] C.D. Bain, P.M. Claesson, D. Langevin, R. Meszaros, T. Nylander, C. Stubenrauch, S. Titmuss, R. von Klitzing, Complexes of surfactants with oppositely charged polymers at surfaces and in bulk, Adv. Colloid Interface Sci. 155 (2010) 32–49. https://doi.org/10.1016/j.cis.2010.01.007

[54] D. Langevin, Complexation of oppositely charged polyelectrolytes and surfactants in aqueous solutions. A review, Adv. Colloid Interface Sci. 147–148 (2009) 170–177. https://doi.org/10.1016/j.cis.2008.08.013

[55]   L. Chiappisi, I. Hoffmann, M. Gradzielski, Complexes of oppositely charged polyelectrolytes and surfactants-recent developments in the field of biologically derived polyelectrolytes, Soft Matter. 9 (2013) 3896-3909. https://doi.org/10.1039/c3sm27698h

[56]   Y.C. Wei, S.M. Hudson, Binding of sodium dodecyl sulfate to a polyelectrolyte based on chitosan, Macromolecules. 26 (1993) 4151–4154. https://doi.org/10.1021/ma00068a013

[57]   M. Foroughi-Dahr, H. Abolghasemi, M. Esmaieli, G. Nazari, B. Rasem, Experimental study on the adsorptive behavior of congo red in cationic surfactant-modified tea waste, Process Saf. Environ. Prot. 95 (2015) 226–236. https://doi.org/10.1016/j.psep.2015.03.005

[58]   L.D.L. Miranda, C.R. Bellato, M.P.F. Fontes, M.F. de Almeida, J.L. Milagres, L.A. Minim, Preparation and evaluation of hydrotalcite-iron oxide magnetic organocomposite intercalated with surfactants for cationic methylene blue dye removal, Chem. Eng. J. 254 (2014) 88–97. https://doi.org/10.1016/j.cej.2014.05.094

[59]   P. Janoš, V. Šmídová, Effects of surfactants on the adsorptive removal of basic dyes from water using an organomineral sorbent—iron humate, J. Colloid Interface Sci. 291 (2005) 19–27. https://doi.org/10.1016/j.jcis.2005.04.065

[60]   A. Adak, M. Bandyopadhyay, A. Pal, Adsorption of anionic surfactant on alumina and reuse of the surfactant-modified alumina for the removal of crystal violet from aquatic environment., J. Environ. Sci. Health. A. Tox. Hazard. Subst. Environ. Eng. 40 (2005) 167–182. https://doi.org/10.1081/ESE-200038392

[61]   A. Adak, A. Pal, M. Bandyopadhyay, Removal of phenol from water environment by surfactant-modified alumina through adsolubilization, Colloids Surfaces A Physicochem. Eng. Asp. 277 (2006) 63–68. https://doi.org/10.1016/j.colsurfa.2005.11.012

[62]   T.D. Pham, M. Kobayashi, Y. Adachi, Adsorption of anionic surfactant sodium dodecyl sulfate onto alpha alumina with small surface area, Colloid Polym. Sci. 293 (2014) 217–227. https://doi.org/10.1007/s00396-014-3409-3

[63]   M.U. Khobragade, A. Pal, Investigation on the adsorption of Mn(II) on surfactant-modified alumina: Batch and column studies, J. Environ. Chem. Eng. 2 (2014) 2295–2305. https://doi.org/10.1016/j.jece.2014.10.008

[64]   M.U. Khobragade, A. Pal, Adsorptive removal of Cu(II) and Ni(II) from single-metal, binary-metal, and industrial wastewater systems by surfactant-modified alumina, J. Environ. Sci. Health A Tox. Hazard Subst. Environ. Eng. 50 (2015) 385-395. https://doi.org/10.1080/10934529.2015.987535

[65] E.P. Azadeh, H. Seyed F, A. Yousefi, Surfactant-modified wheat straw: preparation, characterization and its application for methylene blue adsorption from aqueous solution, Chem. Eng. Process Technol. 6 (2015) 1–9.

[66] T.D. Pham, H.H. Nguyen, N.V. Nguyen, T.T. Vu, T.N.M. Pham, T.H.Y. Doan, M.H. Nguyen, T.M.V. Ngo, Adsorptive removal of copper by using surfactant modified laterite soil, J. Chem. 2017 (2017) 1-10. https://doi.org/10.1155/2017/1986071

[67] T. Mitani, N. Fukumuro, C. Yoshimoto, H. Ishii, Effects of counter ions (SO_4^{2-} and Cl^-) on the adsorption of copper and nickel ions by swollen chitosan beads, Agric. Biol. Chem. 55 (1991) 2419.

[68] A. Adak, A. Pal, M. Bandyopadhyay, Spectrophotometric determination of anionic surfactants in wastewater using acridine orange, Indian J. Chem. Technol. 12 (2005) 145–148.

[69] C. Lin, S. Wang, H. Sun, R. Jiang, Adsorption of anionic dye by anionic surfactant modified chitosan beads: Influence of hydrophobic tail and ionic head-group, J. Dispers. Sci. Technol. 39 (2017) 1–10.

[70] G. Asgari, B. Roshani, G. Ghanizadeh, The investigation of kinetic and isotherm of fluoride adsorption onto functionalize pumice stone, J. Hazard. Mater. 217–218 (2012) 123–132. https://doi.org/10.1016/j.jhazmat.2012.03.003

[71] S. An, X. Liu, L. Yang, L. Zhang, Enhancement removal of crystal violet dye using magnetic calcium ferrite nanoparticle: Study in single- and binary-solute systems, Chem. Eng. Res. Des. 94 (2015) 726–735. https://doi.org/10.1016/j.cherd.2014.10.013

[72] S. Chatterjee, D.S. Lee, M.W. Lee, S.H. Woo, Enhanced adsorption of congo red from aqueous solutions by chitosan hydrogel beads impregnated with cetyl trimethyl ammonium bromide, Bioresour. Technol. 100 (2009) 2803–2809. https://doi.org/10.1016/j.biortech.2008.12.035

[73] M. Seredych, C. Portet, Y. Gogotsi, T.J. Bandosz, Nitrogen modified carbide-derived carbons as adsorbents of hydrogen sulfide, J. Colloid Interface Sci. 330 (2009) 60–66. https://doi.org/10.1016/j.jcis.2008.10.022

[74] S. Chatterjee, T. Chatterjee, S. Lim, S.H. Woo, Adsorption of a cationic dye, methylene blue, on to chitosan hydrogel beads generated by anionic surfactant gelation, Environ. Technol. 32 (2011) 1503–1514. https://doi.org/10.1080/09593330.2010.543157

Chitosan-Based Adsorbents for Wastewater Treatment, Ed. Abu Nasar   Materials Research Forum LLC
Materials Research Foundations **34** (2018) 227-254   doi: http://dx.doi.org/10.21741/9781945291753-10

Chapter 10

Utilization of Chitosan and its Nanocomposites as Adsorbents for Efficient Removal of Dyes

Ufana Riaz*, Jyoti Kashyap

Materials Research Laboratory, Department of Chemistry, Jamia Millia Islamia, New Delhi-110025, India

*ufana2002@yahoo.co.in

Abstract

Chitosan polymer nanocomposites have gained attention as effective adsorbent due to low cost and bioavailability. Many chitosan nanocomposites are obtained through chemical and physical modifications of raw chitosan that include cross-linking, grafting and impregnation of the chitosan backbone. Modification of chitosan backbone enhances the adsorption capability of the nanocomposite for adsorption of different types of dyes. This chapter discusses the different types of chitosan polymer nanocomposites that have been utilized for the removal of various organic dyes. The effects of various parameters like adsorbent dosage, dye concentration, pH, temperature, and adsorption mechanisms have also been highlighted which can help researchers to analyze and identify the future research work in this area.

Keywords

Chitosan, Adsorbent, Nanocomposites, Dye Removal, Kinetics

Contents

1.   Introduction

Water is one of the most vital resources for mankind and water contamination has become a serious issue for human health and other living organisms [1-6]. Textile industries as well as other dyeing industries such as paper, printing, leather, food and plastic have been identified as the major sources of water pollution [7-9]. The presence of dyes in textile wastewater can impose serious health issues as most of the common dyes are carcinogenic in nature [10-13]. Hence, treatment of wastewater is a subject of immense concern and a lot of research has been carried out in this area. There are many water treatment processes applied for the removal of dyes from wastewater. Adsorption is considered to be one of the facile techniques for wastewater remediation [14-15]. Several studies are being undertaken to investigate the use of low-cost adsorbents [16]. Therefore, during the past several years, research groups have focused on the use of various cheaper alternatives that could be used as green adsorbents. The new trends are mainly focused on sustainable resource-based biodegradable polymers [17-20].

2.   Structure and properties of Chitosan

Chitosan, Fig.1 is a well-known derivative of chitin- a natural amino-polysaccharide which is non-toxic and biodegradable in nature exhibiting high binding potential, and anti-bacterial characteristics [21-25]. Chitosan possesses amino groups and hydroxyl groups in its chemical structure which make it a potential adsorbent for dyes [26]. Chitosan, as compared to other commercial adsorbents, has received the immense attention from the researchers due to its cationic characteristics, high adsorption capacity, abundance in nature and low cost [27-30]. Chitosan usually requires modifications, such as crosslinking, grafting that enhances its resistance against acids, but this process reduces its adsorption capacity [31-35]. The major adsorption site of chitosan is the primary amino group which undergoes protonation to form $-NH_3^+$ in acidic medium. Besides this, the adsorption capacity of chitosan against anionic dyes occurs due to the electrostatic attraction between the protonated amino groups and the sulfonic groups of the anionic dyes. The effective electrostatic interaction of the $-NH_3^+$ groups of chitosan with the dye anions has been widely used to explain its mechanism of dye adsorption [36-40].

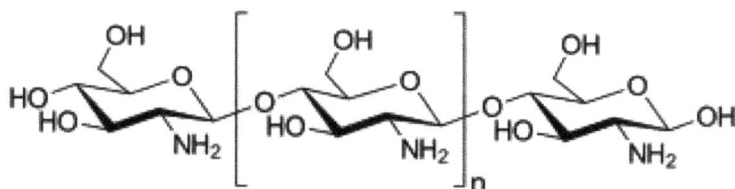

Figure 1 Structure of Chitosan.

3.　Role of Chitosan-based nanocomposite in enhancing absorption

Chitosan-based nanocomposites have been extensively investigated as adsorbents due to their high hydrophilic nature, chemical reactivity, and flexible structure of the polymer chain [41-46]. Modified chitosan nanocomposite based on graphene, silica, copper, cellulose etc. have been reported to show remarkable charge transporting properties, along with remarkable thermal stability as well as adsorption capability [47-50]. The upcoming section describes the synthesis and adsorption properties of chitosan-based nanocomposites.

4.　Synthesis and adsorption properties of some Chitosan nanocomposite

4.1　TLAC/Chitosan composite

An activated carbon that is derived from plant source known as *Typhalatifolia* activated carbon (TLAC) was utilized by *Kumari et al.* [51] for synthesizing novel and low-cost TLAC/Chitosan composites. Crystal violet (CV) was chosen as a model dye for the adsorption studies in a batch process. Results demonstrated that adsorption capacity was influenced by the pH of the dye medium, adsorbent dose, and initial dye concentration. The adsorption capacity was found to increase with an increase in the pH of the medium, showing maximum adsorption of upto 99% at pH 9 in 60 min using 400 mg of the adsorbent dose at 30 ppm dye concentration. Adsorption capacity also revealed an increase with the increasing adsorbent dose due to the increase in the number of activated sites. The porosity of the composite was observed to be 49%. Langmuir as well as Freundlich models revealed that pseudo-second order kinetics model was well fitted for the dye adsorption onto the composite.

4.2   Chitosan/cobalt-silica (Co-MCM) nanocomposites

Chitosan/based cobalt-silica (Co-MCM) nanocomposites were synthesized by *Khan et al.* [52] by mixing 5ml - 25 ml of Co-MCM solution to chitosan solution to formulate chitosan/Co-MCM-5, chitosan/Co-MCM-15 and chitosan/Co-MCM-25, respectively. The nanocomposite was tested for the removal of several dyes such as Methyl Orange (MO), Acridine Orange (AO), Indigo Carmine (IC) and Congo Red(CR). Chitosan/Co-MCM-15 nanocomposites, showed the highest adsorption among all other nanocomposites. Morphology of chitosan/Co-MCM nanocomposite was analyzed using FESEM which revealed cobalt-silica to be embedded in the chitosan structure, Fig.2. Good adsorption of the nanocomposites was observed towards MO, IC and CR dyes, while negligible adsorption was observed towards AO dye.

Highest adsorption tendency was shown by Chitosan/Co-MCM-15 for MO dye as compared to IC dye, Fig.3. Negligible adsorption was noticed for AO dye. The rate of adsorption of IC was observed to be very fast, which was found to gradually decrease with time (Fig.3).

The removal efficiency of MO dye was observed to decrease with the increase in pH of the dye solution. The maximum removal efficiency of 76% was achieved between pH 6-8, while other dyes were adsorbed in lower capacity, Fig.4. The nanocomposites were evaluated for their activity against some gram negative, gram-positive and multidrug-resistant bacteria. Among all the nanocomposites, Chitosan/Co-MCM-15 revealed the strongest activity which was attributed to the optimal loading of Co-MCM in chitosan.

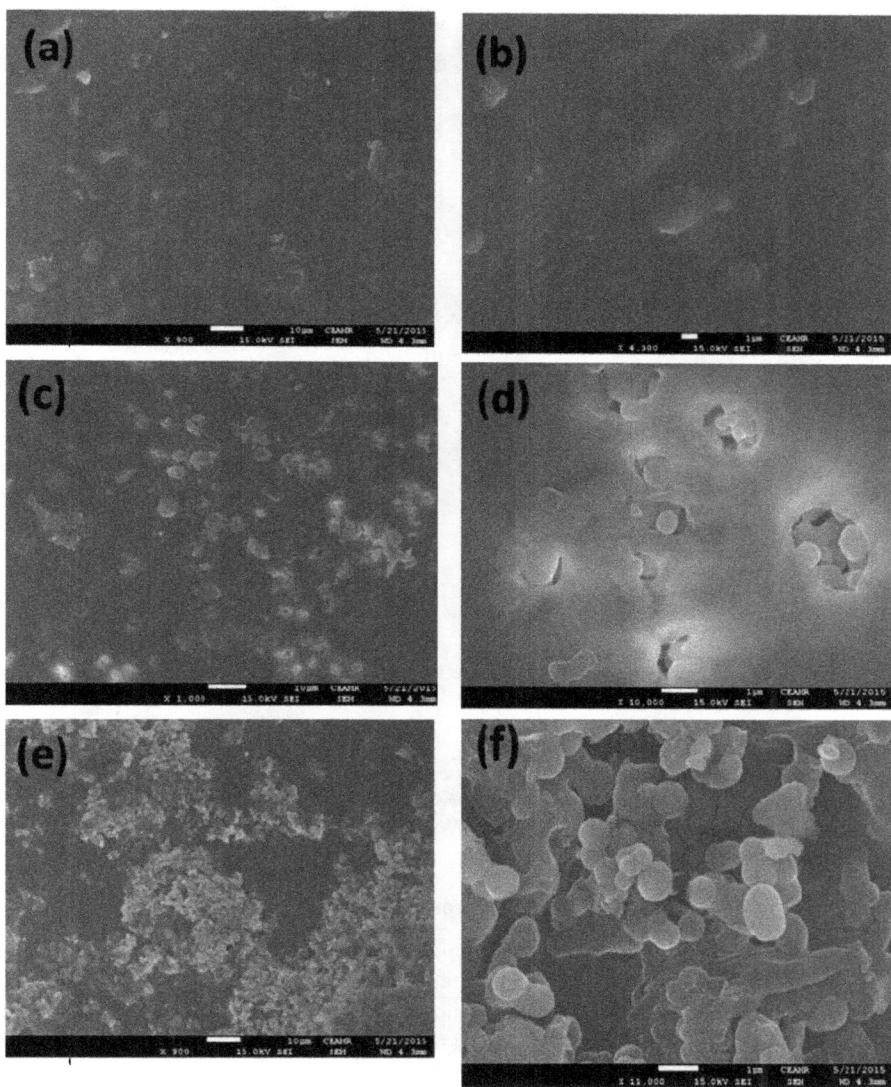

Figure 2   FESEM of   (a,b) chitosan/Co-MCM-5 (c,d), chitosan/Co-MCM-15 (e,f) chitosan/Co-MCM-25. (Reprinted with permission from Elsevier, Khan et al. International Journal of Biological Macromolecules 91 (2016) 744–751)

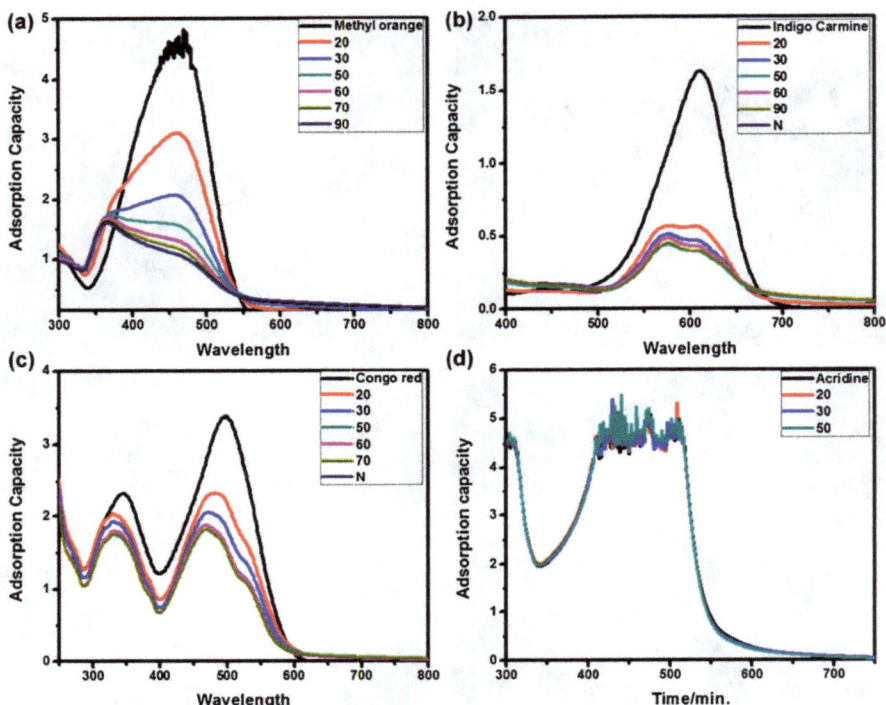

Figure 3   UV spectra of dyes in presence of chitosan/Co-MCM-15 as an adsorbent showing variation in the concentration of (a) MO, (b) IC ,(c) CR and (d) AO dyes. (Reprinted with permission from Elsevier, Khan et al. International Journal of Biological Macromolecules 91 (2016) 744–751)

4.3   Chitosan/copper oxide (CS/CuO) composite

Khan et al.[53] also synthesized chitosan/copper oxide (CS/CuO) composites by mixing of CuO in chitosan solution. The composite revealed a spherical but rough morphology with aggregation of CuO inside the chitosan matrix. Adsorption capability of the composite was investigated for IC, CR and MO dyes and the values were compared with pure chitosan(CS). CS/CuO composite spheres were more selective towards MO adsorption and removal percentage of MO with pure chitosan after 10 min was 30% while the composite showed 25% removal of MO dye. It was reported that with the increase in the contact time, MO concentration decreased at a higher rate (after 180 min)

Chitosan-Based Adsorbents for Wastewater Treatment, Ed. Abu Nasar   Materials Research Forum LLC
Materials Research Foundations **34** (2018) 227-254   doi: http://dx.doi.org/10.21741/9781945291753-10

when the composite was used as an adsorbent, and showed 94% of MO dye adsorption, whereas pure chitosan showed only 76% adsorption of MO dye, Fig.5.The ideal pH for the dye adsorption was found to be ranging between pH 6–8, while further increase in pH above 8 led to a decrease in dye removal efficiency (Fig.6).

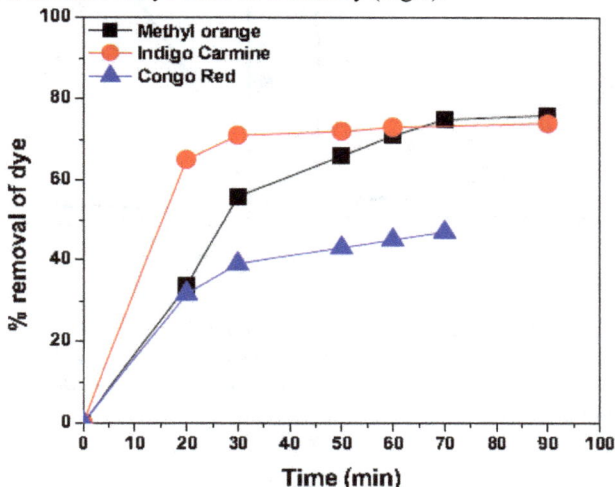

Figure 4   % Removal of MO, IC and CR dyes using chitosan/Co-MCM-15. (Reprinted with permission from Elsevier, Khan et al. International Journal of Biological Macromolecules 91 (2016) 744–751)

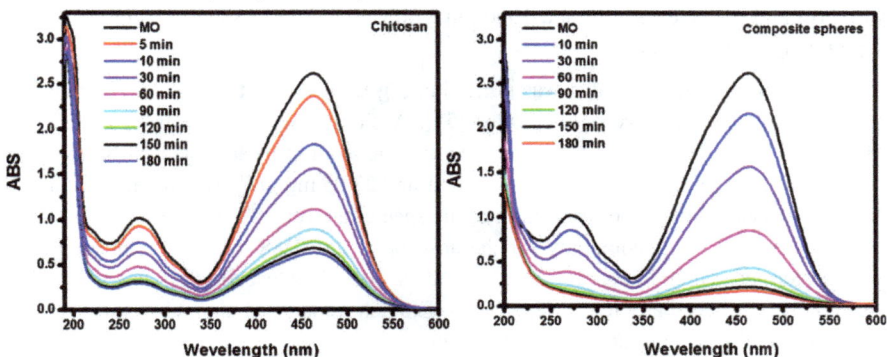

Figure 5   UV spectra of MO dye adsorption using (a) CS as adsorbent and (b) composite spheres as an adsorbent. (Reprinted with permission from Elsevier, Khan et al. International Journal of Biological Macromolecules 88 (2016) 113–119)

Chitosan-Based Adsorbents for Wastewater Treatment, Ed. Abu Nasar   Materials Research Forum LLC
Materials Research Foundations **34** (2018) 227-254   doi: http://dx.doi.org/10.21741/9781945291753-10

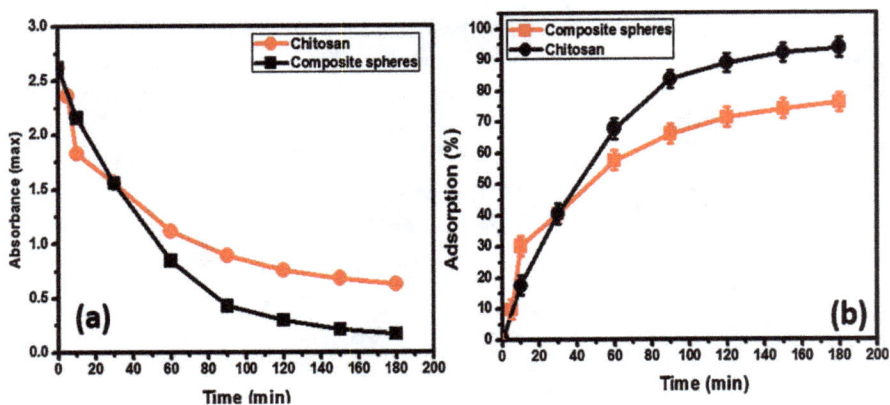

Figure 6(a)   CS and composite spheres as adsorbents showing (a) variation in absorbance vs contact time, (b) adsorption (%) vs contact time for MO dye adsorption. (Reprinted with permission from Elsevier, Kant et al. International Journal of Biological Macromolecules 88 (2016) 113–119)

Soltani et al. [54] explored the efficiency of immobilized nanosized bio-silica (average crystalline size of 20 nm) within chitosan to formulate a nanocomposite as an adsorbent for removing Acid Red (AR) 88 dye. The nanocomposites revealed highly porous structure, Fig.7. A non-uniform size distribution was observed for bio-silica, Fig. 7(a), Fig. 7(c) and (d) taken at 125 and 1000x magnifications which showed a rough surface topography for pure chitosan.

Higher adsorption of AR88 dye was noticed at 3 g L^{-1} adsorbent dosage in 120 min due to the availability of more binding sites [55]. A similar trend was noticed for 10, 25, 50,100, 200 and 400 mg L^{-1} concentrations and the amount of adsorbed AR88 dye was calculated to be 2.65, 6.63, 14.25, 18.82, 21.11 and 23.05 mg gL^{-1}. The pH of the solution also influenced the degree of ionization of the species as well as the surface charge of the adsorbent [56]. The kinetic study for the adsorption of AR88 dye onto nanocomposite was performed using four main kinetic models, including pseudo-first order, pseudo-second order, intra-particle diffusion and Elovich models. Langmuir isotherm and pseudo-second-order kinetic model were all in agreement for adsorption of AR88 dye.

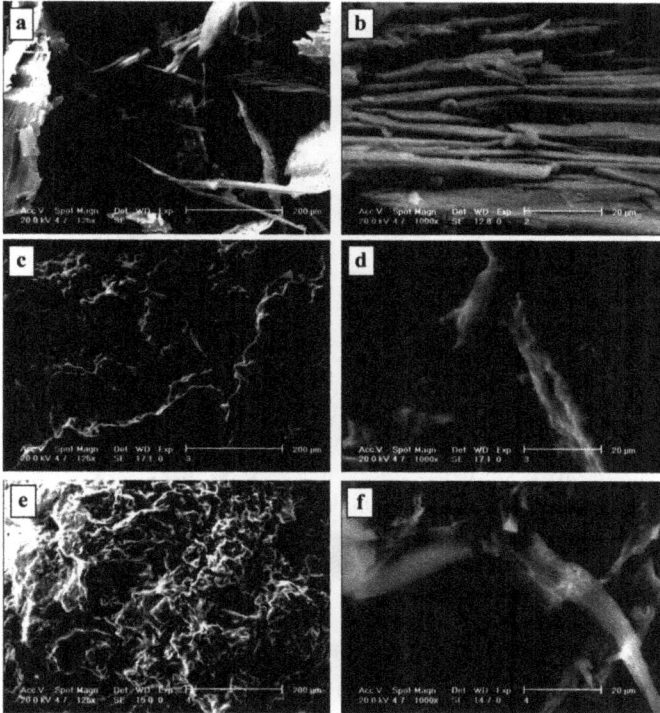

Figure 7   SEM images of pure bio-silica at 125X (a) and 1000×magnification (b), pure chitosan at 125×(c) and 1000×magnification (d), bio-silica/chitosan nanocomposite at 125× (e)and 1000× magnification (f). (Reprinted with permission from Elsevier, Soltani et al. International Biodeterioration & Biodegradation 85 (2013) 383-391)

4.4   Chitosan /Dialdehydes microfibrillated cellulose composite

Tang et al.[57] prepared novel dialdehyde microfibrillated cellulose (DAMFC) based chitosan composite films by the solvent-casting method. Congo red (CR) was selected as a model dye to evaluate the adsorption behavior of DAMFC/chitosan composite films (Fig.8).

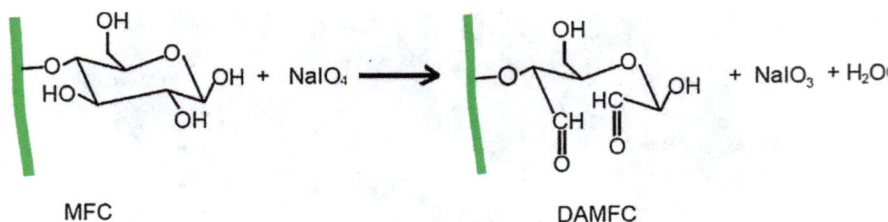

Figure 8   Scheme of DAMFC formation through the periodate oxidation of MFC. (Reprinted with permission from Elsevier, Tang et al. International Journal of Biological Macromolecules 107 (2018) 283–289)

The morphology of the prepared nanocomposites showed a 3-dimensional web-like structure. DAMFC/chitosan films 2/100 composite film efficiently removed the dye and the adsorption capacity of CR dye at equilibrium was reported to be 152.5 mg gL^{-1} with a high removal rate of of 99.95% within 10 min. The pH of the solution also influenced the adsorption capability of DAMFC/chitosan composite and the ideal pH value for CR was noticed to be in the range of pH 3–8. The pseudo-second-order model was more suitable to describe the adsorption of CR onto DAMFC/chitosan composites.

4.5   Graphene-based chitosan nanocomposite

Banerjee et al.[58] explored the adsorption characteristics of magnetic chitosan-graphene oxide nanocomposite (GO-Cs-Nc) synthesized via ultrasonication for adsorption of Acid Yellow 36 (AY) and Acid Blue 74 (AB) dyes. Around 99.58 % and 99.60 % removal was noticed for AY and AB dyes respectively when GO-Cs-Nc adsorbent dosage of 0.50 g^{-1} was used. High adsorption capacity was observed within a short exposure time due to the strong influence of ultrasound irradiation on mass transfer [59]. Adsorption isotherm models such as Langmuir, Freundlich, Tempkin, and Dubinin-Radushkevich were well-fitted and low adsorbent dosage, as well as reduced contact time, indicated energy saving and environment friendly nature of the adsorbent. *Fan et al.* [60] developed magnetic β-cyclodextrin–chitosan/graphene oxide materials (MCCG) via facile chemical route for the reduction of Methylene Blue (MB) dye as shown in Fig. 9.

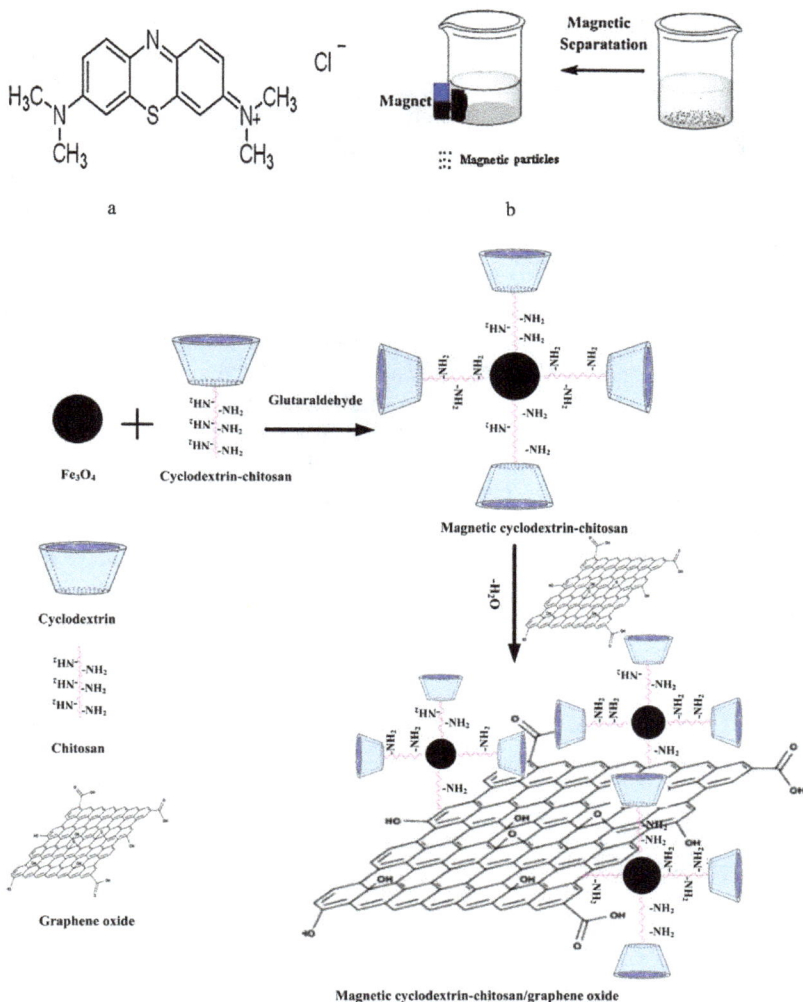

Figure 9   Schematic depiction of the formation of magnetic cyclodextrin-chitosan/graphene oxide; (b) the chemical structure of methylene blue; (c) schematic depiction of magnetic particles separated by a magnet. (Reprinted with permission from Elsevier, Fan et al. Colloids and Surfaces B: Biointerfaces 103 (2013) 601– 607)

Chitosan-Based Adsorbents for Wastewater Treatment, Ed. Abu Nasar   Materials Research Forum LLC
Materials Research Foundations **34** (2018) 227-254   doi: http://dx.doi.org/10.21741/9781945291753-10

The morphology of MCCG composite exhibited the formation of small magnetic chitosan particles on the surface of GO layers. The BET surface area, pore volume and pore size of MCCG were calculated to be 402.1 m^2/g, 0.4152 cm^3/g, and 3.178 nm respectively. The average particle size of MCCG was observed to be 240 nm whereas the adsorption capacity of MB increased with increasing pH of the solution due to the electrostatic attraction between the negatively charged surface of the MCCG and cationic nature of MB dye. The adsorption followed pseudo-second-order kinetics. The equilibrium adsorption isotherms were well suited to the Langmuir isotherm model.

Kamal et al. [61] utilized tetraethyl orthosilicate (TEOS) for the synthesis of a crosslinked nanocomposite composed of chitosan and Graphene oxide (GO). Membrane films of the nanocomposites were prepared and the ones containing GO were designated as CPG whereas the ones without GO were named as CP. SEM micrographs showed that GO was homogeneously mixed with chitosan BET results revealed increase in the surface area upon the inclusion of GO in chitosan. The BET surface area values were shown to be 0.279 m^2/g for CP and 1.809 m^2/g for CPG membrane, whereas the pore sizes were calculated to be 0.498 cm^3/g and 0.522 cm^3/g respectively. Congo Red (CR) was chosen as a model dye and adsorption was carried out at various contact times as shown in Fig. 10(a). 58% dye adsorption was revealed by CPG in 1 min which reached up to 90% in 7 min but CP exhibited only 89% adsorption 30 min. Maximum dye adsorption capacity of CP and CPG were calculated to be 145.2 and 175.9 mg/g for CR, (Fig.10(c) which showed an increase with an increase in the CR dye concentration Fig.10(d).

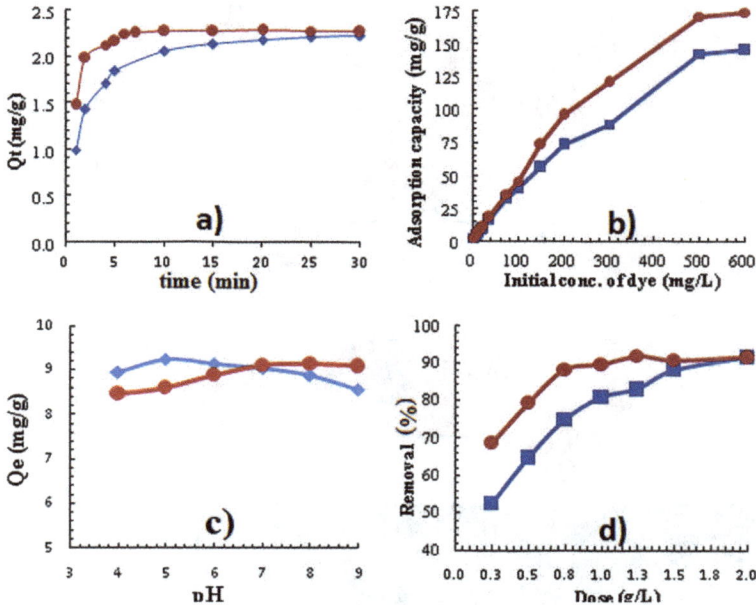

Figure 10 Removal of CR dye using CP and CPG showing the effect of (a) contact time (b) initial concentration (c) pH (d) amount of adsorbent dosage on removal efficiency. (Reprinted with permission from Elsevier, Kamal et al. Reactive and Functional Polymers 110 (2017) 21–29)

Langmuir isotherm model was well-fitted and adsorption kinetic was explained by the pseudo-second-order kinetic model. The increased rate of CR adsorption onto CPG was attributed to the presence of GO with active sites (Fig.11).

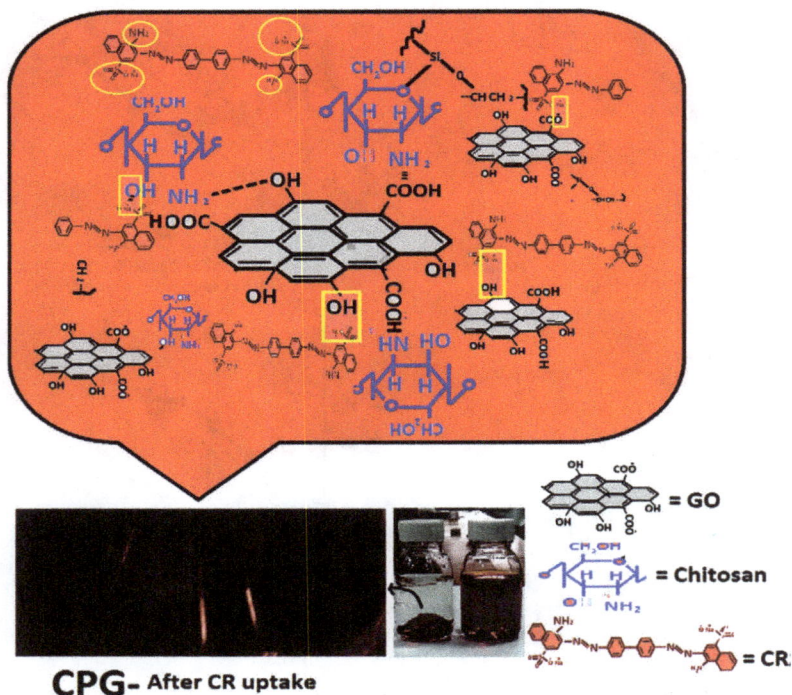

Figure 11 Proposed scheme for CR uptake on CPG membrane. (Reprinted with permission from Elsevier, Kamal et al. Reactive and Functional Polymers 110 (2017) 21–29)

Deliyanniet al. [62] developed graphite oxide magnetic chitosan (GO–Chm) nanocomposite via functionalization of graphite oxide (GO) with magnetic chitosan (Chm) as an adsorbent for the adsorption of Reactive Black 5 (Fig.12).

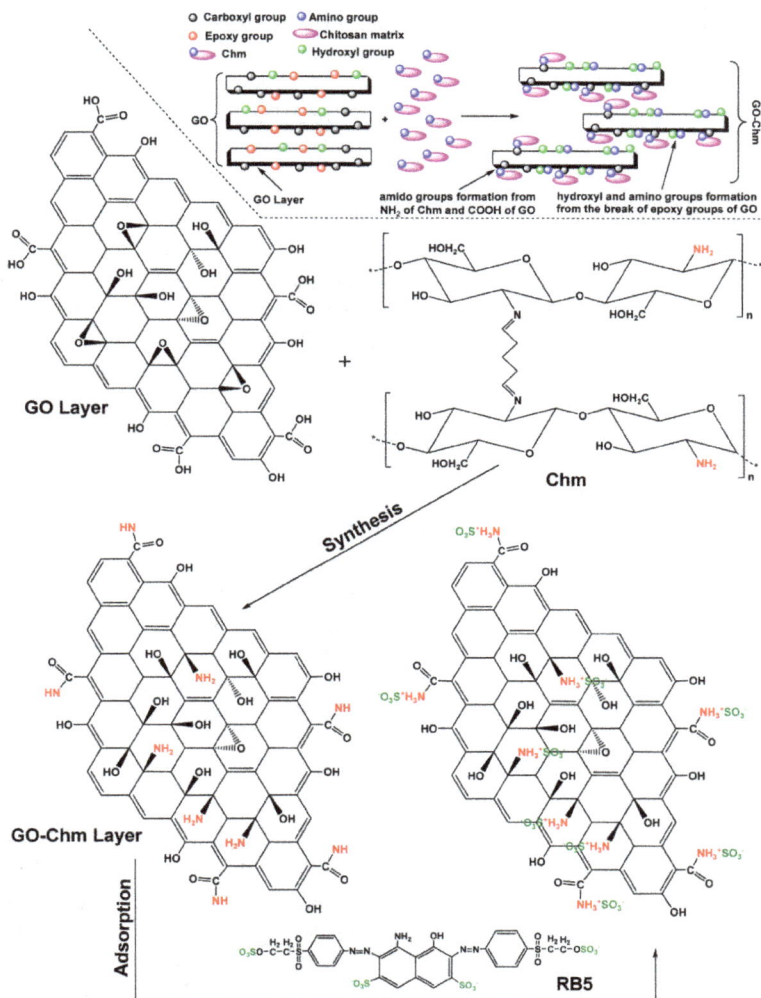

Figure 12 Proposed Synthesis Mechanism of GO−Chm after Functionalization of Chm on GO and Proposed Mechanism of RB5 Adsorption onto the Prepared GO−Chm. (Reprinted with permission from Elsevier, Deliyanni et al. Langmuir 2013, 29, 1657−1668)

Magnetic chitosan with a core–shell was synthesized possessing 1 μm diameter size and the rough surface morphology of GO–Chm nanocomposite was attributed to the accumulation of chitosan nanoparticles on the surface of the GO layers. Adsorption studies showed that the removal efficiency of the reactive dye was 81%, 77%, 70%, 62%, 55%, and 35%, at pH 3, 4, 6, 8, 10, and 12 respectively. Adsorption capability of GO–Chm at pH 12 was relatively higher and the removal efficiency was found to be 35%. The equilibrium adsorption isotherms were in agreement with Langmuir, Freundlich, and Langmuir–Freundlich (L–F) models.

Taher et al. [63] also synthesized magnetic/chitosan/graphene oxide (MCGO) 3-D nanostructure nanocomposites via ultrasonication. Morphology of MCGO revealed that the crystals were homogenously distributed with two diverse phases of iron, hollow nanorods and hexagon. A predictable mechanism for the growth of hollow nanorods goethite into 1D morphology was explained by a scheme as shown in Fig. 13. Disperse Blue 367 (DB367) dye was chosen for the adsorption studies and the experimental results revealed that the adsorbed amount of DB367 dye by MCGO increased rapidly upon decreasing the pH from 8 to 2. Highest adsorption efficiency of 92.9%, was observed at 90 °C at pH 2 in 120 min. The adsorption efficiency of DB367 dye increased as the temperature increased and maximum adsorption efficiency was noted to be 98.4% for 0.15 g MCGO. The adsorption capacity decreased from 298.27 to 16.27 mg g^{-1} as the adsorbent dosage of MCGO increased from 0.01g to 0.20 g. The rapid adsorption process was achieved during the first few minutes and maximum removal efficiency of 99.1% (21.13 mg g^{-1}) was attained in 30 min. Kinetics of DB367 dye adsorption was well-fitted in pseudo-second-order kinetic model.

4.6   Magnetic Chitosan-polyethylenimine (Fe$_3$O$_4$/CS-PEI) polymer composite

You et al.[64] synthesized Fe$_3$O$_4$/CS-PEI polymer composite via crosslinking of chitosan with polyethylenimine, Fig. 14. Chitosan (CS) was cross-linked with polyethylenimine (PEI) in the present of epichlorohydrin as a crosslinker. In alkaline medium, the epoxy group in epichlorohydrin reacted with the amino/hydroxyl groups in CS as well as with amino group of PEI, which resulted in the formation of a cross-linked copolymer. The BET surface area of the Fe$_3$O$_4$/CS-PEI was determined to be 109.2 m^2/g with an average pore width of 15.08 nm while the total pore volume was found to be 0.24 cm^3/g. Adsorption studies confirmed that the synthesized adsorbent exhibited a high removal efficiency of 99.3% for CR dye with a concentration of the dye being 100 mg/L and that adsorbent dose being 1.4 g/L. It was also reported that high temperature was beneficial for adsorption but the adsorption capacity decreased when the pH value increased. Kinetics study suggested that the adsorption mechanism of CR followed the pseudo-

Chitosan-Based Adsorbents for Wastewater Treatment, Ed. Abu Nasar   Materials Research Forum LLC
Materials Research Foundations **34** (2018) 227-254   doi: http://dx.doi.org/10.21741/9781945291753-10

second model and Fe_3O_4/CS-PEI revealed high positive charge, high surface area, multi-level pore distribution and magnetic responsiveness. The Fe3O4/CS-PEI thus showed enhanced capacity (1876 mg/g at 40∘C) for the removal of CR dye from aqueous solutions.

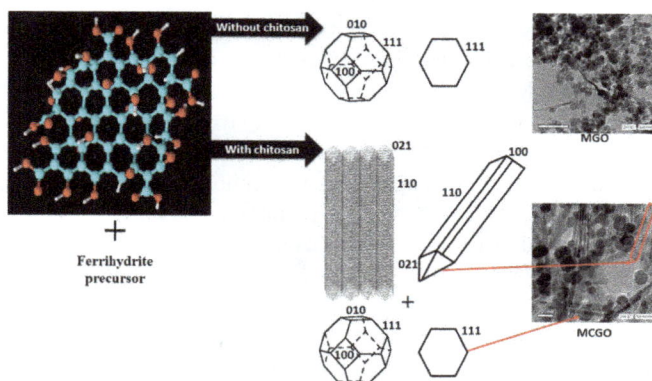

Figure 13 A possible growth mechanism for the formation of MGO nanocomposite and hierarchical MCGO 3D nanostructure. (Reprinted with permission from Elsevier, Taher et al. Materials Research Bulletin 97 (2018) 361–368)

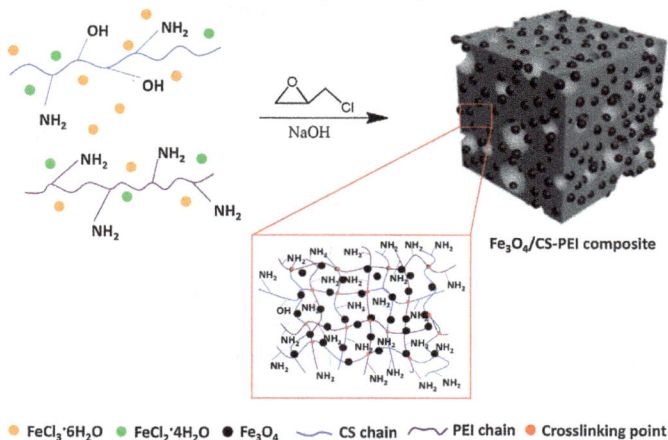

Figure 14 Synthetic procedure of the Fe_3O_4/CS-PEI polymer composite. (Reprinted with permission from Elsevier, You et al. International Journal of Biological Macromolecules 107 (2018) 1620–1628)

Ahmed et al. [65] tailored molecularly imprinted polymer (MIP) chitosan-TiO_2 nanocomposite (CTNC) synthesized via a sol-gel method for the effective removal of Rose Bengal (RB) dye. Morphology of the CTNC-MIP nanoparticles reflected high homogeneity of the prepared nanoparticles. The particles exhibited "spheroid like morphology" showing a large number of cavities that were dispersed through the whole sample matrix. The specific surface area was found to be 95.38 m^2/g. The adsorption studies showed that adsorption capacity increased from 50.84 to 84.8 mg g^{-1} at 30°C for 60 min with increasing RB dye concentration. When pH of the dye solution increased to 5.0, the adsorption capacity increased from 74.13 mg g-1 to 76 mg g-1 but at pH above 5.0, adsorption capacity was greatly reduced. Adsorption efficiency of RB dye removal was reported to be 98.89% as compared to pure chitosan whose adsorption efficiency was 53.73% at pH = 5. Pseudo-first-order, pseudo-second-order Elovich, and Weber-Morris models were used to analyze the kinetic data. The remarkable enhancement of dye removal was attributed to the tailored cavities of LMIP nanoparticles that attracted the RB dye molecules. *Xiao et al.* [66] also prepared MFe_3O_4/CS nanocomposite and brilliant red X-3B (X-3B), was selected as a model dye to evaluate the adsorption characteristics of MFe_3O_4/CS NPs by batch adsorption experiments, Fig.15. Morphology and size of MFe_3O_4/CS NPs were found to be quasi-spherical / ellipsoidal in shape and the diameter was found to be ranging between 5–8 nm. The BET surface area of MFe_3O_4/CS NPs was found to be 106.4 m^2/g and the pore diameters ranged from 5 nm- 10 nm. Adsorption potential of X-3B onto MFe_3O_4/CS was strongly dependent on both initial pH and the adsorbent dosage. The result showed that the adsorption capacity of X-3B on MFe_3O_4/CSNPs increased significantly when pH of the solution decreased to 2 and in case of adsorbent dose, the adsorption capacity decreased with the increase in adsorbent amount, which was due to the presence of unsaturated adsorption active sites [67]. X-3B dye adsorption was analyzed by varying the dosage of MFe_3O_4/CS from 0.1 g/L to 1.3 g/L. The X-3B adsorption increased significantly from 24.6% to 98.5% with the increase in adsorbent amount from 0.1 g/L to 0.6 g/L which was considered as the optimum adsorbent dosage. MFe_3O_4/CS nanocomposite exhibited a reusable property and removal efficiency of X-3B dye was over 90% after five cycles, indicating that MFe_3O_4/CS NPs had good reusability. Langmuir, Freundlich, and Sips isotherm models were fitted for equilibrium adsorption. The results of the kinetic study showed that the adsorption process followed the pseudo-second-order kinetic model.

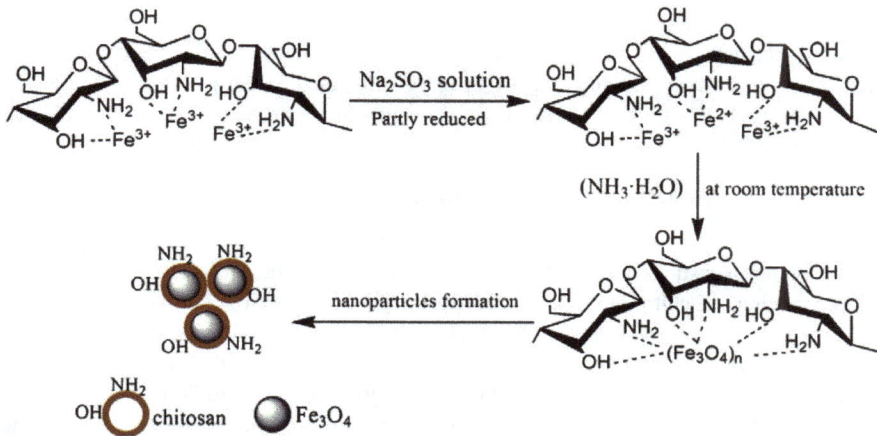

Figure 15 Schematic representation of the formation of MFe_3O_4/CS NPs. (Reprinted with permission from Elsevier, Cao et al. Powder Technology 260 (2014) 90–97)

4.7   Succinyl-grafted Chitosan

Kyzas et al.[68] prepared a succinyl grafted chitosan nanocomposite that showed good adsorption properties and high biodegradability. The grafted chitosan material was used for the efficient removal of basic dye (Remarcyl Red TGL denoted as CR). BET surface area was found to be 4.920 m²/g. The CR removal efficiency was 54%, 71%, 83%, and 97% at pH 2, 3, 4, and 5, respectively. Langmuir-Freundlich model and pseudo-second-order equation. The reusability of this super-adsorbent was found to decrease at 10th cycle(91% adsorption), while the loss until the 20th and 30th cycle was 9% and 10%, respectively.

4.8   Chitosan–Polyaniline/ZnO hybrid composite

Thambidurai et al. [69] synthesized chitosan–polyaniline (CPA) and chitosan–polyaniline/ZnO hybrids using aniline with different content of ZnO for adsorption evaluations of reactive orange 16 (RO16). Morphology of nanocomposite reveal granular like structure and particles size of CPA–ZC/ZO particle was between 29 nm - 40 nm. The specific surface area of the CPA–ZC2.5 was 45.9 m²/g and measured pore volume was for 0.479383 cm3/g CPA–ZC2.5 Experimental results demonstrated that adsorption efficiency was 90.1% of CPA–ZC2.5, and 80.8% of CPA respectively at low concentration of dye (25 mg L^{-1}) but it was 31.5% for CPA–ZC2.5 and 21.9% for CPA at higher concentration of dye (500 mg L^{-1}) the adsorption efficiency was decreased. The increase in adsorbent dosage from 0.1 to 0.5 g/L resulted in the increase of the RO16 adsorption to 41.7% to 99.8% adsorption efficiency of CPA–ZC2. The results indicated that CPA–ZC2.5 hybrids were non-toxic and cost-effective for dye removal process.

4.9   Chitosan/PVA/Na–Titanate/TiO$_2$ composites

Afifi et al. [70] prepared Chitosan/PVA/Na–titanate/TiO$_2$ composite film were prepared via the solution casting method in a different ratio of PVA and TiO$_2$. The schematic way of preparation of chitosan/PVA/Na-titanate/TiO$_2$ (Fig.16). Two anionic dyes, methyl orange, and congo red were selected for adsorption elucidation of the composites. Morphology of prepared composite having a rough surface and finer particles were observed and the diameter of filler particles was varied from 100 to 700 nm. Adsorption of MO was 99.9% using composite having higher weightage of chitosan and crystalline TiO$_2$ phase. CR adsorption was 95.76%. Adsorption percentage decreased at higher loading of PVA due to the deactivation of the amino group of chitosan by making a hydrogen bond with the hydroxyl group of PVA. The adsorption behavior of the composites was explained by pseudo-second-order kinetic model and Lagergren-first-order model for MO and CR dyes, respectively.

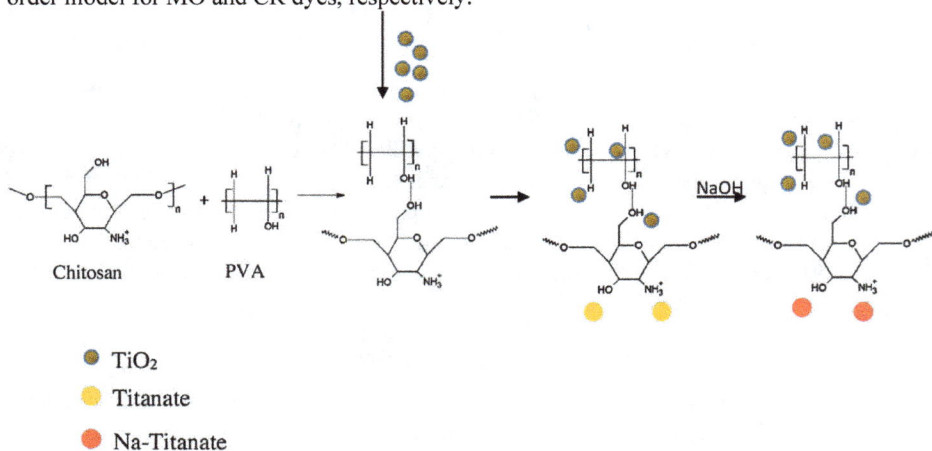

● TiO$_2$

● Titanate

● Na-Titanate

Figure 16   Schematic mechanism for preparation of Chitosan/PVA/Na-titanate/TiO$_2$ composite. (Reprinted with permission from Elsevier, Afifi et al. Carbohydrate Polymers 149 (2016) 317–331)

5. Scope of future work

The present chapter described the recent developments in the field of chitosan-based nanocomposites as adsorbents for the removal of dye from wastewater. Chitosan and its nanocomposite exhibit unique characteristics such as low cost, high adsorption capacity, ease of processibility and biodegradability. To enhance the adsorption performance of chitosan nanocomposite, many groups have employed cross-linking, grafting and impregnation methods. Adsorption using chitosan nanocomposites is becoming a promising alternative to replace conventional adsorbents in removing dyes. Although chitosan nanocomposites have attractive application in wastewater treatment, there are still several gaps which need to be filled. There is some important issue to be summarized below: The method of preparation of chitosan nanocomposite can be more simplified to one-pot synthesis. For the actual wastewater treatment, it's quite necessary to study the adsorption behavior of a multiple solute system, and the influence of co-existing ions in the solution on the target ions. Research till now is only scaled up to batch studies, but pilot-plant studies should also be conducted utilizing chitosan nanocomposite to check their utility on a commercial scale. Multi-purpose, economically feasible magnetic chitosan composites can be developed to remove a variety of pollutants. The study of dynamic adsorption of target contaminant deserves more attention and also a concern with estimating the cost of production and application. Low production cost with higher removal efficiency can make the process economical and efficient. Chitosan significantly shows advantages over currently available commercially expensive activated carbons and, in addition, contributes to an overall waste minimization strategy.

References

[1]   H.H.G. Savenije, Why water is not an ordinary economic good, or why the girl is special. Phys Chem Earth. Phys Chem Earth 27 (2002) 741–4. https://doi.org/10.1016/S1474-7065(02)00060-8

[2]   F. A. Klink, E. P. Moriana, J. S. García. The social construction of scarcity. The case of water in Tenerife (Canary Islands). Ecological Economics. Ecol. Econ. 34 (2000) 233–45. https://doi.org/10.1016/S0921-8009(00)00160-9

[3]   A. Bhatnagar, M. Sillanpää. Utilization of agro-industrial and municipal waste materials as potential adsorbents for water treatment—A review, Chem. Eng. J. 157 (2010) 277–96. https://doi.org/10.1016/j.cej.2010.01.007

[4]   L. Hu, Z. Yang, L. Cui, Y. Li, Y. H. H. Ngo, Q. Wang, H. Wei, L. Ma, B. Yan. Fabrication of hyperbranched polyamine functionalized graphene for high-

efficiency removal of Pb(II) and methylene blue. Chem. Eng. J. 287 (2016) 545–556. https://doi.org/10.1016/j.cej.2015.11.059

[5]   M. Toor, B. Jin, S. Dai, V. Vimonses. Activating natural bentonite as a cost-effective adsorbent for removal of Congo-red in wastewater, J. Ind. Eng. Chem. 21 (2015) 653–661. https://doi.org/10.1016/j.jiec.2014.03.033

[6]   S. Hashemian, A. Foroghimoqhadam. Effect of copper doping on CoTiO$_3$ ilmenite type nanoparticles for removal of congo red from aqueous solution. Chem. Eng. J. 235 (2014) 299–306. https://doi.org/10.1016/j.cej.2013.08.089

[7]   M. Vakili, M. Rafatullah, B. Salamatinia, A.Z. Abdullah, M.H. Ibrahim, K.B. Tan, Z. Gholami, P. Amouzgar. Application of chitosan and its derivatives as adsorbents for dye removal from water and wastewater: A review. Carbohydr. Polym. 113 (2014) 115–130. https://doi.org/10.1016/j.carbpol.2014.07.007

[8]   T. R. Kant. Textile dyeing industry an environmental hazard. Natural Science 4 (2012) 22-26. https://doi.org/10.4236/ns.2012.41004

[9]   C. Zubieta, M.B. Sierra, M.A. Morini, P.C. Schulz, L. Albertengo. The adsorption of dyes used in the textile industry on mesoporous materials. Colloid Poly. Sci. 286 (2008) 377–384. https://doi.org/10.1007/s00396-007-1777-7

[10]   H. Yan, H. Li, H. Yang, A. Li, R. Cheng. Removal of various cationic dyes from aqueous solutions using a kind of fully biodegradable magnetic composite microsphere. Chem. Eng. J. 223 (2013) 402–411. https://doi.org/10.1016/j.cej.2013.02.113

[11]   I. Savin, R. butnaru, wastewater characteristics in textile finishing mills. Environ. Eng. Manag. J. 7 (2008) 859-864.

[12]   S. Ghorai, A.K. Sarkar, A.B. Panda, S. Pal. Effective removal of Congo red dye from aqueous solution using modified xanthan gum/silica hybrid nanocomposite as adsorbent. Biores. Technol. 144 (2013) 485–491. https://doi.org/10.1016/j.biortech.2013.06.108

[13]   A.U. Metin, H. Çiftçi, E. Alver. Efficient Removal of Acidic Dye Using Low-Cost Biocomposite Beads. Ind. Eng. Chem. Res. 52 (2013) 10569–10581. https://doi.org/10.1021/ie400480s

[14]   E. Forgacs, T. Cserháti, G. Oros. Removal of synthetic dyes from wastewaters: a review, Environ. Int. 30 (2004) 953−971. https://doi.org/10.1016/j.envint.2004.02.001

[15] G. Mezohegyi, F. P. van der Zee, J. Font, A. Fortuny, A. Fabregat, Towards advanced aqueous dye removal processes: A short review on the versatile role of activated carbon. J. Environ. Manage. 102 (2012) 148−164. https://doi.org/10.1016/j.jenvman.2012.02.021

[16] Z. Jia, Z. Li, T. Ni, S. Li. Adsorption of low-cost absorption materials based on biomass (*Cortaderia selloana* flower spikes) for dye removal: Kinetics, isotherms and thermodynamic studies. J. Mol. Liq. 229 (2017) 285–292. https://doi.org/10.1016/j.molliq.2016.12.059

[17] R.A.A. Muzzarelli, Potential of chitin/chitosan-bearing materials for uranium recovery: An interdisciplinary review. Carbohydr Polym 84 (2011)54–63. https://doi.org/10.1016/j.carbpol.2010.12.025

[18] A. Bhatnagar, M. Sillanpää, Applications of chitin- and chitosan-derivatives for the detoxification of water and wastewater--a short review. Adv. Colloid. Interface. 159 (2009) 26–38.

[19] G. Crini G, P-M Badot. Application of chitosan, a natural aminopolysaccharide, for dye removal from aqueous solutions by adsorption processes using batch studies: A review of recent literature. Prog. Polym. Sci. 33(2008) 399–447. https://doi.org/10.1016/j.progpolymsci.2007.11.001

[20] L. Pontoni, M. Fabbricino. Use of chitosan and chitosan-derivatives to remove arsenic from aqueous solutions—a mini review. Carbohydr. Res.356 (2012) 86–92. https://doi.org/10.1016/j.carres.2012.03.042

[21] K. Shen, M.A. Gondal. Removal of hazardous Rhodamine dye from water by adsorption onto exhausted coffee ground. J. Saudi Chem. Soc. 21 (2017) S120–S127. https://doi.org/10.1016/j.jscs.2013.11.005

[22] M.N.R. Kumar. A review of chitin and chitosan applications. React. Funct. Poly. 46 (2000) 1–27. https://doi.org/10.1016/S1381-5148(00)00038-9

[23] R. Riva, H. Ragelle, A. Rieux, N. Duhem, C. Jerome, V. Preat, V. Adv. Chitosan and Chitosan Derivatives in Drug Delivery and Tissue Engineering. Polym. Sci. 244 (2011) 244, 19–44.

[24] L. Zhou, J. Jin, Z. Liu, X. Liang, C. Shang Adsorption of acid dyes from aqueous solutions by the ethylenediamine-modified magnetic chitosan nanoparticles.185 (2011) 1045-1052.

[25]   M.-Y. Chang, R.-S. Juang, Adsorption of tannic acid, humic acid, and dyes from water using the composite of chitosan and activated clay. J. Colloid Interface Sci. 278 (2004) 18-25. https://doi.org/10.1016/j.jcis.2004.05.029

[26]   E. Guibal, E. Touraud, J. Roussy. Chitosan Interactions with Metal Ions and Dyes: Dissolved-state vs. Solid-state Application. World J. Microbiol. Biotechnol. 21 (2005) 913–920. https://doi.org/10.1007/s11274-004-6559-5

[27]   R.A.A. Muzzarelli, J. Boudrant, D. Meyer, N. Manno,M. DeMarchis, M.G. Paoletti, Current views on fungal chitin/chitosan, human chitinases, food preservation, glucans, pectins and inulin: A tribute to Henri Braconnot, precursor of the carbohydrate polymers science, on the chitin bicentennial. Carbohydr. Polym. 87 (2012) 995–1012. https://doi.org/10.1016/j.carbpol.2011.09.063

[28]   Wan Ngah, W., Teong, L., & Hanafiah, M. Adsorption of dyes and heavy metal ions by chitosan composites: A review. Carbohydr. Polym. 83(2011) 1446–1456. https://doi.org/10.1016/j.carbpol.2010.11.004

[29]   K.A.G. Gusmão, L.V.A. Gurgel, T.M.S. Melo, L.F. Gil, Application of succinylated sugarcane bagasse as adsorbent to remove methylene blue and gentian violet from aqueous solutions – Kinetic and equilibrium studies. Dyes Pigm. 92 (2012) 967–974. https://doi.org/10.1016/j.dyepig.2011.09.005

[30]   M. Sarkar, P. Majumdar, Application of response surface methodology for optimization of heavy metal biosorption using surfactant modified chitosan bead. Chem. Eng. J. 175 (2011) 376-387. https://doi.org/10.1016/j.cej.2011.09.125

[31]   A.R. Cestari , E.F.S. Vieira, A.G.P. dos Santos, J.A. Mota, V.P. de Almeida. Adsorption of anionic dyes on chitosan beads. 1. The influence of the chemical structures of dyes and temperature on the adsorption kinetics. J. Colloid. Int. Sci. 280 (2004) 380–6. https://doi.org/10.1016/j.jcis.2004.08.007

[32]   G.Gibbs, J.M. Tobin, E. Guibal. Influence of Chitosan Preprotonation on Reactive Black 5 Sorption Isotherms and Kinetics. Ind. Eng. Chem. Res. 43 (2004) 1–11. https://doi.org/10.1021/ie030352p

[33]   A. Hebeish, R. Rafei, A. El-Shafei, Egypt. Crosslinking of chitosan with glutaraldehyde for removal of dyes and heavy metals ions from aqueous solutions. J. Chem. 47 (2004) 65–79. https://doi.org/10.1021/la402778x

[34]   G.Z. Kyzas, P.I. Siafaka, D.A. Lambropoulou, N.K. Lazaridis, D.N. Bikiaris, Poly(itaconic acid)-Grafted Chitosan Adsorbents with Different Cross-Linking for Pb(II) and Cd(II) Uptake Langmuir. 30 (2014) 120–131.

[35]   M.Y. Chan, S. Husseinsyah, S.T. Sam. Chitosan/corn cob biocomposite films by cross-linking with glutaraldehyde. BioResources 8(2013) 2910–2923. https://doi.org/10.15376/biores.8.2.2910-2923

[36]   L. Liu, J. Zhang, R.C. Tang, Adsorption and functional properties of natural lac dye on chitosan fiber, React. Funct. Polym. 73 (2013) 1559–1566. https://doi.org/10.1016/j.reactfunctpolym.2013.08.007

[37]   E. Guibal, P. McCarrick, JM Tobin. Comparison of the Sorption of Anionic Dyes on Activated Carbon and Chitosan Derivatives from Dilute Solutions. Sep Sci Technol. 38 (2003) 3049–73. https://doi.org/10.1081/SS-120022586

[38]   M. Ruiz, AM Sastre, E. Palladium sorption on glutaraldehyde cross-linked chitosan. Guibal. React Funct Polym. 45 (2000) 155–73. https://doi.org/10.1016/S1381-5148(00)00019-5

[39]   M.-S. Chiou, H.-Y. Li. Equilibrium and kinetic modeling of adsorption of reactive dye on cross-linked chitosan beads. J. Hazard. Mater. B 93 (2002) 233–248. https://doi.org/10.1016/S0304-3894(02)00030-4

[40]   W.H. Cheung, Y.S. Szeto, G.McKay. Enhancing the adsorption capacities of acid dyes by chitosan nano particles. Bioresour. Technol. 100 (2009) 1143–1148. https://doi.org/10.1016/j.biortech.2008.07.071

[41]   G. Dotto, J. Moura, T. Cadaval, L. Pinto. Application of chitosan films for the removal of food dyes from aqueous solutions by adsorption. Chem. Eng. J. 214 (2013) 8–16. https://doi.org/10.1016/j.cej.2012.10.027

[42]   D. Balkose, H. Baltacioˇglu, Adsorption of heavy metal cations from aqueous solutions by wool fibers. J. Chem. Technol. Biotechnol. 54 (1992) 393–397. https://doi.org/10.1002/jctb.280540414

[43]   W.W. Ngah, L. Teong, M. Hanafiah. Adsorption of dyes and heavy metal ions by chitosan composites: A review. Carbohydr. Polym. 83 (2011) 1446–1456. https://doi.org/10.1016/j.carbpol.2010.11.004

[44]   M.N.R. Kumar. A review of chitin and chitosan applications. React. Funct. Poly. 46 (2000) 1–27. https://doi.org/10.1016/S1381-5148(00)00038-9

[45]   M. Kaya, F. Dudakli, M. Asan-Ozusaglam, Y.S. Cakmak, T. Baran, A. Mentes, S. Erdogan. Porous and nanofiber α-chitosan obtained from blue crab (*Callinectes sapidus*) tested for antimicrobial and antioxidant activities. LWT- Food Sci. Technol. 65 (2016) 1109–1117.

[46] M. Chiou, M. H. Li. Adsorption behavior of reactive dye in aqueous solution on chemical cross-linked chitosan beads. Chemosphere 50 (2003) 1095–1105. https://doi.org/10.1016/S0045-6535(02)00636-7

[47] A. Shweta, P. Sonia. Pharmaceutical relevance of crosslinked chitosan in microparticulate drug delivery. Int. Res. J. Pharm. 4 (2013) 45–51.

[48] E. Guibal, P. M. Carrick, J.M. Tobin. Comparison of the Sorption of Anionic Dyes on Activated Carbon and Chitosan Derivatives from Dilute Solutions. Sep. Sci. Technol. 38 (2003) 3049–3073. https://doi.org/10.1081/SS-120022586

[49] R. Gaikwad, R., S. Misal. Sorption studies of methylene blue on silica gel. Int. J. Chem.Eng. A. 1(2010) 342–345. https://doi.org/10.7763/IJCEA.2010.V1.59

[50] V.K. Gupta, Suhas, Application of low-cost adsorbents for dye removal – A review J. Environ. Manag. 90 (2009) 2313–2342. https://doi.org/10.1016/j.jenvman.2008.11.017

[51] H.J. Kumari P. Krishnamoorthy. T.K.Arumugam S. Radhakrishnan D.vasudevan. An efficient removal of crystal violet dyes from waste water by adsorption onto TLAC/Chitosan composite: A novel low cost adsorbent. Int. J. Biol. Macromol. 96 (2017) 324-333. https://doi.org/10.1016/j.ijbiomac.2016.11.077

[52] S.Ali Khan, S. B. Khan, T. Kamal, M. Yasir, A. M. Asiri. Antibacterial nanocomposites based on chitosan/Co-MCM as a selective and efficient adsorbent for organic dyes. Int. J. Biol. Macromol. 91 (2016) 744–751. https://doi.org/10.1016/j.ijbiomac.2016.06.018

[53] S. Ali Khan, S. B. Khan, T. Kamal, M. Yasir, A. M. Asiri. CuO embedded chitosan spheres as antibacterial adsorbent for dyes. Int. J. Biol. Macromol. 88 (2016) 113–119. https://doi.org/10.1016/j.ijbiomac.2016.03.026

[54] R. Darvishi, Cheshmeh Soltani, A.R. Khataee , M. Safari , S.W. Joo. Preparation of bio-silica/chitosan nanocomposite for adsorption of a textile dye in aqueous solutions. Int. Biodeter Biodegr J. 85 (2013) 383-391. https://doi.org/10.1016/j.ibiod.2013.09.004

[55] H.Y. Zhu, R. Jiang, Y.Q. Fu, J.H. Jiang, L. Xiao, G.M. Zeng. Preparation, characterization and dye adsorption properties of g-Fe_2O_3/SiO_2/chitosan composite Appl. Surf. Sci. 258 (2011) 1337-1344. https://doi.org/10.1016/j.apsusc.2011.09.045

[56] M.M.F. Silva, M.M. Oliveira, M.C. Avelino, M.G. Fonseca, R.K.S. Almeida, E.C. Silva Filho. Adsorption of an industrial anionic dye by modified-

KSFmontmorillonite: evaluation of the kinetic, thermodynamic and equilibrium data. Chem. Eng. J. 203(2012) 259-268. https://doi.org/10.1016/j.cej.2012.07.009

[57] X. Zheng, X. Li, Jinyang Li, L. Wang, W. Jin, J. liu, Y. Pei, K. Tang. Efficient removal of anionic dye (Congo red) by dialdehyde microfibrillated cellulose/chitosan composite film with significantly improved stability in dye solution. Int. J. Biol. Macromol. 107 (2018) 283–289. https://doi.org/10.1016/j.ijbiomac.2017.08.169

[58] P. Banerjee, S. Roy, B. A. Mukhopadhayay, P. Das. Ultrasound assisted mixed azo dye adsorption by chitosan- graphene oxide nanocomposite. 117 (2017) 43-56.

[59] M. Ghaedi, A.M. Ghaedi, F. Abdi, M. Roosta, R. Sahraei, A. Daneshfar. Principal component analysis-artificial neural network and genetic algorithm optimization for removal of reactive orange 12 by copper sulfide nanoparticles-activated carbon. J. Ind. Eng. Chem. 20 (2014b) 787-795. https://doi.org/10.1016/j.jiec.2013.06.008

[60] L. Fan, C. Luo, M. Sun, H. Qiu, X. Li. Synthesis of magnetic-cyclodextrin–chitosan/graphene oxide as nanoadsorbent and its application in dye adsorption and removal. Colloids Surf B Biointerfaces. 103 (2013) 601– 607. https://doi.org/10.1016/j.colsurfb.2012.11.023

[61]   M.A. Kamal, S. Bibi, S. W. Bokhari, A. H. Siddique, T. Yasin. Synthesis and adsorptive characteristics of novel chitosan/grapheme oxide nanocomposite for dye uptake React. Funct. Polym. 110 (2017) 21–29. https://doi.org/10.1016/j.reactfunctpolym.2016.11.002

[62] N. A. Travlou, G. Z. Kyzas, N. K. Lazaridis, E. A. Deliyanni. Functionalization of Graphite Oxide with Magnetic Chitosan for the Preparation of a Nanocomposite Dye Adsorbent. Langmuir 29 (2013)1657–1668. https://doi.org/10.1021/la304696y

[63] F.A. Taher, F.H. Kamal, N.A. Badawy, A.E Shrshr. Hierarchical magnetic/chitosan/graphene oxide 3D nanostructure as highly effective adsorbent Materials Res. Bull. 97 (2018) 361–368. https://doi.org/10.1016/j.materresbull.2017.09.023

[64] L. You, C. Huang, F. Lu, A. Wang, X. Liu, Q. Zhang. Facile synthesis of high performance porous magnetic chitosan-polyethylenimine polymer composite for Congo red removal. Int. J. Biol. Macromol. 107 (2018) 1620–1628. https://doi.org/10.1016/j.ijbiomac.2017.10.025

[65]   M. A. Ahmed, N. M. Abdelbar, A. A. Mohamed. Molecular imprinted chitosan-TiO2 nanocomposite for the selective removal of Rose Bengal from wastewater. Int. J. Biol. Macromol. 107 (2018) 1046–1053. https://doi.org/10.1016/j.ijbiomac.2017.09.082

[66]   C. Cao, L. Xiao, C. Chen, X. Shi, Q. Cao, L. Gao. In situ preparation of magnetic Fe3O4/chitosan nanoparticles via a novel reduction–precipitation method and their application in adsorption of reactive azo dye Powder Technology. 260 (2014) 90–97. https://doi.org/10.1016/j.powtec.2014.03.025

[67]   H.Y. Zhu, R. Jiang, L. Xiao,W. Li, A novel magnetically separable γ-Fe$_2$O$_3$/crosslinked chitosan adsorbent: Preparation, characterization and adsorption application for removal of hazardous azo dye. J. Hazard. Mater. 179 (2010) 251–257. https://doi.org/10.1016/j.jhazmat.2010.02.087

[68]   G. Z. Kyzas, P. I. Siafaka, E. G. Pavlidou, K. J.Chrissafis, D. N. Bikiaris. Synthesis and adsorption application of succinyl-grafted chitosan for the simultaneous removal of zinc and cationic dye from binary hazardous mixtures. Chem. Eng. J. 259 (2015) 438-448. https://doi.org/10.1016/j.cej.2014.08.019

[69]   K. Pandiselvi, S. Thambidurai. Synthesis of porous chitosan–polyaniline/ZnO hybrid composite and application for removal of reactive orange 16 dye. Colloids Surf. B Biointer. 108 (2013) 229–238. https://doi.org/10.1016/j.colsurfb.2013.03.015

[70]   U. Habiba, M. S. Islam, T. A. Siddique, A. M. Afifi, B. C. Ang. Adsorption and photocatalytic degradation of anionic dyes on Chitosan/PVA/Na–Titanate/TiO$_2$ composites synthesized by solution casting method. Carbohy.Polym.149 (2016) 317–331. https://doi.org/10.1016/j.carbpol.2016.04.127

Chitosan-Based Adsorbents for Wastewater Treatment, Ed. Abu Nasar   Materials Research Forum LLC
Materials Research Foundations **34** (2018) 255-278   doi: http://dx.doi.org/10.21741/9781945291753-11

Chapter 11

Graphene Oxide-Chitosan Furnished Monodisperse Platinum Nanoparticles as Importantly Competent and Reusable Nanosorbents for Methylene Blue Removal

Aysun Savk[1], Betul Sen[1], Buse Demirkan[1], Esra Kuyuldar[1], Aysenur Aygun[1], Mehmet Salih Nas[1,2], Fatih Sen[1,*]

[1]Sen Research Group, Department of Biochemistry, Faculty of Arts and Science, Dumlupınar University, Evliya Çelebi Campus, 43100 Kütahya, Turkey

[2]Department of Environmental Engineering, Faculty of Engineering, Igdir University, 76000 Igdir, Turkey

*fatih.sen@dpu.edu.tr

Abstract

In this study, the microwave assisted methodology was employed to produce uniformly distributed platinum nanoparticles decorated with graphane oxide-chitosan. The capacity of methylene blue removal of these nanohybrids at room temperature was examined via adsorption. Characterizations of these novel nanoadsorbents were accomplished using X-ray diffraction (XRD), X-ray photoelectron spectroscopy (XPS), transmission electron microscopy (TEM) and high-resolution transmission electron microscopy (HRTEM). The X-ray diffractogram of the Pt NPs@CSGO displayed an ordinary structure of face-centered cubic (FCC). Adsorbance measurement results represented significant performance increases for all these novel nanohybrids for methylene blue removal. However, Pt NPs@CSGO hybrid was one of the best nanoadsorbent compared to others produced in this study. Our results presented that the one of highest methylene blue adsorption capacity belongs to the Pt NPs@CSGO, which was 194.6 mg/g, can be considered as an outstanding capacity. Its equilibrium was accomplished in 55 min. Furthermore, all these Pt NPs are reusable materials for the methylene blue removal application because they sustained 74.02 % of the initial efficiency after six successive adsorptions–desorption cycles.

Keywords

Adsorption Capacity, Hybrid Materials, Nanoadsorbents, Graphene Oxide-Chitosan

Contents

1.   Introduction

In numerous industries, dyes are commonly being used; for example, coating, cosmetic, leather, paint, paper, plastic, textile industries, and they have big amounts of effluent water and this effluent water typically contains organic dyes. They are very toxic, and on the other hand, the dyes are visible and seems unaesthetic [1-3]. Therefore, conventional dye removal methods are being used to clean these contaminated effluents. Today, in addition to conventional treatment methods, various techniques are also being utilized for the treatment of wastewaters, which are biological treatment, electrochemical degradation, flocculation, MnO_2 oxidation, photocatalytic degradation, sonochemical degradation, ultrafiltration, and adsorption processes [4-15]. When those techniques have been compared, it can be seen that the adsorption process is the most applicable method for dye removal as it is very efficient, practical, and economic [1, 2, 4, 5]. As adsorption process is a very demandable method for effluent treatments, numerous approaches have been executed to develop various efficient adsorbents. As a result, people have proved that activated carbon, clay materials, chitin, chitosan, peat, silica, and solid waste are good candidates for adsorbents as shown in Table 1. Even though these carry some pros, researchers, however, have revealed their cons which are lack of specificity for their synthesis and treatment, low efficacy and capacity, problems in their recycling and reusing, and longer processing times. Hence, much better adsorbent materials are required for more effective dye removal. Lastly, nanomaterials have been used as efficacious and re-usable materials for many applications [16-23]. They have mechanical flexibility, chemical stability, regulatable pore size, modifiable structures and compositions [24-30]. They can also been used as adsorbents for wastewater treatments.

Thus, for the remediation of dyes from contaminated waters, different novel nanomaterials have been produced, modified and improved by the researchers [9, 11, 31]. Latterly, carbon supported nanoparticles or carbon nanotubes, polyaniline nanotubes, iron oxide nanoparticles, fullerenes, polyurethane foams, polypyrrole/TiO_2 nanocomposites, poly(cyclotri-phosphazene-co-4,4-sulfonyl-diphenol)   (PZS)   and   polydopamine nanospheres and CdS nanostructures have been developed by the researchers as shown in Table 1. These materials and their modified kinds have started to be used in dye removal applications. These nanomaterials have wide surface area. This feature allows better contact, and hence, better dye removal can be obtained. All these create ideal characteristics for demandable nanosorbents. Within all these adsorbents, carbon-based materials (i.e. graphene oxide, reduced graphene oxide) supply many opportunities for remediation of environmental problems, such as pollutant removal in water [32-43]. These makes carbon-based materials an ideal adsorbent to remove cationic dyes like methylene blue. As methylene blue is commonly used for colorizing cotton, silk, or wood, economic adsorbent material developing for methylene blue removal from wastewaters is very important to reduce methylene blue's environmental, health, and esthetical concerns. For successful dye removal in effluent waters, better performing nanocomposites are required, and the research should have a simple adsorbent preparation methodology in addition to the composition of adsorbent. In addition, it is known that homogeneous distribution of NPs on the surface of support materials is a prerequisite to obtaining high performance of adsorbents. Based on very good adsorption properties of chitosan and GO, chitosan-graphene oxide (CSGO) mixture have been thought as biosorbents owing to different surface functional groups of CSGO that make it highly dispersible in water. These makes CSGO an ideal cationic dye adsorbent like methylene blue. Addressed herein, graphene oxide-chitosan (CSGO) material was used as a supporting agent for the nanoparticles. The adsorption capacity and methylene blue removal of this Pt NPs@CSGO that were produced through the microwave-assisted method were investigated. The relationship between contact time and adsorption efficiency and the relationship between the amount of nanomaterials adsorbed per unit weight of methylene blue and dye concentration, as well as the reusability of nanomaterials,   were investigated. X-ray diffraction (XRD), X-ray photoelectron spectroscopy (XPS), transmission electron microscopy (TEM) and high resolution transmission electron microscopy (HRTEM) were used to characterize the Pt NPs@CSGO. The methylene blue removal efficiency was examined via UV-Vis spectrophotometer. It was displayed that the Pt NPs@CSGO has good properties such as high surface area, fast extraction and regeneration periods, ease of operation, and high potential for remediating wastewater for dye removal.

2.   Materials and methods

2.1   Chemicals and techniques

Platinum (IV) chloride ($PtCl_4$ 99%; Alfa), tetrahydrofuran (THF or $(CH_2)_3CH_2O$; 99.5%; Merck), potassium permanganate ($KMnO_4$; Merck), Chitosan (degree of deacetylation: 90 %, Mw = 4.000-6.000), HAc (Aldrich) sodium nitrate ($NaNO_3$; 99.0%; Merck), hydrogen peroxide (H_2O_2; 30%; Merck), hydrochloric acid (HCl; 37%; Merck), methylene blue (MB; Merck) and ethylene glycol (Aldrich) were used in this study. Water was purified using a Millipore filtration system (18 MΩ). Teflon-coated magnetic stir bars and the glassware were washed with aqua regia and distilled water and then dried. TEM images of Pt NPs have been obtained by a JEOL 200 kV TEM instrument. X-ray diffraction (XRD) was performed using a Panalytical Empyrean diffractometer with Ultima + theta-theta high-resolution goniometer, the X-ray generator (Cu K radiation, λ = 1.54056Å) with operation conditions at 45 kV and 40 mA. A Specs spectrometer was used for X-ray photoelectron spectroscopy (XPS) measurements using K lines of Mg (1253.6 eV, 10 mA) as an X-ray source. For the adsorption experiments, after calibration experiments in a different amount of materials by the help of used nanomaterial, to run the batch adsorption experiments, firstly, 25 mg of the nanocomposite was dispersed in water by using an ultrasonic bath for 2 h. Next, the mixture was mixed with 25 mL of methylene blue solution (30 mg/L) and shaken in a water bath (120 rpm) for 24 hours. At the end of 24 h shaking, pH was adjusted to 5.8 using NaOH and HCl solutions. Dye adsorption experiments were done in round bottom flasks at room temperature. After separating the nanocomposite particles by centrifugation (4000 rpm for 10 min), the supernatant solution was analyzed to measure the absorbance at 664 nm, which is the absorption band of methylene blue in water by using a UV-Vis spectrophotometer. Using the calibration curve and the absorbance data, the amount of dye adsorbed was calculated using the equation in the literature [2-6].

For the reusability of the NPs synthesized in this study for methylene blue removal, 15 mg of the nanocomposite was mixed with 25 mL of methylene blue solution. Next, the mixture was sonicated for 30 min at room temperature. After separating the nanocomposite from the mixture by centrifugation, the supernatant was kept for the spectroscopic analyses. Afterward, for desorption, used nanocomposite was washed with 25 mL of ethanol three times at room temperature and then collected by centrifugation. This washed nanocomposite was reused for a next methylene blue adsorption experiment as described above. To figure out the reusability, experiments were repeated six times.

2.2   Preparation of nanosorbents

By using graphite powder and modifying the Hummer's method, graphene oxide (GO) was synthesized [44-52]. The prepared graphene oxide can be easily dispersed in water. Based on this, graphene oxide (GO) powder of 0.20 g was dispersed into 100 mL of ultrapure water by mild ultrasound for 20 min in order to get a homogeneous suspension. Then, 1 ml HAc and 1.5 g chitosan (CS) were added to the suspension. After an hour of stirring at room temperature, the CSGO mixture solutions were prepared. In order to prepare Pt NPs, microwave irradiation method has been performed as shown in Fig. 1. The prepared Pt NPs were mixed with CSGO (1:1 ratio) by using an ultrasonic tip sonicator and then prepared Pt NPs@CSGO was dried at room temperature in a vacuum-drier.

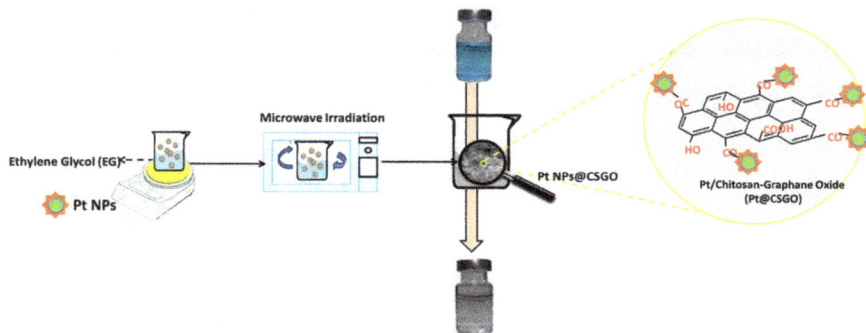

Fig. 1 A schematic illustration for the Pt NPs@CSGO synthesis achieved in this study.

3.   Results and discussion

Characterizations of the Pt NPs@CSGO were performed by XRD, XPS, TEM, and HRTEM. The morphology and structure of the synthesized Pt NPs@CSGO and were analyzed by a TEM and the results were depicted in Fig. 2a. The Pt NPs@CSGO (3.67 ± 0.38 nm) had a uniformly distributed structure on the surface of CSGO (Fig. 2a). Furthermore, the XRD patterns of these nanoparticles were displayed in the Fig. 2b. The peak at around 22.3° is ascribed to the CSGO. Fig. 2b also shows the peaks in face-centered cubic (fcc) Pt crystallites at 2θ of 39.80, 46.50, 67.50, 81.20, and 86.10 which are linked with the Pt (111), Pt (200), Pt (220), Pt (311), and (222) planes, respectively [53-63].

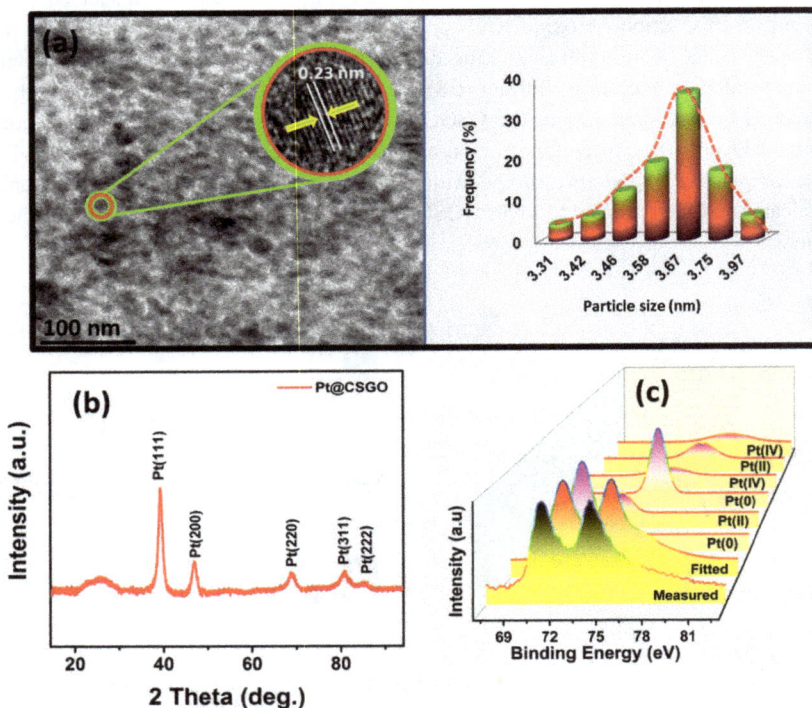

Fig. 2   a) A representative Transmission and High-Resolution Electron Microscopy image and their relevant size distribution graph, b) The X-Ray Diffraction patterns and (c) X-Ray Photoelectron Spectroscopy spectra results for the Pt NPs@CSGO.

X-ray photoelectron spectroscopy was used to examine the oxidation state of the metal in the prepared material [64-85]. Fig. 2c shows the Pt 4f region of the XPS spectrum of Pt NPs@CSGO synthesized by the microwave-assisted method. For this purpose, the Gaussian-Lorentzian method and the Shirley-shaped background was used for fitting of XPS [86-102]. After the fitting process, the intensity ratios of the Pt species were calculated by integrating each peak area. The Pt 4f spectrum of the nanosorbent is composed of two pairs of doublets which are shown in Fig. 2c. The densest doublets at about ·71.0 and 74.4 eV are assigned to zero-valent platinum and this indicating that almost all of the resulting Pt-NPs are at the nanoscale. The second peak shows Pt (II) at 71.8 and 75.9 eV and the third peak shows Pt (IV) at 73.2 and 77.1 eV, respectively, which are derived from PtO_2 and/or $Pt(OH)_4$ derivatives.

In literature, it can be seen that Pt NPs and carbon-chitosane materials have been used in many applications such as solar cells, fuel cells, biofuel cells, dehydrocoupling, some organic reactions, electrochemical sensors, anticancerogen and antimicrobial activities etc [64-76], however, not much study has been performed for the methylene blue (MB) removal from aqueous solutions. On the other hand, as the microwave assisted methodology employed within this work is an appropriate, attractive, rapid, safe, simple and useful method; the synthesis procedure becomes more effective for such a system.

When characterization procedures were performed, adsorption properties of the Pt NPs@CSGO were tested for removal of methylene blue from the water. For this aim, firstly, a calibration curve was made for 2.5, 5, 10, 15, and 30 mg/L methylene blue solutions. For higher methylene blue concentrations, a little dye aggregation was observed therefore, 13 mg/L methylene blue solution was utilized to investigate the relationship between the dye adsorption and contact time at room temperature. The results were presented in Fig. 3a. Our findings presented the equilibrium for methylene blue adsorption in 60 minutes. Current equilibrium time is a relatively short time, hence, it could be said that Pt NPs@CSGO are the efficient adsorbents for methylene blue removal as compared to the other sorbents in Table 1. These results demonstrate that the CSGO is a very good supporting agent for nanoadsorbents. In addition, when the contact time was between 0 and 60 min the methylene blue removal efficiency changed (increased with the contact time) significantly. However, when it reached 60 min, methylene blue removal efficiency started not to change as it can be seen in Fig. 3a as a linear curve. The possible answer for this type of behavior can be the decreasing methylene blue amount during methylene blue adsorption.

Besides, the adsorption isotherm was displayed in Fig. 3b. In this figure, qe, mg/g and Ce, mg/L was shown for the prepared nanomaterials. The results showed that the maximum adsorption capacity was observed in Pt NPs@CSGO as 194.6 mg/g. As shown in Table 1 (previous adsorbent results), we had very good results (Table 1). The possible reasons for this might be the high electrostatic interactions and π-π interactions between CSGO supported and methylene blue dye (due to their opposite charges), and hence the adsorption of methylene blue on the Pt NPs@CSGO was induced. Since the stability of the nanocomposites is very crucial for their practical application as an adsorbent in wastewater treatment processes, we have also tested the durability of the prepared nanomaterials. It is worth to note that no changes were observed in prepared nanomaterials after performing of methylene blue adsorption as shown in Fig. 3c as the good nanosorbents here.

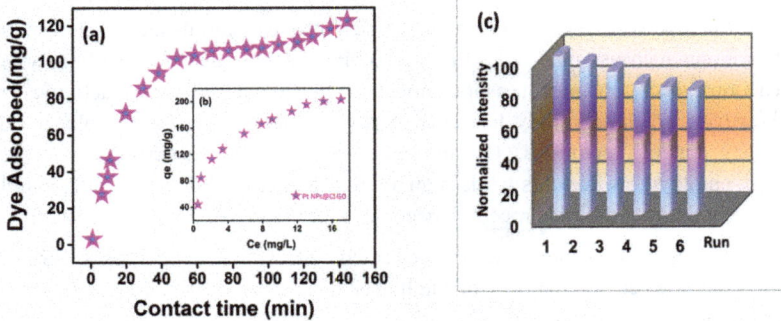

Fig. 3 a) The related graph between adsorption capacity of prepared nanomaterials and various contact time when initial methylene blue concentration was 13 mg/L. b) Isotherm of the methylene blue adsorption of Pt NPs@CSGO. c) The performance of reusability of prepared materials for the methylene blue removal. The experiments were done at 25°C and for 30 min contact time.

On the other hand, a good adsorbent material does not only possess high adsorption capability but also perfect desorption property. Therefore, the reusability of Pt NPs@CSGO was tested in our study. For this aim, 6 successive cycles of adsorption-desorption were done. The results were displayed in Fig. 3c. Although Pt NPs@CGO nanocomposites' adsorption capacity for methylene blue removal showed a decrease, they still had 74.00 % efficiency even after six cycles. Hence, our results showed that the regenerated Pt NPs@CSGO can be used repeatedly as efficient adsorbents for methylene blue removal. Based on our results, the Pt NPs@CSGO is the very good efficient sorbent for methylene blue dye removal from aqueous solutions.

Chitosan-Based Adsorbents for Wastewater Treatment, Ed. Abu Nasar   Materials Research Forum LLC
Materials Research Foundations **34** (2018) 255-278   doi: http://dx.doi.org/10.21741/9781945291753-11

Table 1 Adsorption capacities of different materials for methylene blue removal.

Adsorbent	Adsorption capacity (mg MB/g)	References
Pt NPs@CSGO	194.6	This work
MB-citrus limetta	227.3	[3]
MB-Bacillus subtilis	169.5	[103]
MB-banana leaves	109.9	[104]
MB-cottonseed hull	185.2	[105]
MB-shaddock peel	309.6	[106]
MB-rice hull ash	17.1	[107]
MB-cucumber peels	111.1	[108]
MB-cotton stalk	147.1	[109]
MB-wheat straw	274.1	[110]
GO-Fe_3O_4-SiO_2	111.1	[111]
Graphene	153.85	[112]
GO-Fe_3O_4	190.14	[113]
GO	17.3	[114]
Na-ghassoulite	135	[115]
MWCNTs with Fe_2O_3	42.3	[116]
PZS nanospheres	20	[117]
MPB-AC	163.3	[118]

4.   Conclusions

In summary, in our present study, an effective, simple and practical nanosorbent production method was presented successfully. Our method contains advantageous such as short reaction times, great yields, facile methodology steps and simple work up. Pt NPs@CSGO hybrids indicated remarkable nanosorbent performance together with good dye removal capacity (194.6 mg methylene blue /g nanocomposite) for methylene blue dye in water. The possible explanations are: (i) the uniform distribution of prepared nanomaterial; (ii) higher chemical surface area; (iii) percent Pt contents; and (iv)

Chitosan-Based Adsorbents for Wastewater Treatment, Ed. Abu Nasar   Materials Research Forum LLC
Materials Research Foundations **34** (2018) 255-278   doi: http://dx.doi.org/10.21741/9781945291753-11

electrostatic interactions and π-π interactions between CSGO and methylene blue, which induced the methylene blue adsorption on Pt NPs@CSGO. Additionally, highly stable and reusable absorbents were produced. It was observed that the prepared nanoadsorbents can be used many times for each adsorption-desorption cycles. This facile, straightforward, and controllable method offers a new pathway for the preparation of new electrode materials (CSGO) with high adsorbent performances, which can find extensive applications.

Acknowledgements

This research was supported by Dumlupinar University Research Funding Agency (2014-05 and 2015-35).

References

[1]   M. S. Chiou, P. Y. Ho, H. Y. Li, Adsorption of anionic dyes in acid solutions using chemically cross-linked chitosan beads, Dyes & Pigments. 60 (2004) 69–84. https://doi.org/10.1016/S0143-7208(03)00140-2

[2]   Z. Chen, J. Fu, M. Wang, X. Wang, J. Zhang, Q. Xu, Adsorption of cationic dye (methylene blue) from aqueous solution using poly(cyclotriphosphazene-co- 4,4′-sulfonyldiphenol) nanospheres, Appl. Sur. Sci. 289 (2014) 495-501. https://doi.org/10.1016/j.apsusc.2013.11.022

[3]   Y. Yildiz, T. Onal Okyay, B. Sen, B. Gezer, S. Kuzu, A. Savk, E. Demir, Z. Dasdelen and F. Sen, Highly Monodisperse Pt/Rh Nanoparticles Confined in the Graphene Oxide for Highly Efficient and Reusable Sorbents for Methylene Blue Removal from Aqueous Solutions. Chemistry Select, 2 (2) (2017) 697-70. https://doi.org/10.1002/slct.201601608

[4]   S. Shakoor, A. Nasar, Removal of methylene blue dye from artificially contaminated water using citrus limetta peel waste as a very low cost adsorbent, J. Taiwan Inst. Chem. Eng. 66 (2016) 154–163. https://doi.org/10.1016/j.jtice.2016.06.009

[5]   M. A. Khan, S. H. Lee, S. Kang, K. J. Paeng, G. Lee, S. E. Oh, and B. H. Jeon, Adsorption Studies for the Removal of Methyl tert-Butyl Ether on Various Commercially Available GACs from an Aqueous Medium, Separation Science and Tech. 46 (2011) 1121-1130. https://doi.org/10.1080/01496395.2010.551395

[6]   M. S. Chiou, P. Y. Ho, H. Y. Li, Adsorption of Anionic Dyes in Acid Solutions Using Chemically Cross- Linked Chitosan Beads, Dyes and Pigments. 60 (2004) 69-84. https://doi.org/10.1016/S0143-7208(03)00140-2

[7]   Y. Yıldız, T. Onal Okyay, B. Gezer, Z. Dasdelen, B. Sen, F. Sen, Monodisperse Mw-Pt NPs@VC as highly efficient and reusable adsorbents for methylene blue removal. Journal of Cluster Science, 27 (2016) 1953–1962. https://doi.org/10.1007/s10876-016-1054-3

[7]   S. Shakoor, A. Nasar, Adsorptive treatment of hazardous methylene blue dye from artificially contaminated water using cucumis sativus peel waste as a low-cost adsorbent, Groundw. Sustain. Dev. 5 (2017) 152–159. https://doi.org/10.1016/j.gsd.2017.06.005

[8]   A. Y. Zahrim, C. Tizaoui, N. Hilal, Coagulation with polymers for nanofiltration pre-treatment of highly concentrated dyes: A review, Desalination. 266 (2011) 1–16. https://doi.org/10.1016/j.desal.2010.08.012

[9]   L. Fan, C. Luo, M. Sun, H. Qiu, and X. Li, Synthesis of magnetic-cyclodextrin–chitosan/graphene oxide as nanoadsorbent and its application in dye adsorption and removal, Colloids and Surfaces B: Biointerfaces. 103 (2013) 601-607. https://doi.org/10.1016/j.colsurfb.2012.11.023

[10]   F. Liu, S. Chung, G. Oh, T. S. Seo, Three-Dimensional Graphene Oxide Nanostructure for Fast and Efficient Water-Soluble Dye Removal, ACS Applied Materials & Interfaces. 4 (2012) 922-927. https://doi.org/10.1021/am201590z

[11]   Qamruzzaman, A. Nasar, Degradation of tricyclazole by colloidal manganese dioxide in the absence and presence of surfactants, J. Ind. Eng. Chem. 20 (2014) 897–902. https://doi.org/10.1016/j.jiec.2013.06.020

[12]   Qamruzzaman, A. Nasar, Kinetics of metribuzin degradation by colloidal manganese dioxide in absence and presence of surfactants, Chem. Pap. 68 (2014). https://doi.org/10.2478/s11696-013-0424-7.

[13]   Qamruzzaman, A. Nasar, Degradation of acephate by colloidal manganese dioxide in the absence and presence of surfactants, Desalin. Water Treat. 55 (2015) 2155–2164. https://doi.org/10.1080/19443994.2014.937752

[14]   Qamruzzaman, A. Nasar, Treatment of acetamiprid insecticide from artificially contaminated water by colloidal manganese dioxide in the absence and presence of surfactants, RSC Adv. (2014). https://doi.org/10.1039/C4RA09685A

[15]   J. P. Giraldo, M. P. Landry, S. M. Faltermeier et.al., A Nanobionic Approach to Augment Plant Photosynthesis and Biochemical Sensing Using Targeted Nanoparticles. Nature Materials, 13 (2014) 400–408. https://doi.org/10.1038/nmat3890

[16]   F. Sen, Z. W. Ulissi, X. Gong et.al., Spatiotemporal Intracellular Nitric Oxide Signaling Captured Using Internalized, Near-Infrared Fluorescent Carbon Nanotube Nanosensors. Nano Letters, 14 (8) (2014) 4887–4894. https://doi.org/10.1021/nl502338y

[17]   F. Sen, A. A. Boghossian, S. Sen, et.al. Application of Nanoparticle Antioxidants to Enable Hyperstable Chloroplasts for Solar Energy Harvesting. Advanced Energy Materials, 3 (7) (2013) 881–893. https://doi.org/10.1002/aenm.201201014

[18]   S. Sen, F. Sen, A. A. Boghossian, et al. The Effect of Reductive Dithiothreitol and Trolox on Nitric Oxide Quenching of Single Walled Carbon Nanotubes, Journal of Physical Chemistry C. 117 (1) (2013) 593-602. https://doi.org/10.1021/jp307175f

[19]   J.T. Abrahamson, F. Sen, B. Sempere, et. al. Excess Thermopower and the Theory of Thermopower Waves, ACS Nano.7 (8) (2013) 6533–6544. https://doi.org/10.1021/nn402411k

[20]   N. M. Iverson, P. W. Barone, Mia Shandell, et. al. In vivo biosensing via tissue-localizable near- infrared-fluorescent single-walled carbon nanotubes, Nature Nanotechnology. 8 (11) (2013) 873-880. https://doi.org/10.1038/nnano.2013.222

[21]   F. Sen, A A. Boghossian, S. Sen, et.al. Observation of Oscillatory Surface Reactions of Riboflavin, Trolox, and Singlet Oxygen Using Single Carbon Nanotube Fluorescence Spectroscopy, ACS Nano. 6 (12) (2012) 10632-10645. https://doi.org/10.1021/nn303716n

[22]   J. Zhang, M. P. Landry, P. W. Barone,et.al. Molecular recognition using corona phase complexes made of synthetic polymers adsorbed on carbon nanotubes, Nature Nanotechology. 8 (12) (2013) 959-968. https://doi.org/10.1038/nnano.2013.236

[23]   B. Şen, E. H. Akdere, A. Şavk, E. Gültekin, H. Göksu and F. Şen, A novel thiocarbamide functionalized graphene oxide supported bimetallic monodisperse Rh-Pt nanoparticles (RhPt/TC@GO NPs) for Knoevenagel condensation of aryl aldehydes together with malononitrile, Applied Catalysis B: Environmental. 225 5 (2018) 148-153. https://doi.org/10.1016/j.apcatb.2017.11.067

[24]   B. Şen, N. Lolak, M. Koca, A. Şavk, S. Akocak, F. Şen, Bimetallic
       PdRu/graphene oxide-based Catalysts for one-pot three-component synthesis of 2-
       amino-4H-chromene derivatives. Nano-Structures & Nano-Objects, 12 (2017) 33-
       40. https://doi.org/10.1016/j.nanoso.2017.08.013

[25]   R. Ayranci, G. Baskaya, M. Guzel, S. Bozkurt, M. Ak, A. Savk, F. Sen, Enhanced
       optical and electrical properties of PEDOT via nanostructured carbon materials: A
       Comparative investigation. Nano-Structures and Nano-Objects, 11 (2017) 13–19.
       https://doi.org/10.1016/j.nanoso.2017.05.008

[26]   B. Şahin, E. Demir, A. Aygün et.al. Investigation of the Effect of Pomegranate
       Extract and Monodisperse Silver Nanoparticle Combination on MCF-7 Cell Line.
       Journal of Biotechnology 260C (2017) 79-83.
       https://doi.org/10.1016/j.jbiotec.2017.09.012

[27]   H. Göksu, B. Kilbas and F. Sen, Recent Advances in the Reduction of Nitro
       Compounds by Heterogenous Catalysts, Current Organic Chemistry. 21 (9) (2017)
       794-820. https://doi.org/10.2174/1385272820666160525123907

[28]   B. Şahin, A. Aygün, H. Gündüz, K. Şahin, E. Demir, S. Akocak, F. Şen, Cytotoxic
       Effects of Platinum Nanoparticles Obtained from Pomegranate Extract by the
       Green Synthesis Method on the MCF-7 Cell Line, Colloids and Surfaces B:
       Biointerfaces. 163 (2018) 119–124. https://doi.org/10.1016/j.colsurfb.2017.12.042

[29]   S. Akocak, B. Şen, N. Lolak, A. Şavk, M. Koca, S. Kuzu, F. Şen, One-pot three-
       component synthesis of 2-Amino-4H-Chromene derivatives by using
       monodisperse Pd nanomaterials anchored graphene oxide as highly efficient and
       recyclable catalyst. Nano-Structures & Nano-Objects, 11 (2017) 25–31.
       https://doi.org/10.1016/j.nanoso.2017.06.002

[30]   Y. Yıldız, T. Onal Okyay, B. Sen, B. Gezer, S. Bozkurt, G. Başkaya and F. Sen,
       Activated Carbon Furnished Monodisperse Pt nanocomposites as a superior
       adsorbent for methylene blue removal from aqueous solutions, Journal of
       Nanoscience and Nanotechnology. 17 (2017) 4799–4804.
       https://doi.org/10.1166/jnn.2017.13776

[31]   T. Demirci, B. Çelik, Y. Yıldız, S. Eriş, M. Arslan, B. Kilbas and F. Sen, One-Pot
       Synthesis of Hantzsch Dihydropyridines Using Highly Efficient and Stable
       PdRuNi@GO Catalyst. RSC Advances, 6 (2016) 76948 – 76956.
       https://doi.org/10.1039/C6RA13142E

[32]   E. Demir, A. Savk, B. Sen, F. Sen, A novel monodisperse metal nanoparticles anchored graphene oxide as Counter Electrode for Dye-Sensitized Solar Cells. Nano-Structures & Nano-Objects, 12 (2017) 41-45. https://doi.org/10.1016/j.nanoso.2017.08.018

[33]   D. R. Dreyer, S. Park, C. W. Bielawski and R. S. Ruoff, The chemistry of graphene oxide, Chemical Society Reviews. 39 (2010) 228-240. https://doi.org/10.1039/B917103G

[34]   F. Ahmed and D. F. Rodrigues, Investigation of acute effects of graphene oxide on wastewater microbial community: A case study, J. Hazard. Mater. 256–257 (2013) 33-39. https://doi.org/10.1016/j.jhazmat.2013.03.064

[35]   B. Şen, N. Lolak, Ö. Paralı, M. Koca, A. Şavk, S. Akocak, F. Şen, Bimetallic PdRu/graphene oxide based Catalysts for one-pot three-component synthesis of 2-amino-4H-chromene derivatives, Nano-Structures & Nano-Objects. 12 (2017) 33-40. https://doi.org/10.1016/j.nanoso.2017.08.013

[36]   İ. Esirden, E. Erken, M. Kaya and F. Sen, Monodisperse Pt NPs@rGO as highly efficient and reusable heterogeneous catalysts for the synthesis of 5-substituted 1H-tetrazole derivatives. Catal. Sci. Technol., 5 (2015) 4452-4457. https://doi.org/10.1039/C5CY00864F

[37]   H. Pamuk, B. Aday, M. Kaya, Fatih Şen, Pt Nps@GO as Highly Efficient and Reusable Catalyst for One-Pot Synthesis of Acridinedione Derivatives. RSC Advances, 5 (2015) 49295-49300. https://doi.org/10.1039/C5RA06441D

[38]   H. Goksu, Y. Yıldız, B. Celik, M. Yazıcı, B. Kılbas and F. Sen, Highly Efficient and Monodisperse Graphene Oxide Furnished Ru/Pd Nanoparticles for the Dehalogenation of Aryl Halides via Ammonia Borane, Chemistry Select. 1 (2016) 953-958. https://doi.org/10.1002/slct.201600207

[39]   J. Xu, H. Lv, S. T. Yang and J. Luo, Preparation of graphene adsorbents and their applications in water purification, Reviews in Inorganic Chem. 33 (2013) 139-160. https://doi.org/10.1515/revic-2013-0007

[40]   Y. Yildiz, E. Erken, H. Pamuk and F. Sen, Monodisperse Pt Nanoparticles Assembled on Reduced Graphene Oxide: Highly Efficient and Reusable Catalyst for Methanol Oxidation and Dehydrocoupling of Dimethylamine-Borane (DMAB) J. Nanosci. Nanotechnol. 16 (2016) 5951-5958. https://doi.org/10.1166/jnn.2016.11710

[41]   S. Akocak, B. Şen, N. Lolak, A. Şavk, M. Koca, S. Kuzu, F. Şen, One-pot three-component synthesis of 2-Amino-4H-Chromene derivatives by using monodisperse Pd nanomaterials anchored graphene oxide as highly efficient and recyclable catalyst, Nano-Structures & Nano-Objects. 11 (2017) 25–31. https://doi.org/10.1016/j.nanoso.2017.06.002

[42]   S. Eigler, A. Hirsch, Chemistry with Graphene and Graphene Oxide—Challenges for Synthetic Chemists, Angewandte Chemie International Edition. 53 (2014) 7720-7738.

[43]   B. Khodadadi, M. Bordbar, M. Nasrollahzadeh, Facile and green solvothermal synthesis of palladium nanoparticle-nanodiamond-graphene oxide material with improved bifunctional catalytic properties, Journal of the Iranian Chemical Society. 14 (2017) 2503-2512.

[44]   Z. Dasdelen, Y. Yıldız, S. Eriş, F. Şen, Enhanced electrocatalytic activity and durability of Pt nanoparticles decorated on GO-PVP hybride material for methanol oxidation reaction. Applied Catalysis B: Environmental 219C (2017) 511-516. https://doi.org/10.1016/j.apcatb.2017.08.014

[45]   H. Goksu, Y. Yıldız, B. Çelik, M. Yazici, B. Kilbas, and F. Sen, Eco-friendly hydrogenation of aromatic aldehyde compounds by tandem dehydrogenation of dimethylamine-borane in the presence of reduced graphene oxide furnished platinum nanocatalyst. Catalysis Science and Technology, 6 (2016) 2318 – 2324. https://doi.org/10.1039/C5CY01462J

[46]   H. Goksu, Y. Yıldız, B. Celik, M. Yazıcı, B. Kılbas and F. Sen, Highly Efficient and Monodisperse Graphene Oxide Furnished Ru/Pd Nanoparticles for the Dehalogenation of Aryl Halides via Ammonia Borane. Chemistry Select, 1 (5) (2016) 953-958. https://doi.org/10.1002/slct.201600207

[47]   B. Aday, Y. Yıldız, R. Ulus, S. Eriş, M. Kaya, and F. Sen, One-Pot, Efficient and Green Synthesis of Acridinedione Derivatives using Highly Monodisperse Platinum Nanoparticles Supported with Reduced Graphene Oxide. New Journal of Chemistry, 40 (2016) 748 – 754. https://doi.org/10.1039/C5NJ02098K

[48]   S. Bozkurt, B. Tosun, B. Sen, S. Akocak, A. Savk, M. F. Ebeoğlugil, F. Sen, A hydrogen peroxide sensor based on TNM functionalized reduced graphene oxide grafted with highly monodisperse Pd nanoparticles. Analytica Chimica Acta 989C (2017) 88-94. https://doi.org/10.1016/j.aca.2017.07.051

[49]   B. Aday, H. Pamuk, M. Kaya, and F. Sen, Graphene Oxide as Highly Effective and Readily Recyclable Catalyst Using for the One-Pot Synthesis of 1,8-Dioxoacridine Derivatives, J. Nanosci. Nanotechnol. 16 (2016) 6498-6504. https://doi.org/10.1166/jnn.2016.12432

[50]   R. Ayranci, G. Baskaya, M. Guzel, S. Bozkurt, M. Ak, A. Savk, F. Sen, Carbon based Nanomaterials for High Performance Optoelectrochemical Systems. Chemistry Select, 2 (4) (2017) 1548-1555. https://doi.org/10.1002/slct.201601632

[51]   B. Çelik, G. Başkaya, Ö. Karatepe, E. Erken, F. Şen, Monodisperse Pt(0)/DPA@GO nanoparticles as highly active catalysts for alcohol oxidation and dehydrogenation of DMAB. International Journal of Hydrogen Energy, 41 (2016) 5661-5669. https://doi.org/10.1016/j.ijhydene.2016.02.061

[51]   F. Sen, and G. Gokagac, Different sized platinum nanoparticles supported on carbon: An XPS study on these methanol oxidation catalysts. Journal of Physical Chemistry C, 111 (2007) 5715-5720. https://doi.org/10.1021/jp068381b

[52]   F. Sen, Z. Ozturk, S. Sen, G. Gokagac, The preparation and characterization of nano-sized Pt-Pd alloy catalysts and comparison of their superior catalytic activities for methanol and ethanol oxidation, Journal of Materials Science. 47 (2012) 8134–8144. https://doi.org/10.1007/s10853-012-6709-3

[53]   F. Sen, and G. Gokagac, The activity of carbon supported platinum nanoparticles towards methanol oxidation reaction – role of metal precursor and a new surfactant, tert-octanethiol. Journal of Physical Chemistry C, 111 (2007) 1467-1473. https://doi.org/10.1021/jp065809y

[54]   F. Sen, and G. Gokagac, Improving Catalytic Efficiency in the Methanol Oxidation Reaction by Inserting Ru in Face-Centered Cubic Pt Nanoparticles Prepared by a New Surfactant, tert-Octanethiol. Energy & Fuels, 22 (3) (2008) 1858- 1864. https://doi.org/10.1021/ef700575t

[55]   F. Sen, Z. Ozturk, S. Sen, G. Gokagac, The preparation and characterization of nano-sized Pt-Pd alloy catalysts and comparison of their superior catalytic activities for methanol and ethanol oxidation. Journal of Materials Science, 47 (2012) 8134–8144. https://doi.org/10.1007/s10853-012-6709-3

[56]   S. Sen, F. Sen, G. Gokagac, Preparation and characterization of nano-sized Pt–Ru/C catalysts and their superior catalytic activities for methanol and ethanol oxidation. Phys. Chem. Chem. Phys., 13 (2011) 6784-6792. https://doi.org/10.1039/c1cp20064j

[57]   F. Sen, S. Sen, G. Gokagac, Efficiency enhancement in the methanol/ethanol oxidation reactions on Pt nanoparticles prepared by a new surfactant, 1,1-dimethyl heptanethiol, and surface morphology by AFM. Phys. Chem. Chem. Phys., 13 (2011) 1676-1684. https://doi.org/10.1039/C0CP01212B

[58]   J. Fan, K. Qi, L. Zhang, S. Yu, X. Cui, Engineering Pt/Pd Interfacial Electronic Structures for Highly Efficient Hydrogen Evolution and Alcohol Oxidation, ACS Applied Materials and Interfaces. 9 (2017) 18008-18014. https://doi.org/10.1021/acsami.7b05290

[59]   P. Qiu, S. Lian, G. Yang, S. Yang, Halide ion-induced formation of single crystalline mesoporous PtPd bimetallic nanoparticles with hollow interiors for electrochemical methanol and ethanol oxidation reaction, Nano Research. 10 (2017) 1064-1077. https://doi.org/10.1007/s12274-016-1367-4

[60]   F. Sen, S. Ertan, S. Sen, G. Gokagac, Platinum nanocatalysts prepared with different surfactants for C1 to C3 alcohol oxidations and their surface morphologies by AFM. Journal of Nanoparticle Research, 14 (2012) 922-26. https://doi.org/10.1007/s11051-012-0922-5

[61]   F. Sen, S. Sen, G. Gokagac, High performance Pt nanoparticles prepared by new surfactants for C1 to C3 alcohol oxidation reactions. Journal of Nanoparticle Research, 15 (2013) 1979. https://doi.org/10.1007/s11051-013-1979-5

[62]   H. Klug, L. Alexander, X-Ray Diffraction Procedures: For Polycrystalline and Amorphous Materials, 2nd Edition, Wiley, New York, 1954.

[63]   E. Erken, İ. Esirden, M. Kaya and F. Sen, A Rapid and Novel Method for the Synthesis of 5-Substituted 1H-tetrazole Catalyzed by Exceptional Reusable Monodisperse Pt NPs@AC under the Microwave Irradiation. RSC Advances, 5 (2015) 68558-68564. https://doi.org/10.1039/C5RA11426H

[64]   Ö. Karatepe, Y. Yıldız, H. Pamuk, S. Eriş, Z. Dasdelen and F. Şen, Enhanced electro catalytic activity and durability of highly mono disperse Pt@PPy-PANI nanocomposites as a novel catalyst for electro-oxidation of methanol. RSC Advances, 6 (2016) 50851 – 50857. https://doi.org/10.1039/C6RA06210E

[65]   E. Erken, H. Pamuk, Ö. Karatepe, G. Başkaya, H. Sert, O. M. Kalfa, F. Şen, New Pt(0) Nanoparticles as Highly Active and Reusable Catalysts in the C1–C3 Alcohol Oxidation and the Room Temperature Dehydrocoupling of Dimethylamine-Borane (DMAB). Journal of Cluster Science, (2016) 27: 9. https://doi.org/10.1007/s10876-015-0892-8

[66] F. Sen, and G. Gokagac, Pt Nanoparticles Synthesized with New Surfactans: Improvement in C1-C3 Alcohol Oxidation Catalytic Activity. Journal of Applied Electrochemistry, 44(1) (2014) 199 – 207. https://doi.org/10.1007/s10800-013-0631-5

[67] F. Şen, Y. Karataş, M. Gülcan, M. Zahmakıran, Amylamine stabilized platinum (0) nanoparticles: active and reusable nanocatalyst in the room temperature dehydrogenation of dimethylamine- borane. RSC Advances, 4 (4) (2014) 1526-1531. https://doi.org/10.1039/C3RA43701A

[68] S. Eris, Z. Daşdelen, Y. Yıldız, F. Sen, Nanostructured Polyaniline-rGO decorated platinum catalyst with enhanced activity and durability for Methanol oxidation, International Journal of Hydrogen Energy. 43 (3) 2018 1337–1343. https://doi.org/10.1016/j.ijhydene.2017.11.051

[69] Y. Yıldız, S. Kuzu, B. Sen, A. Savk, S. Akocak, F. Şen, Different ligand based monodispersed metal nanoparticles decorated with rGO as highly active and reusable catalysts for the methanol oxidation. International Journal of Hydrogen Energy, 42 (18) 2017 13061-13069. https://doi.org/10.1016/j.ijhydene.2017.03.230

[70] Y. Yıldız, H. Pamuk, Ö. Karatepe, Z. Dasdelen and F.Sen, Carbon black hybride material furnished monodisperse Platinum nanoparticles as highly efficient and reusable electrocatalysts for formic acid electro-oxidation. RSC Advances, 6 (2016) 32858 – 32862. https://doi.org/10.1039/C6RA00232C

[71] E. Erken, Y. Yildiz, B. Kilbas, and F. Sen, Synthesis and Characterization of Nearly Monodisperse Pt Nanoparticles for C1 to C3 Alcohol Oxidation and Dehydrogenation of Dimethylamine-borane (DMAB). J. Nanosci. Nanotechnol. 16 (2016) 5944-5950. https://doi.org/10.1166/jnn.2016.11683

[72] B. Çelik, E. Erken, S. Eriş, Y. Yıldız, B. Şahin, H. Pamuk and F. Sen, Highly monodisperse Pt(0)@AC NPs as highly efficient and reusable catalysts: the effect of the surfactant on their catalytic activities in room temperature dehydrocoupling of DMAB. Catalysis Science and Technology, 6 (2016) 1685 – 1692. https://doi.org/10.1039/C5CY01371B

[73] B. Çelik, S. Kuzu, E. Erken, Y. Koskun, F. Sen, Nearly Monodisperse Carbon Nanotube Furnished Nanocatalysts as Highly Efficient and Reusable Catalyst for Dehydrocoupling of DMAB and C1 to C3 Alcohol Oxidation. International

Journal of Hydrogen Energy, 41 (2016) 3093-3101.
https://doi.org/10.1016/j.ijhydene.2015.12.138

[74]   G. Baskaya, I. Esirden, E. Erken, F. Sen, and M. Kaya, Synthesis of 5-Substituted-
1H-Tetrazole Derivatives Using Monodisperse Carbon Black Decorated Pt
Nanoparticles as Heterogeneous Nanocatalysts. J. Nanosci. Nanotechnol. 17
(2017) 1992-1999. https://doi.org/10.1166/jnn.2017.12867

[75]   B. Sen, S. Kuzu, E. Demir et.al., Polymer-Graphene hybride decorated Pt
Nanoparticles as highly eficient and reusable catalyst for the Dehydrogenation of
Dimethylamine-borane at room temperature. International Journal of Hydrogen
Energy, 42 (36) (2017) 23284-23291.
https://doi.org/10.1016/j.ijhydene.2017.05.112

[76]   E. Demir, B. Sen, F. Sen, Highly efficient Pt nanoparticles and f-MWCNT
nanocomposites based counter electrodes for dye-sensitized solar cells. Nano-
Structures & Nano-Objects (Invited), 11 (2017) 39-45.
https://doi.org/10.1016/j.nanoso.2017.06.003

[77]   S. Eris, Z. Daşdelen, F. Sen, Investigation of electrocatalytic activity and stability
of Pt@f-VC catalyst prepared by in-situ synthesis for Methanol electrooxidation,
International Journal of Hydrogen Energy. 43 (1) (2018) 385-390.
https://doi.org/10.1016/j.ijhydene.2017.11.063

[78]   J.M. Sieben, A.E. Alvarez, V. Comignani, M.M.E. Duarte, Methanol and ethanol
oxidation on carbon supported nanostructured Cu core Pt-Pd shell electrocatalysts
synthesized via redox displacement, International Journal of Hydrogen Energy. 39
(2014) 11547-11556. https://doi.org/10.1016/j.ijhydene.2014.05.123

[79]   S. Eris, Z. Daşdelen, F. Sen, Enhanced electrocatalytic activity and stability of
monodisperse Pt nanocomposites for direct methanol fuel cells, Journal of Colloid
and Interface Science. 513 (2018) 767–773.
https://doi.org/10.1016/j.jcis.2017.11.085

[80]   Saipanya, S., S. Lapanantnoppakhun, T. Sarakonsri, Electrochemical deposition of
platinum and palladium on gold nanoparticles loaded carbon nanotube support for
oxidation reactions in fuel cell, Journal of Chemistry. (2014)
https://dx.doi.org/10.1155/2014/104514.

[81]   G. Yang, Y. Zhou, H.B. Pan, (...), Zhu, J. J, Y. Lin, Ultrasonic-assisted synthesis
of Pd-Pt/carbon nanotubes nanocomposites for enhanced electro-oxidation of

ethanol and methanol in alkaline medium, Ultrasonics Sonochemistry. 28 (2016) 192-198. https://doi.org/10.1016/j.ultsonch.2015.07.021

[82]   Y. Gao, F. Wang, Y. Wu, R. Naidu, Z. Chen, Comparison of degradation mechanisms of microcystin-LR using nanoscale zero-valent iron (nZVI) and bimetallic Fe/Ni and Fe/Pd nanoparticles, Chemical Engineering Journal. 285 (2016) 459-466. https://doi.org/10.1016/j.cej.2015.09.078

[83]   B. Sen, S. Kuzu, E. Demir, T. Onal Okyay, F. Sen, Hydrogen liberation from the dehydrocoupling of dimethylamine-borane at room temperature by using novel and highly monodispersed RuPtNi nanocatalysts decorated with graphene oxide. International Journal of Hydrogen Energy, 42 (36) (2017) 23299-23306. https://doi.org/10.1016/j.ijhydene.2017.04.213

[84]   B. Celik, Y. Yildiz, H. Sert, E. Erken, Y. Koskun, F. Sen, Monodispersed palladium–cobalt alloy nanoparticles assembled on poly(N-vinyl-pyrrolidone) (PVP) as a highly effective catalyst for dimethylamine borane (DMAB) dehydrocoupling, RSC Adv. 6 (2016) 24097 – 24102. https://doi.org/10.1039/C6RA00536E

[85]   B. Celik, S. Kuzu, E. Demir, E. Yıldırır, F. Sen, Highly Efficient Catalytic Dehydrogenation of Dimethly Ammonia Borane via Monodisperse Palladium-Nickel Alloy Nanoparticles Assembled on PEDOT, Int. J. Hydrogen Energy. 42 (2017) 23307-23314. https://doi.org/10.1016/j.ijhydene.2017.05.115

[86]   T. Kim, X. Fu, D. Warther, M.J. Sailor, Green synthesis of Pd nanoparticles at Apricot kernel shell substrate using Salvia hydrangea extract: Catalytic activity for reduction of organic dyes, ACS Nano. 11 (2017) 2773-2784. https://doi.org/10.1021/acsnano.6b07820

[87]   A. Zhang, Y. Xiao, F. Gong, L. Zhang, Y. Zhang, Solid-state synthesis, formation mechanism and enhanced electrocatalytic properties of Pd nanoparticles supported on reduced graphene oxide, ECS Journal of Solid State Science and Technology. 6 (2017) M13-M18. https://doi.org/10.1149/2.0271701jss

[88]   L.L. Carvalho, F. Colmati, A.A. Tanaka, Nickel–palladium electrocatalysts for methanol, ethanol, and glycerol oxidation reactions, International Journal of Hydrogen Energy. 42 (2011) 16118-16116. https://doi.org/10.1016/j.ijhydene.2017.05.124

[89]   N.R. Elezovic, P. Zabinski, P. Ercius, U.Č. Lačnjevac, N.V. Krstajic, High surface area Pd nanocatalyst on core-shell tungsten based support as a beneficial catalyst

for low temperature fuel cells application, Electrochimica Acta. 247 (2017) 674-684. https://doi.org/10.1016/j.electacta.2017.07.066

[90] J. Zhang, S. Lu, Y. Xiang, J. Liu, S.P. Jiang, Carbon-Nanotubes-Supported Pd Nanoparticles for Alcohol Oxidations in Fuel Cells: Effect of Number of Nanotube Walls on Activity, ChemSusChem. 8 (2015) 2956-2966. https://doi.org/10.1002/cssc.201500107

[91] B. Çelik, Y. Yildiz, E. Erken and Y. Koskun, F. Sen, Monodisperse Palladium-Cobalt Alloy Nanoparticles Assembled on Poly (N-vinyl-pyrrolidone) (PVP) as Highly Effective Catalyst for the Dimethylammine Borane (DMAB) dehydrocoupling. RSC Advances, 6 (2016) 24097 – 24102. https://doi.org/10.1039/C6RA00536E

[92] H. Göksu, B. Çelik, Y. Yıldız, B. Kılbaş and F. Şen, Superior monodisperse CNT-Supported CoPd (CoPd@CNT) nanoparticles for selective reduction of nitro compounds to primary amines with NaBH4 in aqueous medium. Chemistry Select, 1 (10) (2016) 2366-2372. https://doi.org/10.1002/slct.201600509

[93] Y. Yıldız, İ. Esirden, E. Erken, E. Demir, M. Kaya and F. Şen, Microwave (Mw)-assisted Synthesis of 5-Substituted 1H-Tetrazoles via [3+2] Cycloaddition Catalyzed by Mw-Pd/Co Nanoparticles Decorated on Multi-Walled Carbon Nanotubes. Chemistry Select, 1 (8) (2016) 1695-1701. https://doi.org/10.1002/slct.201600265

[94] G. Baskaya, Y. Yıldız, A. Savk, T. Onal Okyay, S. Eris, F. Sen, Rapid, Sensitive, and Reusable Detection of Glucose by Highly Monodisperse Nickel nanoparticles decorated functionalized multi-walled carbon nanotubes. Biosensors and Bioelectronics, 91 (2017) 728–733. https://doi.org/10.1016/j.bios.2017.01.045

[95] B. Sen, S. Kuzu, E. Demir, S. Akocak, F. Sen, Monodisperse Palladium-Nickel Alloy Nanoparticles Assembled on Graphene Oxide with the High Catalytic Activity and Reusability in the Dehydrogenation of Dimethylamine-Borane. International Journal of Hydrogen Energy, 42 (36) (2017) 23276-23283. https://doi.org/10.1016/j.ijhydene.2017.05.113

[96] B. Sen, S. Kuzu, E. Demir, E. Yıldırır, F. Sen, Highly Efficient Catalytic Dehydrogenation of Dimethly Ammonia Borane via Monodisperse Palladium-Nickel Alloy Nanoparticles Assembled on PEDOT. International Journal of Hydrogen Energy, 42 (36) 2017 23307-23314. https://doi.org/10.1016/j.ijhydene.2017.05.115

[97] B. Sen, S. Kuzu, E. Demir, et.al, Highly Monodisperse RuCo Nanoparticles decorated on Functionalized Multiwalled Carbon Nanotube with the Highest Observed Catalytic Activity in the Dehydrogenation of Dimethylamine Borane. International Journal of Hydrogen Energy, 42 (36) (2017) 23292-23298. https://doi.org/10.1016/j.ijhydene.2017.06.032

[98] D.H. Nagaraju, S. Devaraj, P. Balaya, Palladium nanoparticles anchored on graphene nanosheets: Methanol, ethanol oxidation reactions and their kinetic studies, Materials Research Bulletin. 60 (2014) 150-157. https://doi.org/10.1016/j.materresbull.2014.08.027

[99] K. Mishra, N. Basavegowda, Y.R. Lee, Biosynthesis of Fe, Pd, and Fe-Pd bimetallic nanoparticles and their application as recyclable catalysts for [3 + 2] cycloaddition reaction: A comparative approach, Catalysis Science and Technology. 5 (2015) 2612-2621. https://doi.org/10.1039/C5CY00099H

[100] Y. Yıldız, R. Ulus, S. Eris, B. Aday, M. Kaya and F. Sen, Functionalized multi-walled carbon nanotubes (f-MWCNT) as Highly Efficient and Reusable Heterogeneous Catalysts for the Synthesis of Acridinedione Derivatives. Chemistry Select, 1 (13) (2016) 3861–3865.

[101] A. Ayla , A. Cavus , Y. Bulut , Z. Baysal , C. Aytekin , Removal of methylene blue from aqueous solutions onto Bacillus subtilis: determination of kinetic and equilibrium parameters. Desalination 51 (2013) 7596–603. https://doi.org/10.1080/19443994.2013.791780

[102] Q. Zhou , W. Gong , C. Xie , X. Yuan , Y. Li , C. Bai , et al. Biosorption of methylene blue from aqueous solution on spent cottonseed hull substrate for pleurotus ostreatus cultivation. Desalination Water Treat 29 (2011) 317–25. https://doi.org/10.5004/dwt.2011.2238

[103] RR. Krishni , KY. Foo, BH. Hameed, Adsorptive removal of methylene blue using the natural adsorbent-banana leaves. Desalination Water Treat 52 (2014) 6104–12. https://doi.org/10.1080/19443994.2013.815687

[104] J. Liang , J .Wu , P. Li , X. Wang , B. Yang , Shaddock peel as a novel low-cost adsor- bent for removal of methylene blue from dye wastewater. Desalination Water Treat 39 (2012) 70–5. https://doi.org/10.1080/19443994.2012.669160

[105] X. Chen , S. Lv, S. Liu, P. Zhang , A. Zhang , J. Sun , et al. Adsorption of methylene blue by rice hull ash. Sep Sci Technol 47 (2012) 147–56. https://doi.org/10.1080/01496395.2011.606865

[106] G. Akkaya , F. Guzel , Application of some domestic wastes as new low-cost biosorbents for removal of methylene blue: kinetic and equilibrium studies. Chem Eng Commun 201 (2014) 557–78. https://doi.org/10.1080/00986445.2013.780166

[107] H. Deng, J. Lu , G. Li , G. Zhang , X. Wang, Adsorption of methylene blue on adsor- bent materials produced from cotton stalk. Chem Eng J 172 (2011) 326–34. https://doi.org/10.1016/j.cej.2011.06.013

[108] W. Zhang , H. Yan, H. Li , Z. Jiang, L. Dong, X. Kan , et al. Removal of dyes from aqueous solutions by straw based adsorbents: Batch and column studies. Chem Eng J, 168 (2011) 1120–7. https://doi.org/10.1016/j.cej.2011.01.094

[109] Y. Yao, S. Miao, S. Yu, LP. Ma, H. Sun, S.J. Wang, Fabrication of Fe3O4/SiO2 core/shell nanoparticles attached to graphene oxide and its use as an adsorbent. Colloid Interface Sci, 379 (2012) 20. https://doi.org/10.1016/j.jcis.2012.04.030

[110] T. Liu, Y. Li, Q. Du, J. Sun, Y. Jiao, G. Yang, et al. Adsorption of methylene blue from aqueous solution by graphene. Colloids Surf B, 90 (2012) 197. https://doi.org/10.1016/j.colsurfb.2011.10.019

[111] F. He, JT. Fan, D. Ma, L. Zhang, C. Leung, HL. Chan. The attachment of Fe3O4 nanoparticles to graphene oxide by covalent bonding. Carbon, 48 (2010) 3139. https://doi.org/10.1016/j.carbon.2010.04.052

[112] GK. Ramesha, AV. Kumara, H.B Muralidhara, S. Sampath, Graphene and graphene oxide as effective adsorbents toward anionic and cationic dyes. Journal of Colloid Interface Science, 361 (2011) 270. https://doi.org/10.1016/j.jcis.2011.05.050

[113] Y. El Mouzdahir, A. Elmchaouri, R. Mahboub, A. Gil, and S. Korili, Adsorption of Methylene Blue from Aqueous Solutions on a Moroccan Clay. Journal of Chemical & Engineering Data, 52 (2007) 1621-1625. https://doi.org/10.1021/je700008g

[114] S. Qu, F. Huang, S. Yu, G. Chen, and J. Kong, Magnetic removal of dyes from aqueous solution using multi-walled carbon nanotubes filled with Fe2O3 particles. Journal of Hazardous Materials, 160 (2008) 643-647. https://doi.org/10.1016/j.jhazmat.2008.03.037

[115] Z. Chen, J. Fu, M. Wang, X. Wang, J. Zhang, Q. Xu, and R. A. Lemons, Self-assembly fabrication of microencapsulated n-octadecane with natural silk fibroin

shell for thermal-regulating textiles, Appl. Surf. Sci., 289 (2014) 495–501. https://doi.org/10.1016/j.apsusc.2013.11.022

[116] K. T. Wong, N. C. Eu, S. Ibrahim, H. Kim, Y. Yoon, and M. Jang, Recyclable magnetite-loaded palm shell-waste based activated carbon for the effective removal of methylene blue from aqueous solution. Journal of Cleaner Production, 115(2016) 337-342. https://doi.org/10.1016/j.jclepro.2015.12.063

Keyword Index

Chitosan-Based Adsorbents for Wastewater Treatment, Ed. Abu Nasar   Materials Research Forum LLC
Materials Research Foundations **34** (2018)   doi: http://dx.doi.org/10.21741/9781945291753

About the Editor

Dr. Abu Nasar is presently working as an Associate Professor at the Department of Applied Chemistry, Z.H. College of Engineering & Technology, Aligarh Muslim University.

He has received his Ph.D. and done postdoctoral work at the prestigious organization, Indian Institute of Technology, Banaras Hindu University. He has published 42 research papers in high quality journals of international repute, 5 book chapters, and edited 1 book.

His areas of interest include physical chemistry, environmental chemistry, and materials science.

A research paper presented by him at the International Conference on "Advanced Semiconductor Devices and Microsystems" held at Smolenice, Slovakia during Oct. 20-24, 1996 was selected as one of the best papers of the Conference. In the recognition of his work, Dr. Nasar has received the **Young Scientists'** and **Young Metallurgists' Awards** by the Indian Science Congress Association and the Ministry of Steel, Government of India, respectively.

www.ingramcontent.com/pod-product-compliance
Lightning Source LLC
Chambersburg PA
CBHW071333210326
41597CB00015B/1434